Next Generation Wireless LANs

If you've been searching for a way to get up to speed quickly on IEEE 802.11n without having to wade through the entire standard, then look no further. This comprehensive overview describes the underlying principles, implementation details, and key enhancing features of 802.11n. For many of these features, the authors outline the motivation and history behind their adoption into the standard. A detailed discussion of the key throughput, robustness, and reliability enhancing features (such as MIMO, 40 MHz channels, and packet aggregation) is given, in addition to a clear summary of the issues surrounding legacy interoperability and coexistence. Advanced topics such as beamforming and fast link adaption are also covered. With numerous MAC and physical layer examples and simulation results included to highlight the benefits of the new features, this is an ideal reference for designers of WLAN equipment, and network managers whose systems adopt the new standard. It is also a useful distillation of 802.11n technology for graduate students and researchers in the field of wireless communication.

Eldad Perahia is a member of the Wireless Standards and Technology group at Intel Corporation, Chair of the IEEE 802.11 Very High Throughput Study Group, and the IEEE 802.11 liaison to IEEE 802.19. Prior to joining Intel, Dr. Perahia was the 802.11n lead for Cisco Systems. He was awarded his Ph.D. in Electrical Engineering from the University of California, Los Angeles, and has fourteen patents in various areas of wireless communications.

Robert Stacey is a member of the Wireless Standards and Technology group at Intel Corporation. He was a member of the IEEE 802.11 High Throughput Task Group (TGn) and a key contributor to the various proposals culminating in the final joint proposal submission that became the basis for the 802.11n draft standard, and has numerous patents filed in the field of wireless communications.

Next Generation Wireless LANs

Throughput, Robustness, and Reliability in 802.11n

ELDAD PERAHIA AND ROBERT STACEY

CAMBRIDGE
UNIVERSITY PRESS

CAMBRIDGE UNIVERSITY PRESS
Cambridge, New York, Melbourne, Madrid, Cape Town, Singapore, São Paulo, Delhi

Cambridge University Press
The Edinburgh Building, Cambridge CB2 8RU, UK

Published in the United States of America by Cambridge University Press, New York

www.cambridge.org
Information on this title: www.cambridge.org/9780521885843

First published 2008

Printed in the United Kingdom at the University Press, Cambridge

A catalog record for this publication is available from the British Library

Library of Congress Cataloging in Publication data
Perahia, Eldad, 1967–
Next generation wireless LANs : throughput, robustness, and reliability in 802.11n / Eldad Perahia and
Robert Stacey.
    p.   cm.
Includes index.
ISBN 978-0-521-88584-3 (hbk.)
1. Wireless LANS.   I. Stacey, Robert.   II. Title.
TK5105.78.P47   2008
621.384–dc22   2008023551

ISBN 978-0-521-88584-3 hardback

To my wife Sarah and our son Nathan
– Eldad Perahia

To my father, who nurtured and guided an inquiring mind
– Robert Stacey

Brief contents

Contents

Part II Medium access control layer

7   Medium access control   181

Foreword

The first version of the 802.11 standard was ratified in 1997 after seven long years of development. However, initial adoption of this new technology was slow, partly because of the low penetration of devices that needed the "freedom of wireless."

The real opportunity for 802.11 came with the increased popularity of laptop computers just a few years later. This popularity brought a rapidly growing user base wanting network connectivity not only while connected to an Ethernet cable at home or at work, but also in between: in hotels, airports, conference centers, restaurants, parks, etc. 802.11 provided a cheap and easy way to make laptop mobility a reality for anyone who wanted it.

However, technology by itself is rarely sufficient, particularly in the networking space, where interoperability of devices from multiple vendors is almost always the key to market success. Having been formed as WECA in 1999, the Wi-Fi Alliance was ready to provide certification of multi-vendor interoperability.

With the right technology from the IEEE 802.11 Working Group, certified interoperability from the Wi-Fi Alliance, and a real market need based on a growing installed base of laptops, the conditions were ripe for the Wi-Fi market to take off, and indeed it did. By 2007 virtually every new laptop contains Wi-Fi as standard equipment. More importantly, and unlike some other "successful" wireless technologies, many of these devices are used regularly. With this wide use came a growing understanding of the power of cheap, easy-to-deploy, and easy-to-manage interoperable Wi-Fi networks.

The natural next step was for people to ask, "What else can we use Wi-Fi for?" The answer is increasingly becoming "everything, everywhere!" Not just laptops, but now almost anything mobile and even many fixed devices contain Wi-Fi, and they are used in a phenomenal range of applications, including data, voice, games, music, video, location, public safety, vehicular, etc. In 2007, more than 300 million Wi-Fi devices were shipped. By 2012, some analysts are forecasting that more than one billion Wi-Fi devices will be shipped every year.

The 2.4 GHz 802.11b 11 Mb/s DSSS/CCK PHY and the basic 802.11 contention-based MAC provided the basis for a great industry. However, the rapid growth of the Wi-Fi market challenged the capabilities of the technology. It was not long before better security (802.11i certified by the Wi-Fi Alliance as WPA/WPA2^{TM}) and better Quality of Service (802.11e certified by the Wi-Fi Alliance as WMM^{TM} and WMM Power Save) were defined, certified, and deployed.

It was also not long before higher data rates were demanded for greater data density and to support the many new and exciting devices and applications. 802.11a, providing 54 Mbps based on OFDM in the 5 GHz band, failed to garner significant support because two radios were required to maintain backward compatibility with 2.4 GHz 802.11b devices; the cost of two radios was often too high. The real success story was 802.11g, which provided 54 Mbps based on OFDM in the 2.4 GHz band in a way that was backward-compatible with 802.11b.

The success of 802.11g drove the use of Wi-Fi to new heights and expanded the demands on the technology yet again; everyone wanted more. Fortunately, the technology continued to develop and in 2002 the IEEE 802.11 Working Group started defining the next generation of PHY and MAC features as part of 802.11n. 802.11n will define mechanisms to provide users some combination of greater throughput, longer range and increased reliability, using mandatory and optional features in the PHY (including MIMO technology and 40 MHz channels) and the MAC (including more efficient data aggregation and acknowledgments).

Interestingly, 802.11n operates in both the 2.4 GHz and 5 GHz bands. It is expected that 5 GHz operation will be more popular than when 802.11a was introduced, because 2.4 GHz is now more congested, the number of available channels in the 5 GHz band has been expanded with the introduction of DFS and TPC technology, there is more need for high throughput 40 MHz channels, and the cost of dual-band radios has decreased.

The 802.11n standard is not yet complete, and is unlikely to be ratified by the IEEE until at least mid 2009. Until August 2006, the Wi-Fi Alliance had a policy to not certify 802.11n products until the standard was ratified. However, some vendors decided the market could not wait for ratification of the 802.11n standard and started releasing pre-standard products. These products were often not interoperable at the expected performance levels because they were not based on a common interpretation of the draft 802.11n specification. The problem for the Wi-Fi Alliance was that these products were adversely affecting the reputation of Wi-Fi. The Wi-Fi Alliance decided the only way forward was to certify the basic features of 802.11n from a pre-standard draft. Such a decision is not without precedent. In 2003, certification of WPA started before the 802.11i standard was ratified and in 2004 certification of WMM started before 802.11e was ratified. The Wi-Fi Alliance commenced certification of 802.11n draft 2.0 on 26 June 2007.

The decision has turned out to be the right one for the industry and for users. The Wi-Fi CERTIFIED 802.11n draft 2.0 programme has been remarkably successful, with more than 150 products certified in less than five months. This represents a significantly higher number of certified products than for the 802.11g programme during a similar period after launch. The Wi-Fi Alliance's certification program has helped ensure interoperability for the many products that will be released before the ratification of the 802.11n standard. This is particularly important given that the likely ratification date of the 802.11n standard has been extended by more than a year since the decision to start a certification program was announced by the Wi-Fi Alliance. The next challenge for the Wi-Fi Alliance is to ensure a backward-compatible transition path from the 802.11n draft 2.0 as certified by the Wi-Fi Alliance to the final ratified standard.

Standards are never the most accessible of documents. The 802.11 standard is particularly difficult to understand because it has been amended so many times by different groups and editors over a long period. A draft amendment to the standard, such as 802.11n D2.0, is even harder to interpret because many clauses are still being refined and the refinement process often has technical and political aspects that are only visible to those participating full time in the IEEE 802.11 Working Group.

Books like this one are invaluable because they provide the details and the background that allow readers to answer the questions, "What is likely to be in the final standard and how does it work?" Eldad and Robert should be congratulated on taking up the challenge.

Dr. Andrew Myles
Chairman of the BoD
Wi-Fi Alliance
6 December 2007

Preface

Having worked on the development of the 802.11n standard for some time, we presented a full day tutorial on the 802.11n physical layer (PHY) and medium access control (MAC) layer at the IEEE Globecom conference held in San Francisco in December 2006. Our objective was to provide a high level overview of the draft standard since, at the time, there was very little information on the details of the 802.11n standard available to those not intimately involved in its development. After the tutorial, we were approached by Phil Meyler of Cambridge University Press and asked to consider expanding the tutorial into a book.

Writing a book describing the standard was an intriguing prospect. We felt that a book would provide more opportunity to present the technical details in the standard than was possible with the tutorial. It would fill the gap we saw in the market for a detailed description of what is destined to be one of the most widely implemented wireless technologies. While the standard itself conveys details on what is needed for interoperability, it lacks the background on why particular options should be implemented, where particular aspects came from, the constraints under which they were designed, or the benefit they provide. All this we hoped to capture in the book. The benefits various features provide, particularly in the physical layer, are quantified by simulation results. We wanted to provide enough information to enable the reader to model the physical layer and benchmark their simulation against our results. Finally, with the amended standard now approaching 2500 pages, we hoped to provide an accessible window into the most important aspects, focusing on the throughput and robustness enhancements and the foundations on which these are built.

The book we came up with is divided into three parts. The first part covers the physical layer (PHY), and begins with background information on the 802.11a/g OFDM PHY on which the 802.11n PHY is based and interoperates, and proceeds with an overview of spatial multiplexing, the key throughput enhancing technology in 802.11n. This is followed by details on exactly how high throughput is achieved in 802.11n using spatial multiplexing and wider, 40 MHz channels. This in turn is followed by details on robustness enhancing features such as receive diversity, spatial expansion, space-time block codes, and low density parity check codes.

The second part covers the medium access control (MAC) layer. This part provides background on the original 802.11 MAC as well as the 802.11e quality of service (QoS) enhancements. It gives an overview of why changes were needed in the MAC to achieve higher throughput, followed by details on each of the new features introduced. Given the large installed base of 802.11 devices, coexistence and interoperability are considered crucial to the smooth adoption of the standard. To this end, the book provides a detailed discussion on features supporting coexistence and interoperability.

In the third part we provide details on two of the more advanced aspects of the standard, transmit beamforming and link adaptation. These topics are covered in a section of their own, covering both the PHY and the MAC.

Writing this book would not have been possible without help from our friends and colleagues. We would like to thank Thomas (Tom) Kenney and Brian Hart for reviewing the PHY portion of the book and Solomon Trainin, Tom Kenney, and Michelle Gong for reviewing the MAC portion of the book. They provided insightful comments, suggestions, and corrections that significantly improved the quality of the book.

One of the goals of this book is to provide the reader with a quantitative feel of the benefit of the PHY features in the 802.11n standard. This would have been impossible without the extensive simulation support provided to us by Tom Kenney. He developed an 802.11n PHY simulation platform that includes most of the 802.11n PHY features and is also capable of modeling legacy 802.11a/g. The simulation includes all the 802.11n channel models. Furthermore, Tom modeled receiver functionality such as synchronization, channel estimation, and phase tracking. The simulation also included impairments such as power amplifier non-linearity and phase noise to provide a more realistic measure of performance.

The simulation supports both 20 MHz and 40 MHz channel widths. With the 40 MHz simulation capability, Tom generated the results given in Figure 5.8 in Section 5.1.5 modeling MCS 32 and Figure 5.9 in Section 5.1.7 which illustrates the range and throughput improvement of 40 MHz modes. With the MIMO/SDM capability of the simulation in both AWGN channel and 802.11n channel models, Tom produced the results for Figures 5.12–5.15 in Section 5.3. By designing the simulation with the flexibility to set the transmitter and receiver to different modes, he also produced the results given in Figure 5.18 in Section 5.4 modeling the behavior of a legacy 802.11a/g device receiving a GF transmission. Tom also incorporated short guard interval into the simulation with which the results for sensitivity to time synchronization error in Figures 5.20–5.22 in Section 5.5 were generated.

Tom designed the simulation with the ability to select an arbitrary number of transmitter and receiver antennas independent from the number of spatial streams. Using this capability he produced the results for receive diversity gain in Figures 6.2–6.4 in Section 6.1 and spatial expansion performance in Figures 6.5 and 6.6 in Section 6.2. Tom also incorporated space-time block coding and low density parity check coding into the simulation and generated the results given in Figures 6.8, 6.9, 6.14, 6.15, and 6.16 in Section 6.3 and Figure 6.24 in Section 6.4.

To accurately model the performance of a transmit beamforming system, it is important to include aspects like measurement of the channel state information, beamforming weight computation, and link adaptation. Tom incorporated all of these functions into the simulation to generate the waterfall curves in Figures 12.11–12.16 and the throughput curves in Figures 12.17 and 12.18 in Section 12.18.

We sincerely hope our book provides you with insight and a deeper understanding of the 802.11n standard and the technology upon which it is built.

Abbreviations

μs	microseconds
2G	second generation (cellular)
3G	third generation (cellular)
AC	access category
ACK	acknowledgement
ADC	analog-to-digital converter
ADDBA	add block acknowledgement
ADDTS	add traffic stream
AGC	automatic gain control
AID	association identifier
AIFS	arbitration inter-frame space
A-MPDU	aggregate MAC protocol data unit
A-MSDU	aggregate MAC service data unit
AoA	angle of arrival
AoD	angle of departure
AP	access point
APSD	automatic power save delivery
A-PSDU	aggregate PHY service data unit
AS	angular spectrum
ASEL	antenna selection
AWGN	additive white Gaussian noise
BA	block acknowledgement
BAR	block acknowledgement request
BCC	binary convolution code
BF	beamforming
BICM	bit interleaved coded modulation
bps	bits-per-second
BPSCS	coded bits per single carrier for each spatial stream
BPSK	binary phase shift keying
BSS	basic service set
BSSID	BSS identifier
BW	bandwidth
CBPS	coded bits per symbol
CBPSS	coded bits per spatial stream

CBW	channel bandwidth
CCA	clear channel assessment
CCDF	complementary cumulative distribution function
CCK	complementary code keying
CFP	contention free period
CP	contention period
CRC	cyclic redundancy code
CS	carrier sense
CSD	cyclic shift diversity
CSI	channel state information
CSMA	carrier sense multiple access
CSMA/CA	carrier sense multiple access with collision avoidance
CSMA/CD	carrier sense multiple access with collision detection
CTS	clear to send
CW	contention window
DA	destination address
DAC	digital-to-analog converter
dB	decibels
dBc	decibels relative to carrier
dBi	decibels isotropic relative to an antenna
dBm	decibel of measured power referenced to one milliwatt
DBPS	data bits per OFDM symbol
dBr	dB (relative)
DC	direct current
DCF	distributed coordination function
DELBA	delete block acknowledgement
DIFS	DCF inter-frame space
DLS	direct link session
DS	distribution system
DSL	digital subscriber line
DSSS	direct sequence spread spectrum
DTIM	delivery traffic indication message
DVD	digital versatile disc
EDCA	enhanced distributed channel access
EIFS	extended inter-frame space
ERP	enhanced rate PHY
ESS	extended service set
ETSI	European Telecommunications Standards Institute
EVM	error vector magnitude
EWC	Enhanced Wireless Consortium
FCC	Federal Communications Commission
FCS	frame check sequence
FEC	forward error correction
FFT	fast Fourier transform

FHSS	frequency hopped spread spectrum
FS	free space
FTP	file transfer protocol
GF	Greenfield
GF-HT-STF	Greenfield High Throughput Short Training field
GHz	gigahertz
GI	guard interval
GIF	graphics interchange format
GPS	global positioning system
HC	hybrid coordinator
HCCA	HCF controlled channel access
HCF	hybrid coordination function
HEMM	HCCA, EDCA mixed mode
HT	high throughput
HTC	high throughput control
HT-DATA	High Throughput Data field
HT-LTF	High Throughput Long Training field
HTSG	High Throughput Study Group
HT-SIG	High Throughput Signal field
HT-STF	High Throughput Short Training field
HTTP	hypertext transfer protocol
Hz	Hertz
IBSS	independent basic service set
IC	integrated circuit
IDFT	inverse discrete Fourier transform
IEEE	Institute of Electrical and Electronic Engineers
IFFT	inverse fast Fourier transform
IFS	inter-frame space
IP	Internet Protocol
IPv6	Internet Protocol version 6
IR	infrared
ISI	inter-symbol interference
ISM	industrial, scientific, and medical
JPEG	Joint Photographic Experts Group
kHz	kilohertz
km/h	kilometers per hour
LAN	local area networking
LDPC	low density parity check
LLC	logical link control
L-LTF	Non-HT (Legacy) Long Training field
LNA	low noise amplifier
LOS	line-of-sight
LSB	least significant bit
L-SIG	Non-HT (Legacy) Signal field

L-STF	Non-HT (Legacy) Short Training field
LTF	Long Training field
m	meters
MAC	medium access control
MAI	MRQ or ASEL indication
MAN	metropolitan area networking
Mbps	megabit per second
MCS	modulation and coding scheme
MF	mixed format
MFB	MCS feedback
MFSI	MCS feedback sequence indication
MHz	megahertz
MIB	management information base
MIMO	multiple-input multiple-output
ML	maximum likelihood
MMPDU	MAC management protocol data unit
MMSE	minimum mean-square-error
MPDU	MAC protocol data unit
MPEG	Moving Picture Experts Group
MRC	maximal-ratio combining
MRQ	MCS request
Msample/s	mega-samples per second
MSB	most significant bit
MSDU	MAC service data unit
MSE	mean-square-error
MSFI	MCS feedback sequence identifier
MSI	MCS request sequence identifier
NAV	network allocation vector
NDP	null data packet
NF	noise figure
NLOS	non-line-of-sight
nsec	nanosecond
OBO	output back-off
OBSS	overlapping BSS
OFDM	orthogonal frequency division multiplexing
OSI	open systems interconnection
PA	power amplifier
PAR	project authorization request
PAS	power angular spectrum
PC	point coordinator
PCF	point coordination function
PCO	phased coexistence operation
PDU	protocol data unit
PER	packet error rate

PHY	physical layer
PIFS	PCF inter-frame space
PLCP	physical layer convergence procedure
PPDU	PLCP protocol data unit
ppm	parts per million
PSD	power spectral density
PSDU	PLCP service data unit
PSMP	power-save multi-poll
PSMP-DTT	PSMP downlink transmission time
PSMP-UTT	PSMP uplink transmission time
QAM	quadrature amplitude modulation
QoS	quality of service
QPSK	quadrature phase shift keying
R	code rate
RA	receiver address
RD	reverse direction
RDG	reverse direction grant
RF	radio frequency
RIFS	reduced inter-frame space
RMS	root-mean-square
RSSI	received signal strength indication
RTS	request to send
Rx	receive
SA	source address
SAP	service access point
SCP	secure copy protocol
SDM	spatial division multiplexing
SDU	service data unit
SE	spatial expansion
SIFS	short inter-frame space
SIG	Signal field
SIMO	single-input, multiple-output
SISO	single-input, single-output
SMTP	simple mail transfer protocol
SNR	signal-to-noise ratio
SOHO	small-office, home-office
SS	spatial stream
SSC	starting sequence control
SSID	service set identifier
SSN	starting sequence number
STA	station
STBC	space-time block coding
STF	Short Training field
STS	space-time stream

SVD	singular value decomposition
SYM	symbol
TA	transmitter address
TBTT	target beacon transmission time
TC	traffic category
TCLAS	traffic classification
TCM	trellis coded modulation
TCP	transmission control protocol
TDD	time division duplexing
TGn	Task Group n
TGy	Task Group y
TID	traffic identifier
TIFF	tagged image file format
TRQ	training request
TS	traffic stream
TSID	traffic stream identifier
TSPEC	traffic specification
TV	television
Tx	transmit
TxBF	transmit beamforming
TXOP	transmit opportunity
TXTIME	transmit time
UDP	user datagram protocol
USA	United States of America
VoIP	voice over IP
VPN	virtual private network
WEP	wired equivalent privacy
WFA	Wi-Fi Alliance
WLAN	wireless local area network
WM	wireless medium
WNG SC	Wireless Next Generation Standing Committee
WWiSE	world wide spectral efficiency
XOR	exclusive-or
ZF	zero-forcing
ZIP	ZIP file format

1 Introduction

Wireless local area networking has experienced tremendous growth in the last ten years with the proliferation of IEEE 802.11 devices. Its beginnings date back to Hertz's discovery of radio waves in 1888, followed by Marconi's initial experimentation with transmission and reception of radio waves over long distances in 1894. In the following century, radio communication and radar proved to be invaluable to the military, which included the development of spread spectrum technology. The first packet-based wireless network, ALOHANET, was created by researchers at the University of Hawaii in 1971. Seven computers were deployed over four islands communicating with a central computer in a bi-directional star topology.

A milestone event for commercial wireless local area networks (WLANs) came about in 1985 when the United States Federal Communications Commission (FCC) allowed the use of the experimental industrial, scientific, and medical (ISM) radio bands for the commercial application of spread spectrum technology. Several generations of proprietary WLAN devices were developed to use these bands, including WaveLAN by Bell Labs. These initial systems were expensive and deployment was only feasible when running cable was difficult.

Advances in semiconductor technology and WLAN standardization with IEEE 802.11 led to a dramatic reduction in cost and the increased adoption of WLAN technology. With the increasing commercial interest, the Wi-Fi Alliance (WFA) was formed in 1999 to certify interoperability between IEEE 802.11 devices from different manufacturers through rigorous testing. Since 2000, shipments of Wi-Fi certified integrated circuits (IC) reached 200 million per year in 2006 (ABIresearch, 2007). Shipments are expected to exceed a billion units per year by 2012 (ABIresearch, 2007), as illustrated in Figure 1.1.

Such large and sustained growth is due to the benefits WLANs offer over wired networking. In existing homes or enterprises, deploying cables for network access may involve tearing up walls, floors, or ceilings, which is both inconvenient and costly. In contrast, providing wireless network connectivity in these environments is often as simple as installing a single wireless access point. Perhaps more importantly though, the proliferation of laptops and handheld devices has meant that people desire connectivity wherever they are located, not just where the network connection is located. Network connectivity in a conference room or while seated on the sofa in the living room are just two examples of the flexibility afforded by WLANs.

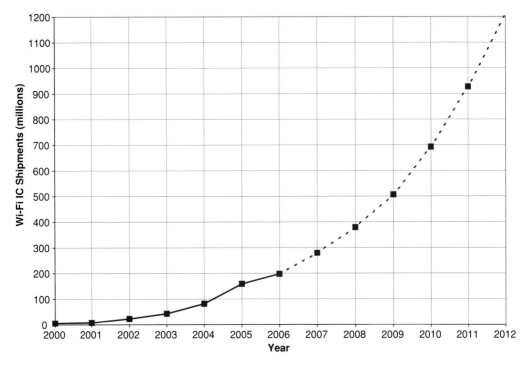

Figure 1.1 Wi-Fi IC shipments. Source: ABIresearch (2007).

Building on the convenience of mobility from the cellular world, WLANs are now enabling Internet access at little or no cost in public wireless networks. In 2005, Google offered to deploy a free Wi-Fi service covering San Francisco at no cost to the city. There has also been a proliferation of small scale deployments providing Internet access in coffee shops, airports, hotels, etc., which have come to be known as hotspots. Additionally, when these networks are used in conjunction with virtual private network (VPN) technology, employees can securely access corporate networks from almost anywhere.

The vast majority of WLAN products and systems today are based on the 802.11b, 802.11g, and 802.11a standard amendments, which provide throughput enhancements over the original 802.11 PHYs. Progress in WLAN technology continues with the development of 802.11n. Increased data rates are achieved with the multiple-input multiple-output (MIMO) concept, with its origins by Foschini (1996) at Bell Labs. In 2004, Atheros demonstrated that 40 MHz devices could be produced at almost the same cost as 20 MHz devices. During a similar time frame, the FCC and ETSI adopted new regulations in the 5 GHz band that added an additional 400 MHz of unlicensed spectrum for use by commercial WLANs. These events paved the way for the broad acceptance of 40 MHz operating modes in 802.11n. When spectrum is free, increasing the channel bandwidth is the most cost effective way to increase the data rate.

Typically product development lags standardization efforts and products are released after the publication of the standard. An interesting event occurred in 2003 when Broadcom released a chipset based on a draft version of the 802.11g amendment, prior to final publication. This set a precedent for the flurry of "pre-n" or "draft-n" products in 2005 and 2006, as industry players rushed to be first to market. Most of these products were either proprietary implementations of MIMO, or based on draft 1.0 of 802.11n, and thus unlikely to be compliant with the final standard.

Through early 2007, major improvements and clarifications were made to the 802.11n draft resulting in IEEE 802.11n draft 2.0. To continue the market momentum and forestall interoperability issues, the IEEE took the unusual step of releasing 802.11n D2.0 to the public while work continued toward the final standard. This allowed the WFA to begin interoperability testing and certification of devices based on a subset of the 802.11n D2.0 features in May 2007. WFA certified 802.11n D2.0 products provide consumers the assurance of interoperability between manufacturers that was not guaranteed by previous "pre-n" or "draft-n" products. These were major steps in speeding up the standardization and certification process of new technology.

1.1 History of IEEE 802.11

The IEEE 802.11 working group began development of a common medium access control (MAC) layer for multiple physical layers (PHY) to standardize wireless local area networking. As a member of the IEEE 802 family of local area networking (LAN) and metropolitan area networking (MAN) standards, 802.11 interfaces with 802.1 architecture, management, and interworking, and 802.2 logical link control (LLC). The combination of 802.2 LLC and 802.11 MAC and PHY make up the data link and physical layers of the Open Systems Interconnection (OSI) reference model, as described in Table 1.1.

Table 1.1 OSI reference model (Zimmerman, 1980; Teare, 1999)

OSI Reference Model layers	Description	Examples	Layer categories
Application	Interacts with software applications that implement a communicating component	Telnet, FTP, SMTP	Application
Presentation	Coding and conversion functions that are applied to application layer data	QuickTime, MPEG, GIF, JPEG, TIFF	Application
Session	Establishes, manages, and terminates communication sessions	ZIP, AppleTalk, SCP, DECnet Phase IV	
Transport	Accepts data from the session layer and segments the data for transport across the network	TCP, UDP	Data transport
Network	Defines the network address	IP, IPv6	Data transport
Data link	Transit of data across a physical network link	802.2 LLC	
Physical	Electrical, mechanical, procedural, and functional specifications	802.11 PHY	

The initial version of the 802.11 standard was completed in 1997. Influenced by the huge market success of Ethernet (standardized as IEEE 802.3), the 802.11 MAC adopted the same simple distributed access protocol, carrier sense multiple access (CSMA). With CSMA, a station wishing to transmit first listens to the medium for a predetermined period. If the medium is sensed to be "idle" during this period then the station is permitted to transmit. If the medium is sensed to be "busy," the station has to defer its transmission. The original (shared medium) Ethernet used a variation called CSMA/CD or carrier sense multiple access with collision detection. After determining that the medium is "idle" and transmitting, the station is able to receive its own transmission and detect collisions. If a collision is detected, the two colliding stations backoff for a random period before transmitting again. The random backoff period reduces the probability of a second collision.

With wireless it is not possible to detect a collision with one's own transmission directly in this way: thus 802.11 uses a variation called CSMA/CA or carrier sense multiple access with collision avoidance. With CSMA/CA, if the station detects that the medium is busy, it defers its transmission for a random period following the medium going "idle" again. This approach of always backing off for a random period following another station's transmission improves performance since the penalty for a collision is much higher on a wireless LAN than on a wired LAN. On a wired LAN collisions are detected electrically and thus almost immediately, while on wireless LAN collisions are inferred through the lack of an acknowledgement or other response from the remote station once the complete frame has been transmitted.

There is no doubt that the simplicity of this distributed access protocol, which enables consistent implementation across all nodes, significantly contributed to Ethernet's rapid adoption as the industry LAN standard. Likewise, the adoption by the industry of 802.11 as the wireless LAN standard is in no small part due to the simplicity of this access protocol, its similarity to Ethernet, and again the consistent implementation across all nodes that has allowed 802.11 to beat out the more complex, centrally coordinated access protocols of competing WLAN technologies such as HyperLAN.

The original (1997) 802.11 standard included three PHYs: infrared (IR), 2.4 GHz frequency hopped spread spectrum (FHSS), and 2.4 GHz direct sequence spread spectrum (DSSS). This was followed by two standard amendments in 1999: 802.11b built upon DSSS to increase the data rate in 2.4 GHz and 802.11a to create a new PHY in 5 GHz. 802.11b enhanced DSSS with complementary code keying (CCK), increasing the data rate to 11 Mbps. With higher data rates, IEEE 802.11b devices achieved significant market success, and markets for IR and FHSS PHYs did not materialize.

The development of 802.11a introduced orthogonal frequency division multiplexing (OFDM) to 802.11. Even though 802.11a introduced data rates of up to 54 Mbps, it is confined to the 5 GHz band and, as a result, adoption has been slow. New devices wishing to take advantage of the higher rates provided by 802.11a but retain backward compatibility with the huge installed base of 802.11b devices would need to implement two radios, one to operate using 802.11b in the 2.4 GHz band and one to operate using 802.11a in the 5 GHz band. Furthermore, international frequency regulations in the 2.4 GHz band uniformly allowed commercial use, whereas in 1999 and 2000 the non-military use of the 5 GHz band was limited to select channels in the United States.

Table 1.2 Overview of 802.11 PHYs

	802.11	802.11b	802.11a	802.11g	802.11n
PHY technology	DSSS	DSSS/CCK	OFDM	OFDM DSSS/CCK	SDM/OFDM
Data rates	1, 2 Mbps	5.5, 11 Mbps	6–54 Mbps	1–54 Mbps	6–600 Mbps
Frequency band	2.4 GHz	2.4 GHz	5 GHz	2.4 GHz	2.4 and 5 GHz
Channel spacing	25 MHz	25 MHz	20 MHz	25 MHz	20 and 40 MHz

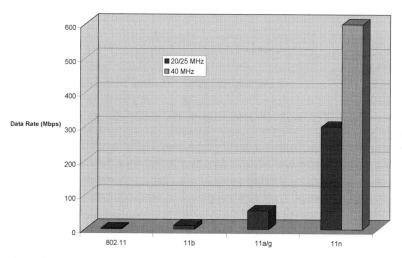

Figure 1.2 Increase in 802.11 PHY data rate.

In 2001, the FCC permitted the use of OFDM in the 2.4 GHz band. Subsequently, the 802.11 working group developed the 802.11g amendment, which incorporates the 802.11a OFDM PHY in the 2.4 GHz band, and adopted it as part of the standard in 2003. In addition, backward compatibility and interoperability is maintained between 802.11g and the older 802.11b devices. This allows for new 802.11g client cards to work in existing 802.11b hotspots, or older 802.11b embedded client devices to connect with a new 802.11g access point (AP). Because of this and new data rates of up to 54 Mbps, 802.11g has experienced large market success. A summary of the high level features of each PHY is given in Table 1.2.

With the adoption of each new PHY, 802.11 has experienced a five-fold increase in data rate. This rate of increase continues with 802.11n with a data rate of 300 Mbps in 20 MHz and 600 Mbps in 40 MHz. The exponential rate of increase in data rate is illustrated in Figure 1.2.

1.2   History of high throughput and 802.11n

1.2.1   The High Throughput Study Group

Interest in a high data rate extension to 802.11a began with a presentation to the Wireless Next Generation Standing Committee (WNG SC) of IEEE 802.11 in January 2002. Market drivers were outlined, such as increasing data rates of wired Ethernet, more data rate intensive applications, non-standard 100+ Mbps products entering the market, and the need for higher capacity WLAN networks (Jones, 2002). The presentation mentioned techniques such as spatial multiplexing and doubling the bandwidth as potential approaches to study for increasing data rate.

After many additional presentations, the High Throughput Study Group (HTSG) was formed with its first meeting in September 2002. The primary objective of HTSG was to complete two documents necessary for the creation of the High Throughput Task Group (TGn). These are the project authorization request (PAR) form and five criteria form. The PAR defined the scope and purpose of the task group as follows:

The scope of this project is to define an amendment that shall define standardized modifications to both the 802.11 physical layers (PHY) and the 802.11 medium access control layer (MAC) so that modes of operation can be enabled that are capable of much higher throughputs, with a maximum throughput of at least 100 Mbps, as measured at the MAC data service access point (SAP). IEEE (2006)

By this statement, the standard amendment developed by TGn must contain modes of operation that are capable of achieving at least 100 Mbps *throughput*. Throughput is the measure of "useful" information delivered by the system and by using throughput as the metric, both MAC and PHY overhead must be considered. 802.11a/g systems typically achieve a maximum throughput of around 25 Mbps; thus this statement required at least a four fold increase in throughput. Meeting this requirement would in essence mandate PHY data rates well in excess of 100 Mbps as well as significant enhancements to MAC efficiency.

Additional explanatory notes were included with the PAR outlining many evaluation metrics. These include throughput at the MAC SAP, range, aggregate network capacity, power consumption, spectral flexibility, cost complexity flexibility, backward compatibility, and coexistence (IEEE, 2006).

The five criteria form requires that the study group demonstrate the necessity of creating an amendment to the standard. The criteria include (1) broad market potential, (2) compatibility with existing IEEE 802.1 architecture, (3) distinct identity from other IEEE 802 standards, (4) technical feasibility, and (5) economic feasibility (Rosdahl, 2003). The goal is to create a standard amendment which results in marketable products, but that will also be differentiated from other potentially similar products.

In addition to completing the PAR and five criteria forms, HTSG also began development of new multipath fading MIMO channel models (Erceg *et al.*, 2004) and usage models (Stephens *et al.*, 2004). The channel models and usage models were used to create a common framework for simulations by different participants in the standard development process.

1.2.2   Formation of the High Throughput Task Group (TGn)

The PAR was accepted and approved by the 802 working group, creating Task Group n (TGn) with the first meeting of the task group held in September 2003. The standard amendment developed by the task group would be proposal driven, meaning that members of the task group would make partial or complete technical proposals, with the complete proposals proceeding through a down-selection process culminating in a single proposal upon which the standard amendment would be based. Partial proposals would be informative and could be incorporated in a complete proposal along the way. To that end, the task group began development of the functional requirements (Stephens, 2005) and comparison criteria (Stephens, 2004) documents. These two documents would provide, respectively, the technical requirements complete proposals must meet and criteria by which complete proposals would be compared.

The task group began with nine functional requirements. One of the functional requirements was a catch-all, requiring that proposals meet the PAR and five criteria. A second requirement was a reiteration of the PAR requirement to achieve 100 Mbps throughput at the top of the MAC. Furthermore, since it was expected that not all regulatory domains would allow a single device to use multiple 20 MHz channels (an easy way to achieve the throughput objective), the second requirement added a restriction that 100 Mbps throughput be achieved in a single 20 MHz channel. To enforce efficient use of spectrum, another requirement was added for a mode of operation with a spectral efficiency of at least 3 bps/Hz.

Four functional requirements addressed operational bands and backward compatibility. One of these requirements was that the protocol should support operation in the 5 GHz band due to the large availability of spectrum there. Another requirement was that at least some modes of operation be backward compatible with 802.11a systems. Noteworthy was the fact that there was no requirement to support operation in the 2.4 GHz band. However, if a proposal did support 2.4 GHz band operation, it was required that there be modes of operation that were backward compatible with 802.11g systems. In this context, some flexibility was given, allowing an 802.11n AP to be configured to accept or reject associations from legacy stations.

The 802.11e amendment to the standard, nearing completion at the time, added many features for improving the quality of service (QoS) in 802.11 systems. Many of the perceived applications for 802.11n involved real time voice and video which necessitate QoS. Therefore a functional requirement was included which mandated that a proposal allow for the implementation of 802.11e features within an 802.11n station.

The comparison criteria in Stephens (2004) outlined metrics and required disclosure of results which would allow for comparison between proposals under the same simulation setup and assumptions. The comparison criteria incorporated the simulation scenarios and usage models defined in Stephens et al. (2004). During the development of the comparison criteria, the task group realized that members of the task group did not always share the same definitions for common terms. Therefore definitions for goodput, backward compatibility, and signal-to-noise ratio (SNR) were provided. The comparison criteria covered four main categories: marketability, backward compatibility and coexistence with legacy devices, MAC related criteria, and PHY related criteria.

Under marketability, the proposal must provide goodput results for residential, enterprise, and hotspot simulation scenarios. Goodput is defined by totaling the number of bits in the MAC service data units (MSDU) indicated at the MAC service access point

(SAP), and dividing by the simulation duration (Stephens, 2004). Two optional criteria included describing the PHY and MAC complexity. The PHY complexity was to be given relative to 802.11a.

To ensure backward compatibility and coexistence with legacy devices, a proposal was required to provide a summary of the means used to achieve backward compatibility with 802.11a and, if operating in 2.4 GHz, 802.11g. Simulation results demonstrating interoperability were also required. The goodput of a legacy device in an 802.11n network and the impact of a legacy device on the goodput of 802.11n devices were also to be reported.

The MAC related criteria included performance measurements and changes that were made to the MAC. In the residential, enterprise, and hotspot simulation scenarios a number of different metrics were to be captured and reported. These included the ability to support the service requirements of various applications, including QoS requirements. Measurements of aggregate goodput of the entire simulation scenario were required to indicate network capacity. MAC efficiency was to be provided, which is defined as the aggregate goodput divided by the average PHY data rate. To ensure reasonable range for the new modes of operation, throughput versus range curves were also to be provided.

The PHY related criteria included PHY rates and preambles, channelization, spectral efficiency, PHY performance, and PHY changes. In addition, the comparison criteria also defined PHY impairments to be used in combination with channel models for PHY simulations. Each proposal was required to generate simulation results for both additive white Gaussian noise (AWGN) and non-AWGN channels. Furthermore, simulation conditions to analyze the impact on packet error rate (PER) of carrier frequency offset and symbol clock offset were also defined.

1.2.3   Call for proposals

The TGn call for proposals was issued on May 17, 2004, with the first proposals presented in September 2004. Over the course of the process two main proposal teams emerged, TGn Sync and WWiSE (world wide spectral efficiency). The TGn Sync proposal team was founded by Intel, Cisco, Agere, and Sony with the objective of covering the broad range of markets these companies were involved in, including the personal computer (PC), enterprise, and consumer electronics markets. The WWiSE proposal team was formed by Broadcom, Conexant, and Texas Instruments. These semiconductor companies were interested in a simple upgrade to 802.11a for fast time to market. Many other companies were involved in the proposal process and most ended up joining one of these two proposal teams.

The key features of all the proposals were similar, including spatial division multiplexing and 40 MHz channels for increased data rate, and frame aggregation for improved MAC efficiency. The proposals differed in scope (TGn Sync proposed numerous minor improvements to the MAC while WWiSE proposed limiting changes) and support for advanced features such as transmit beamforming (initially absent from the WWiSE proposal).

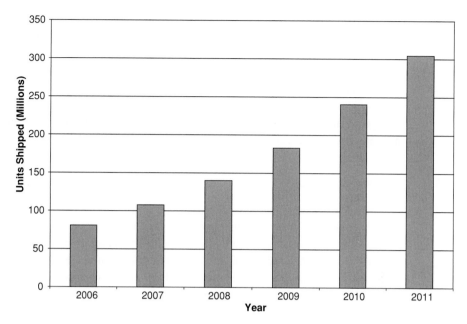

Figure 1.3 Worldwide converged mobile device shipments. Source: IDC (2007).

A series of proposal down-selection and confirmation votes took place between September 2004 and May 2005. During that time, mergers between proposals and enhancements to proposals took place. The TGn Sync proposal won the final down-selection vote between it and WWiSE, but failed the confirmation vote in May 2005.

1.2.4 Handheld devices

During this period interest arose in a new emerging market of converged Wi-Fi and mobile handsets. The shipment of dual mode Wi-Fi/cellular handsets had grown significantly from 2005 to 2006. Some participants in the proposal process believed that handsets would be the dominant Wi-Fi platform within a few years (de Courville *et al.*, 2005). A projected world wide growth of converged mobile devices was given in IDC (2007) and is illustrated in Figure 1.3.

A contentious issue for handheld proponents was the high throughput requirement for 100 Mbps throughput. This, in essence, would force all 802.11n devices to have multiple antennas. This is a difficult requirement for converged mobile devices, since they already contain radios and antennas for cellular 2G, 3G, Bluetooth, and in some cases GPS. Concern was raised that mandating 802.11n devices to have multiple antennas would force handset manufacturers to continue to incorporate single antenna 802.11a/g into handsets and not upgrade to 802.11n. Not only does this diminish the capabilities of the handset device, it burdens all future 802.11n deployments with continued coexistence with 802.11a/g embedded in these new handset devices.

For this reason an ad hoc group was formed to create functional requirements supporting single antenna devices. Two new requirements were added to the functional requirements document in July 2005. The first requirement mandated that a proposal define single antenna modes of operation supporting at least 50 Mbps throughput in a 20 MHz channel. The second requirement dictated that an 802.11n AP or station interoperate with client devices that comply with 802.11n requirements but incorporate only a single antenna. This requirement resulted in 802.11n making mandatory at least two antennas in an AP, but only one antenna in a non-AP device.

1.2.5   Merging of proposals

After the failed confirmation vote, a joint proposal effort was started within the task group to merge the two competing proposals. Due to entrenched positions and the large membership of the group, the joint proposal effort proceeded very slowly. As a result, Intel and Broadcom formed the Enhanced Wireless Consortium (EWC) in October 2005 to produce a specification outside the IEEE that would bring products to market faster. With much of the task group membership ultimately joining the EWC, this effort had the effect of breaking the deadlock within the IEEE, and the EWC specification, which was essentially a merger of the TGn Sync and WWiSE proposals, was adopted as the joint proposal and submitted for confirmation to TGn where it was unanimously passed in January 2006.

1.2.6   802.11n amendment drafts

The joint proposal was converted to a draft 802.11 standard amendment for higher throughput (TGn Draft 1.0), and entered letter ballot. In letter ballot, IEEE 802.11 working group members (not just task group members) vote to either adopt the draft as is or reject it with comments detailing changes needed. The draft requires at least a 75% affirmative vote within the IEEE 802.11 working group in order to proceed to sponsor ballot where it is considered for adoption by the broader IEEE standards association. TGn Draft 1.0 entered letter ballot in March 2006 and, not unusually, failed to achieve the 75% threshold for adoption. Comment resolution began May 2006 on the roughly 6000 unique technical and editorial comments submitted along with the votes.

With resolution of the TGn Draft 1.0 comments, TGn Draft 2.0 went out for letter ballot vote in February 2007 and this time passed with 83% of the votes. However, there were still 3000 unique technical and editorial comments accompanying the letter ballot votes. It is typical for the task group to continue comment resolution until a minimum number of negative votes are received; thus comment resolution for TGn Draft 2.0 continued between March 2007 and September 2007, resulting in TGn Draft 3.0. Since TGn Draft 2.0 passed, TGn Draft 3.0 and possible later drafts only require a recirculation ballot in which comments may only address clauses that changed between the drafts.

At the time this book went to press, the standard amendment was in recirculation ballot and would continue there until a minimum number of negative votes and comments were received. It will then proceed to sponsor ballot. Whereas letter ballot includes only voting members in IEEE 802.11, the sponsor ballot pool may include members from all of the IEEE 802 standard association, providing a broader review of the draft.

The IEEE 802.11n standard amendment is expected to be completed with final IEEE 802 working group and executive committee approval in March 2009. The IEEE standards board is expected to approve the amendment in June 2009. Publication of the amendment would occur shortly after.[1]

The submissions contributed by the task group participants in the development of the standard are publicly available on the World Wide Web.[2] The drafts of the standard are only available to voting members of 802.11, but, draft 2.0 of 802.11n was released to the public. Draft 3.0 is currently available for purchase from the IEEE.[3] Once approved, the final standard amendment would be available there as well.

Of great value to someone investigating the 802.11n PHY is the transmit waveform generator developed by Metalink. The description of the generator, developed in MATLAB®, is given in Anholt and Livshitz (2006). The actual source code is publicly available and is included in Anholt and Livshitz (2007). Most, if not all, transmit waveform features are supported by the generator.

1.3   Environments and applications for 802.11n

The basic service set (BSS) is the basic building block of an 802.11 LAN. Stations that remain within a certain coverage area and form some sort of association form a BSS. The most basic form of association is where stations communicate directly with one another in an ad-hoc network, referred to as an independent BSS or IBSS. This is illustrated as BSS 1 in Figure 1.4.

More typically though, stations associate with a central station dedicated to managing the BSS and referred to as an access point (AP). A BSS built around an AP is called an infrastructure BSS and is illustrated by BSS 2 and BSS 3 in Figure 1.4. Infrastructure BSSs may be interconnected via their APs through a distribution system (DS).

The BSSs interconnected by a DS form an extended service set (ESS). A key concept of the ESS is that stations within the ESS can address each other directly at the MAC layer. The ESS, being an 802.11 concept, encompasses only the 802.11 devices and does not dictate the nature of the DS. In practice, however, the DS is typically an Ethernet (802.3) LAN and the AP functions as an Ethernet bridge. As such, stations in a BSS can also directly address stations on the LAN at the MAC layer.

In the development of 802.11n, three primary environments were considered for studying system performance and capabilities: residential, enterprise, and hotspot. Within each of these environments a different mix of existing and new applications was envisioned. Use cases were defined (Stephens et al., 2004) that describe how an end user uses an

---

[1] The reader is referred to http://grouper.ieee.org/groups/802/11/Reports/802.11_Timelines.htm for the latest update on the timeline of 802.11n.

[2] http://grouper.ieee.org/groups/802/11/

[3] http://standards.ieee.org/getieee802/

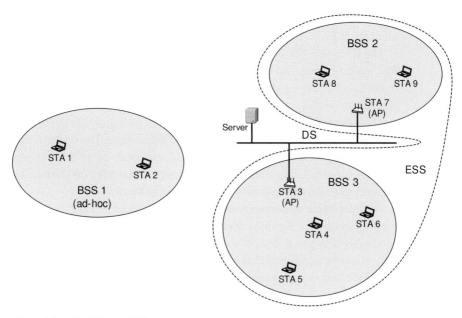

Figure 1.4 BSS, DS, and ESS concepts.

application in a specific WLAN environment. Examples include watching television remotely from the cable or set-top box within the home, or talking on the telephone remotely from one's desk at work. Additionally, usage models were developed for each environment which combined multiple use cases and applications.

Finally, a simulation scenario was created for each usage model and environment. These simulation scenarios were used to stress MAC capability when comparing proposals. Each simulation scenario includes a channel model associated with the particular environment. In addition, the location of the AP and stations were defined, giving a spatial component to the usage model in terms of the distance between the AP and the stations. For each application, system parameters were defined such as packet size, maximum packet loss rate tolerated by the application, maximum delay, network-layer protocol running (e.g. UDP or TCP), and offered load.

The residential usage model consists of a single BSS, as illustrated in Figure 1.5. This model typically includes only one AP and many client stations. In a typical AP-station configuration, applications include Internet access and streaming audio and video. Furthermore, user experience with applications like intra-networking for local file transfer, backups, and printing is enhanced with higher data rates. New applications such as voice over IP (VoIP) and video phones were also incorporated into the residential usage model.

The high throughput task group envisioned an AP that could also take the form of a wireless home media gateway. Such a device would distribute audio and video content throughout the home, such as DVD and standard and high definition TV. Other residential applications benefiting from higher wireless data rates include content download from

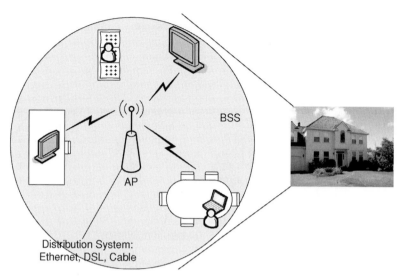

Figure 1.5 Residential usage model.

a video camera or photo camera. Interactive gaming has recently begun to incorporate wireless technology. Gamers benefit from the freedom of not being tethered by wires when the connections between the controller and the console, the console and the display, and the console to internet access are made wireless.

The usage model for an enterprise environment emphasizes network connectivity supported by multiple BSSs to cover larger buildings and floor plans, as illustrated in Figure 1.6. The BSSs are interconnected via the distribution system, typically Ethernet, creating an extended service set (ESS). As in a cellular deployment, each additional AP increases the total network coverage and capacity.

Networking applications such as file transfer and disk backup will benefit greatly from the higher data rates of 802.11n. Higher data rates will increase network capacity providing support for a larger number of clients. Higher throughputs will also enable new applications such as remote display via a wireless connection between a laptop and projector, simplifying presentations in conference rooms. Additionally, wireless video conferencing and VoIP may be supported (Stephens *et al.*, 2004).

The hotspot model envisions locations such as an airport lounge (illustrated in Figure 1.7), coffee shop, library, hotel, or convention center. Some municipalities have also blanketed downtown areas with Wi-Fi coverage. A hotspot could be located either indoors or outdoors and could cover a large open area. Therefore the propagation model could be substantially different from either residential or enterprise. In a hotspot, most traffic goes through the Internet and a session is typically limited to less than two hours (Stephens *et al.*, 2004). Applications include web browsing, Internet file transfer, and email. Also, new hotspot applications are envisioned such as the ability to watch a TV program or movie on a laptop or other display. This would involve the streaming of audio

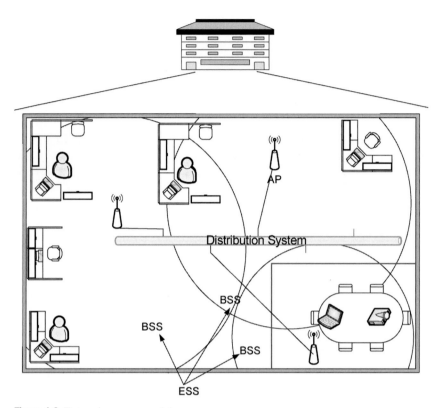

Figure 1.6 Enterprise usage model.

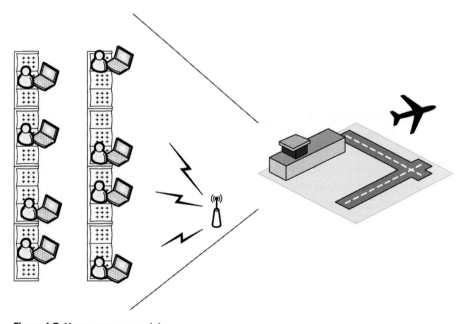

Figure 1.7 Hotspot usage model.

and video content over the Internet or the redistribution of standard or high definition TV signals.

1.4   Major features of 802.11n

PHY data rates in 802.11n are significantly improved over 802.11a and 802.11g primarily through the use of spatial multiplexing using MIMO and 40 MHz operation. To take advantage of the much higher data rates provided by these techniques, MAC efficiency is also improved through the use of frame aggregation and enhancements to the block acknowledgment protocol. These features together provide the bulk of the throughput enhancement over that achievable with 802.11a and 802.11g.

Robustness is improved inherently through the increased spatial diversity provided by the use of multiple antennas. Space-time block coding (STBC) as an option in the PHY further improves robustness, as does fast link adaptation, a mechanism for rapidly tracking changing channel conditions. More robust channel codes are adopted in the form of low density parity check (LDPC) codes. The standard amendment also introduces transmit beamforming, with both PHY and MAC enhancements to further improve robustness.

A number of other enhancements provide further gains. In the PHY, these include a shorter guard interval, which may be used under certain channel conditions. The PHY also includes a Greenfield preamble, which is shorter than the mandatory mixed format preamble. However, unlike the mixed format, it is not backward compatible with existing 802.11a and 802.11g devices without MAC protection. In the MAC, the reverse direction protocol provides a performance improvement for certain traffic patterns, by allowing a station to sublease the otherwise unused portion of its allocated transmit opportunity to its remote peer and thus reducing overall channel access overhead. A reduced interframe space (RIFS) used when transmitting a burst of frames reduces overhead in comparison to the existing short interframe space (SIFS).

An overview of the mandatory and optional features of the 802.11n PHY is given in Figure 1.8. At the time this book went to press, two generations of so called pre-n or draft 2.0 products have been released. The first generation of products typically operate in the 2.4 GHz band only, with up to two spatial streams and 40 MHz channel width. In this book the term spatial streams is used to refer to one or more independent data streams transmitted from the antennas. A device requires at least as many antennas as spatial streams. When using the short guard interval, these initial products are able to achieve a PHY data rate of 300 Mbps. With second generation products, we begin to see dual-band 2.4 GHz and 5 GHz products. These products also achieve 300 Mbps, but several incorporate an extra receive antenna chain for additional receive diversity. Some products also support the Greenfield preamble format. We expect that third generation devices will add another transmit antenna chain to support three spatial streams and

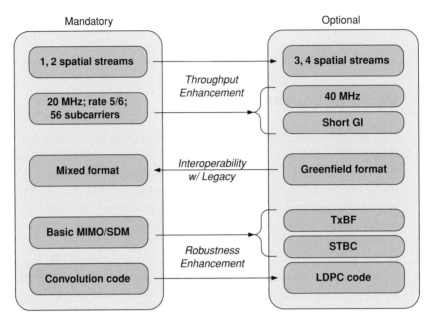

Figure 1.8 Mandatory and optional 802.11n PHY features.

450 Mbps. For robustness, these devices may begin employing STBC and transmit beamforming (TxBF).

An overview of the features added to the MAC in 802.11n is given in Figure 1.9. In addition to the throughput and robustness enhancing features already mentioned, the MAC is extended in a number of other areas.

The numerous optional features in 802.11n mean that extensive signaling of device capability is required to ensure coexistence and interoperability. For example, whether a device supports certain PHY features such as the Greenfield format preamble or MAC features such as the ability to participate in a reverse direction protocol exchange.

The existence of 40 MHz operation also creates a number of coexistence issues. The AP needs to manage the 40 MHz BSS so that 40 MHz and 20 MHz devices, both legacy and high throughput, are able to associate with the BSS and operate. Because 40 MHz operation uses two 20 MHz channels, mechanisms are needed to mitigate the effect this might have on neighboring 20 MHz BSSs operating independently on those two channels. Coexistence is primarily achieved through careful channel selection, i.e. choosing a pair of channels that have little or no active neighborhood traffic. To this end, the amendment adds scanning requirements to detect the presence of active neighborhood BSSs as well as the ability to actively move the BSS to another pair of channels should a neighboring 20 MHz BSS become active. If neighboring BSSs cannot be avoided then a fallback technique called phased coexistence operation (PCO) may be used. This allows the BSS to alternate between 20 MHz and 40 MHz phases of operation, with the 40 MHz phase entered after a frame exchange on the two 20 MHz channels has silenced devices operating there.

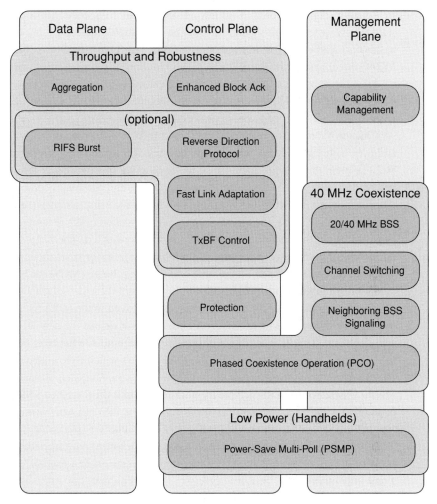

Figure 1.9 Summary of 802.11n MAC enhancements.

Finally, in recognition of the growing importance of handheld devices, a channel access scheduling technique called power-save multi-poll (PSMP) has been added to efficiently support a large number of stations.

1.5   Outline of chapters

The book is divided into three parts. The first part, Chapters 2–6, covers the PHY and provides a comprehensive review of all mandatory and optional PHY features in 802.11n. The second part, Chapters 7–11, covers the MAC, an overview of existing MAC features, and a detailed review of the new features introduced in 802.11n. Transmit beamforming is presented in the last part, Chapter 12, with a description of both PHY and MAC aspects.

Chapter 2 gives a brief overview of orthogonal frequency division multiplexing (OFDM). Chapter 3 begins with a description of multiple-input multiple-output (MIMO) basics and spatial division multiplexing (SDM). This is followed by a discussion of the MIMO environment and the 802.11n MIMO multipath fading channel and propagation models. The chapter concludes with an explanation of linear receiver design and highlights of maximum likelihood estimation. Included in the discussion of MIMO and receiver design are capacity based performance curves.

Chapter 4 details the design of the mixed format (MF) preamble used for interoperability with legacy 802.11a/g OFDM devices. The chapter begins with a review of 802.11a preamble design. Included in the review are illustrations of the waveform. A description of the 802.11a/g packet encoding process and receive procedure is presented, which includes a receiver block diagram. This leads into a discussion of the legacy part of the MF preamble. Next, the high throughput (HT) portion of the preamble is described. Following this, the encoding of the Data field is presented. The chapter ends with a discussion of the receive procedure and a block diagram for basic modes of operation. Tables of parameters of the modulation and coding scheme (MCS) for basic modes of operation are given in an appendix.

Chapter 5 outlines all the PHY techniques employed in the 802.11n specification to increase the data rate. The first section of the chapter details the new 40 MHz channel and waveform. This includes a plot illustrating throughput as a function of range. This is followed by a brief discussion of the extra subcarriers which were added to the 802.11n 20 MHz waveform. The next part of the chapter gives the MCS definitions. This includes several waterfall curves illustrating the packet error rate (PER) versus SNR performance in additive white Gaussian noise (AWGN) and in the 802.11n MIMO multipath fading channel models. A description of the shorter Greenfield (GF) preamble is also provided in this chapter. This also includes discussion on the debate of how much the GF preamble actually improves performance. The last topic covered in this chapter is on the short guard interval (GI).

Chapter 6 covers the subject of improving the robustness of the system. Four techniques are described. The first method is receive diversity, where PER versus SNR waterfall curves are provided along with throughput curves to demonstrate the gain achieved from receive diversity in a MIMO system. The next technique is a straightforward one involving spatial expansion (SE), which provides a small amount of transmit diversity gain. Waterfall curves are provided for SE as well. This is followed by a detailed description of space-time block coding (STBC). Transmit antenna configurations are presented, along with an approach for implementing a receiver and equalizer. Again performance curves are presented that illustrate which system configurations benefit the most from STBC. In the last part of the chapter, low density parity check (LDPC) codes are discussed. The specific characteristics of the LDPC encoding process in the 802.11n standard amendment are detailed. Waterfall curves for LDPC are provided to compare performance with the mandatory binary convolutional code.

The MAC section begins in Chapter 7 with a functional description of the 802.11 MAC as background for the remaining chapters. This chapter covers the basic contention-based access protocol including the 802.11e quality of service (QoS) extensions, channel access timing, the concept of a transmit opportunity, and the basic acknowledgement and block acknowledgement protocols.

Chapter 8 describes why changes are necessary in the MAC to improve throughput and then details the two key throughput enhancing features: aggregation and enhancements to the block acknowledgment protocol.

Beyond the basic contention-based access protocol, the 802.11 MAC includes additional channel access mechanisms. Chapter 9 provides an overview of these mechanisms, including the point coordination function (PCF) from the original 802.11 specification and the hybrid coordinated channel access (HCCA) function from the 802.11e amendment. The chapter then provides details on the power-save multi-poll (PSMP) channel access technique and the reverse direction protocol, both of which are new in 802.11n.

Coexistence and interoperability is a critical issue with 802.11n and Chapter 10 provides details on this broad topic. The chapter covers capability signaling and BSS control. The chapter then covers 40 MHz operation, managing 40 MHz BSS operation, and maintaining interoperability with legacy 20 MHz devices. The critical topic of 40 MHz coexistence with neighboring 20 MHz BSSs is also discussed. Finally, the chapter covers protection mechanisms.

To round out the MAC section, Chapter 11 provides details on MAC frame formats. This chapter is intended as a reference for the discussions in the other chapters.

The final part of the book deals with the complex topics of transmit beamforming and fast link adaptation. Chapter 12 provides details on both the PHY and MAC aspects of this topic.

References

ABIresearch (2007). *Wi-Fi IC Market Data*. 2007-01-31.

Anholt, M. and Livshitz, M. (2006). *Waveform Generator*, IEEE 802.11-06/1714r1.

Anholt, M. and Livshitz, M. (2007). *Waveform Generator Source Code*, IEEE 802.11-07/0106r0.

de Courville, M., Muck, M., van Waes, N., *et al.* (2005). *Handset Requirements for TGn*, IEEE 802.11-05/0433r0.

Erceg, V., Schumacher, L., Kyritsi, P., *et al.* (2004). *TGn Channel Models*, IEEE 802.11-03/940r4.

Foschini, G. J. (1996). Layered space-time architecture for wireless communication in a fading environment when using multi-element antennas. *Bell Labs Technical Journal*, Autumn, 41–59.

IDC (2007). *Worldwide Converged Mobile Device 2007–2001 Forecast and Analysis*, IDC #206585. 2007-05-01.

IEEE (2006) IEEE 802.11n Project Authorization Request, 26 May 2006 available at: http://standards.ieee.org/board/nes/projects/802-11n.pdf.

Jones, V. K, De Vegt, R., and Terry, J. (2002), *Interest for HDR Extension to 802.11a*, IEEE 802.11-02/081r0.

Rosdahl, J. (2003). *Criteria for Standards Development*, IEEE 802.11-02/799r6.

Stephens, A. (2004), *IEEE 802.11 TGn Comparison Criteria*, IEEE 802.11-03/814r31.

Stephens, A. (2005). *802.11 TGn Functional Requirements*, IEEE 802.11-03/813r13.

Stephens, A., Bjerke, B., Jechoux, B., *et al.* (2004). *Usage Models*, IEEE 802.11-03/802r23.

Teare, D. (1999). *Designing Cisco Networks*. Indianapolis, IN: Cisco Press.

Zimmerman, H. (1980). OSI reference model – The ISO model of architecture for open systems interconnection. *IEEE Transactions on Communications*, **COM-28**(4), 425–32.

Part I

Physical layer

2 Orthogonal frequency division multiplexing

2.1 Background

The 802.11n physical layer builds upon the 802.11a orthogonal frequency division multiplexing (OFDM) structure. OFDM is well suited to wideband systems in frequency selective fading environments. Only a few subcarriers are impacted by a deep fade or narrow band interference, which can be protected by forward error correction. Important in a high data rate system, OFDM is tolerant of time synchronization errors. In addition, OFDM is bandwidth efficient, since a nearly square power spectrum can be created with narrow subcarriers with each subcarrier supporting a constellation with many bits per symbol.

With frequency division multiplexing, signals are transmitted simultaneously on different subcarriers. Figure 2.1 illustrates an example with four subcarriers. The top two graphs in the figure depict the time domain waveform of each subcarrier individually for the real and imaginary part of each subcarrier. The bottom graph gives the composite time domain waveform.

Figure 2.2 illustrates the power spectrum of each subcarrier individually. If the subcarriers are separated in frequency (Δ_F) by the inverse of the symbol period (T), the nulls of adjacent subcarriers coincide with the peak of the main lobes of the subcarriers. With the construction of the waveform in this manner, the subcarriers are orthogonal.

A baseband OFDM waveform is constructed as an inverse Fourier transform of a set of coefficients X_k,

$$r(t) = \frac{1}{N} \sum_k X_k \exp\left(j2\pi k \Delta_F t\right) \quad 0 \le t < T \tag{2.1}$$

where Δ_F is the subcarrier frequency spacing and T is the inverse Fourier transform symbol period, with Δ_F equal to $1/T$, and N is the number of samples in the inverse Fourier transform. A set of modulated symbols is transmitted on subcarriers as the coefficients X_k. The inverse Fourier transform is commonly implemented by an IFFT.

In 802.11a, the fundamental sampling rate is 20 MHz, with a 64-point FFT/IFFT. The Fourier transform symbol period, T, is 3.2 µs in duration and Δ_F is 312.5 kHz. Of the 64 subcarriers in 802.11a, there are 52 populated subcarriers. Numbering the subcarrier locations as $-32, -31, \ldots, -1, 0, 1, \ldots, 31$, the populated subcarriers are located from $-26, -25, \ldots, -2, -1, 1, 2, \ldots, 25, 26$. That is, the lowest six subcarriers, the DC subcarrier, and the highest five subcarriers are not used.

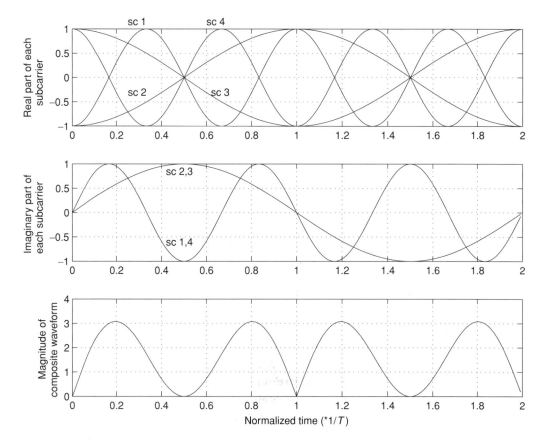

Figure 2.1 Frequency division multiplexing waveform.

Inter-symbol interference (ISI) between OFDM symbols that are adjacent in time degrades the orthogonality between the subcarriers and impairs performance. ISI may be caused by delay spread in the channel and filtering. To minimize the impact of ISI, a guard interval (GI) is added in between adjacent OFDM symbols. Therefore in 802.11a, the total symbol time is 4 μs. Whereby, the first 0.8 μs is the GI consisting of an exact copy of the last 0.8 μs of the OFDM symbol (otherwise known as cyclic extension), followed by the 3.2 μs OFDM symbol.

To extract the information from the received waveform, a Fourier transform is performed on the received signal, as given in Eq. (2.2):

$$X_k = \sum_n r(n) \exp\left(-j2\pi k \cdot {}^n\!/_N\right) \qquad (2.2)$$

Note the Fourier transform is represented in the discrete time domain, whereas the inverse Fourier transform in Eq. (2.1) is represented in the continuous time domain. We have assumed that the received waveform has been sampled prior to the Fourier transform.

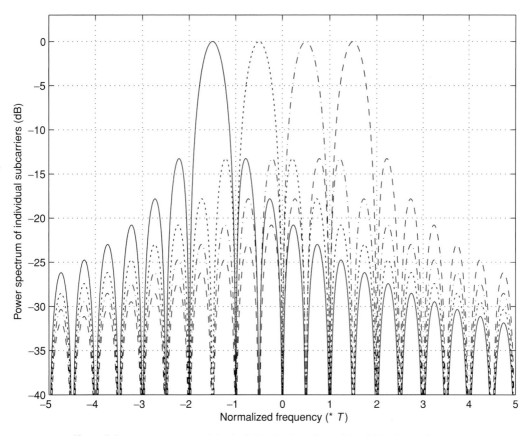

Figure 2.2 Power spectrum of the individual subcarriers of the OFDM waveform.

For an 802.11a OFDM waveform, a 4 μs block of time domain samples is processed at a time. A 64-point FFT (assuming the data is sampled at a 20 MHz sampling rate) is performed on a 3.2 μs subset of the time domain samples. The subset is selected to avoid ISI. Section 4.1 details the 802.11a waveform design and receive procedure.

Further general information on OFDM is provided in Halford (2001) and van Nee and Prasad (2000).

2.2 Comparison to single carrier modulation

With single carrier modulation, data bits are modulated and the pulses are transmitted sequentially in time. As described in the previous section, with OFDM, blocks of data bits are modulated onto subcarriers across the frequency band. An illustrative comparison between single carrier and OFDM is given in Figure 2.3.

An OFDM signal experiences large variation in magnitude when the individual subcarriers are combined into a single time domain waveform. The variation results in a high peak-to-average ratio, whereas the magnitude of the signal of a phase modulated

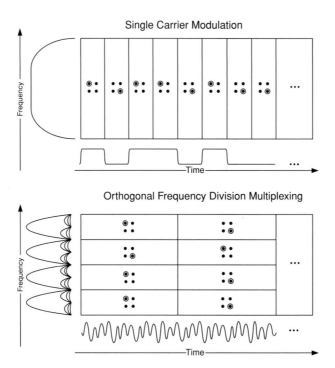

Figure 2.3 Comparison between single carrier modulation and orthogonal frequency division multiplexing.

single carrier waveform theoretically has a constant envelope. A properly designed single carrier system has little to no variation of the magnitude of the signal and a very small peak-to-average ratio.

This is one of the primary limitations of OFDM as compared to single carrier modulation. Due to the non-linear nature of a typical transmitter power amplifier, signals with a high peak-to-average ratio are distorted unless the amplifier is operated with enough power backoff to remain linear and avoid distortion. As a result, OFDM systems may have lower output transmit power or devices may require a larger power amplifier.

However, to achieve high data rates with limited bandwidth, designs of single carrier modulation systems inevitably utilize high order modulation with both amplitude and phase modulation. This results in the peak-to-average ratio of a single carrier waveform approaching that of an OFDM waveform. As such, typically there is only a small difference in the peak-to-average ratio of single carrier systems and OFDM.

The second limitation of OFDM when comparing to single carrier is higher sensitivity to carrier frequency offset and phase noise. In both OFDM and single carrier modulation, carrier frequency offset and phase noise causes unwanted phase variation in the modulated symbols. This requires phase tracking loops in either system to mitigate the degradation.

In addition, in OFDM, carrier frequency offset and phase noise cause the subcarriers to deviate from the $1/T$ spacing required for orthogonality, causing inter-carrier interference. In Pollet *et al.* (1995) an approximation for degradation due to carrier frequency offset (ε_f) is given for OFDM and single carrier, as follows:

$$
D \approx
\begin{cases}
\dfrac{10}{\ln 10}\dfrac{1}{3}(\pi\varepsilon_f T)^2 & \text{single carrier} \\[3mm]
\dfrac{10}{\ln 10}\dfrac{1}{3}(\pi\varepsilon_f T)^2\dfrac{E_s}{N_0} & \text{OFDM}
\end{cases}
\tag{2.3}
$$

Unlike single carrier, with OFDM the degradation is proportional to the signal-to-noise ratio E_s/N_0. For phase noise, an approximation for degradation is given in Pollet *et al.* (1995) as a function of the oscillator linewidth (β), as follows:

$$
D \approx
\begin{cases}
\dfrac{10}{\ln 10}\dfrac{1}{60}(4\pi\beta T)\dfrac{E_s}{N_0} & \text{single carrier} \\[3mm]
\dfrac{10}{\ln 10}\dfrac{11}{60}(4\pi\beta T)\dfrac{E_s}{N_0} & \text{OFDM}
\end{cases}
\tag{2.4}
$$

OFDM may be shown to be several orders of magnitude more sensitive to carrier frequency offset and phase noise than single carrier modulation. This is in part due to the symbol time (represented by T in Eqs. (2.3) and (2.4)) which for OFDM is much longer than for single carrier modulation, as illustrated in Figure 2.3.

The additional sensitivity of OFDM to carrier frequency offset and phase noise may appear to be problematic. However, the proliferation of low cost 802.11 OFDM devices demonstrates that the necessity of a more accurate oscillator does not significantly impact the overall cost of an OFDM device at these frequency bands.

As described in the previous section, OFDM is intrinsically resistant to frequency selective multipath fading. To combat frequency selectivity due to channels with long delay spread, a single carrier system requires an equalizer with a long tap-delay line. The implementation complexity based on number of multiplies, between the FFT in OFDM and the tap-delay line equalizer in single carrier, is shown in van Nee and Prasad (2000) to be ten times more complex for single carrier. In Clark (1998), a reduced complexity frequency domain equalizer for single carrier is proposed, as opposed to a tap-delay line equalizer. This technique requires both an FFT and IFFT as part of the receiver, which has twice the complexity of an OFDM receiver.

References

Clark, M. V., (1998). Adaptive frequency-domain equalization and diversity combining for broadband wireless communications. *IEEE Journal on Selected Areas in Communications*, **16**(8), 1385–95.

Halford, S. (2001). Implementing OFDM in wireless designs. *Communications Design Conference*, 2001-10-01.

Pollet, T., van Bladel, M., and Moeneclaey, M. (1995). BER sensitivity of OFDM systems to carrier frequency offset and Wiener phase noise. *IEEE Transactions on Communications*, **43**(2/3/4), 191–3.

van Nee, R. and Prasad, R. (2000). *OFDM for Wireless Multimedia Communications*, Boston, MA: Artech House.

3   MIMO/SDM basics

3.1   SISO (802.11a/g) background

A basic communication system is described as

$$y = \sqrt{\rho} \cdot h \cdot x + z \qquad (3.1)$$

where x is the transmitted data with unity mean expected power, h is the channel fading coefficient, z is independent, complex additive white Gaussian noise (AWGN) with zero mean and unit variance, ρ is the average signal-to-noise ratio (SNR), and y is the received signal. Typically h is modeled as Rayleigh fading and is defined as complex, zero mean, unit variance Gaussian distribution: Normal $(0, 1/\sqrt{2}) + \sqrt{-1} \cdot$ Normal $(0, 1/\sqrt{2})$. The notation Normal (x, y) defines a Gaussian distributed random variable with a mean of x and a variance of y.

In this model, both the transmitter and receiver are configured with one antenna. This is termed single-input, single-output (SISO), with reference to a single input to the environment and single output from the environment, as illustrated in Figure 3.1.

The capacity for a general SISO system is given by the Shannon capacity formula in Eq. (3.2) (Foschini and Gans, 1998):

$$C\,(\text{bps/Hz}) = \log_2(1 + \rho \cdot |h|^2) \qquad (3.2)$$

In the frequency domain (after the FFT), each subcarrier may be described by Eq. (3.1). Subsequently, the receiver extracts the information by equalizing the received signal, as follows:

$$\begin{aligned} \hat{x} &= (\sqrt{\rho} \cdot h)^{-1} \cdot y \\ &= x + (\sqrt{\rho} \cdot h)^{-1} \cdot z \end{aligned} \qquad (3.3)$$

where $\hat{x}$ is the noisy estimate of the transmitted signal x.

3.2   MIMO basics

Multiple-input, multiple-output (MIMO) describes a system with a transmitter with multiple antennas transmitting through the propagation environment to a receiver with multiple receive antennas, as illustrated in Figure 3.2.

Figure 3.1 SISO system.

Figure 3.2 MIMO system.

In such a system the transmitter may use transmit beamforming (or an adaptive antenna array) to improve the quality of the link (Perahia and Pottie, 1996). An example of a system with a two antenna array transmitter and a single antenna receiver is described mathematically, and in vector notation, in Eq. (3.4):

$$
\begin{aligned}
y &= \sqrt{\rho} \cdot h_1 \cdot w_1 \cdot x + \sqrt{\rho} \cdot h_2 \cdot w_2 \cdot x + z \\
&= \sqrt{\rho} \, [h_1 \quad h_2] \begin{bmatrix} w_1 \\ w_2 \end{bmatrix} x + z
\end{aligned}
\tag{3.4}
$$

This example is more accurately described as multiple-input, single-output. In this example, the received signal is the composite of signals transmitted by the two antennas, with the weight w_1 used to manipulate the data x from the first antenna, and w_2 used to manipulate the data x from the second antenna. The beamforming weights, w_1 and w_2, are applied at the transmitter before transmission. The transmitted signal from the first antenna passes through the environment modeled by h_1, and the transmitted signal from the second antenna passes through the environment modeled by h_2.

One may also improve the robustness of the link by adding extra antennas to the receiver with diversity combining, termed single-input, multiple-output. An example of a two receive antenna array system is described mathematically and in vector notation in Eq. (3.5):

$$
\begin{aligned}
y_1 &= \sqrt{\rho} \cdot h_1 \cdot x + z_1 \\
y_2 &= \sqrt{\rho} \cdot h_2 \cdot x + z_2 \\
\begin{bmatrix} y_1 \\ y_2 \end{bmatrix} &= \sqrt{\rho} \begin{bmatrix} h_1 \\ h_2 \end{bmatrix} x + \begin{bmatrix} z_1 \\ z_2 \end{bmatrix}
\end{aligned}
\tag{3.5}
$$

With this type of system a maximal-ratio combining (MRC) receiver could be used to optimally combine the received signals (Jakes, 1974). The output of the maximal ratio combiner is given in Eq. (3.6). Note that perfect knowledge of the channel has been assumed.

$$r = \begin{bmatrix} h_1^* & h_2^* \end{bmatrix} \begin{bmatrix} y_1 \\ y_2 \end{bmatrix}$$

$$= \sqrt{\rho} \cdot (|h_1|^2 + |h_2|^2) \cdot x + \begin{bmatrix} h_1^* & h_2^* \end{bmatrix} \begin{bmatrix} z_1 \\ z_2 \end{bmatrix} \tag{3.6}$$

After the MRC function, the received signal may be equalized as follows:

$$\hat{x} = \frac{1}{\sqrt{\rho}} \cdot \frac{1}{|h_1|^2 + |h_2|^2} \cdot r$$

$$= x + \frac{1}{\sqrt{\rho}} \cdot \frac{1}{|h_1|^2 + |h_2|^2} \cdot \begin{bmatrix} h_1^* & h_2^* \end{bmatrix} \begin{bmatrix} z_1 \\ z_2 \end{bmatrix} \tag{3.7}$$

These examples of transmit beamforming and receive diversity combining illustrate two antennas. However, these techniques are not limited to two antennas and extension to three, four, or more antennas is straightforward.

Using the combination of transmit beamforming and receive diversity combining improves robustness of the link in what some call a MIMO system. These techniques have been used with IEEE 802.11a/g to increase the range of a given data rate. However, they do not increase the maximum data rate beyond 54 Mbps, the maximum data rate in 802.11a/g.

3.3   SDM basics

A MIMO system may be used to transmit independent data streams (or spatial streams) on different antennas. We define spatial streams as streams of bits transmitted over separate spatial dimensions. When multiple spatial streams are used with MIMO this is termed spatial division multiplexing (SDM), and is illustrated in Figure 3.3. In the previous example of Figure 3.2 note that each antenna transmits the same sequence x, whereas in Figure 3.3 x_1 and x_2 represent independent streams.

Thus, with MIMO/SDM, the maximum data rate of the system increases as a function of the number of independent data streams. The system must contain the same number or more transmit (Tx) antennas as data streams. With a linear receiver, the system must

Figure 3.3 MIMO/SDM system.

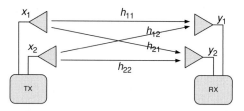

Figure 3.4 Mathematical model of the MIMO/SDM system.

contain the same number or more receive (Rx) antennas as data streams. In other words the data rate of the system increases by min(Tx antennas, Rx antennas, data streams).

A more detailed graphical view of a MIMO/SDM system with two transmit antennas (and two spatial streams) and two receive antennas is given in Figure 3.4. The notation used to describe such systems is "number of Tx antennas" × "number of Rx antennas;" thus Figures 3.3 and 3.4 represent a 2 × 2 MIMO/SDM system.

An M × N MIMO/SDM system is represented more generally by Eq. (3.8). Here it has been assumed that the total transmit power is equally divided over the M transmit antennas:

$$
\begin{aligned}
y_1 &= \sqrt{P/M} \cdot h_{11} \cdot x_1 + \sqrt{P/M} \cdot h_{12} \cdot x_2 + \cdots + \sqrt{P/M} \cdot h_{1M} \cdot x_M + z_1 \\
y_2 &= \sqrt{P/M} \cdot h_{21} \cdot x_1 + \sqrt{P/M} \cdot h_{22} \cdot x_2 + \cdots + \sqrt{P/M} \cdot h_{2M} \cdot x_M + z_2 \\
&\;\;\vdots \quad \vdots \hspace{8cm} \vdots \\
y_N &= \sqrt{P/M} \cdot h_{N1} \cdot x_1 + \sqrt{P/M} \cdot h_{N2} \cdot x_2 + \cdots + \sqrt{P/M} \cdot h_{NM} \cdot x_M + z_N
\end{aligned}
$$

$$(3.8)$$

The set of equations in Eq. (3.8) may be converted to vector form, as shown in Eq. (3.9):

$$
\begin{bmatrix} y_1 \\ y_2 \\ \vdots \\ y_N \end{bmatrix} = \sqrt{P/M} \begin{bmatrix} h_{11} & h_{12} & \cdots & h_{1M} \\ h_{21} & h_{22} & \cdots & h_{2M} \\ \vdots & \vdots & \ddots & \vdots \\ h_{N1} & h_{N2} & \cdots & h_{NM} \end{bmatrix} \begin{bmatrix} x_1 \\ x_2 \\ \vdots \\ x_M \end{bmatrix} + \begin{bmatrix} z_1 \\ z_2 \\ \vdots \\ z_N \end{bmatrix}
$$

$$(3.9)$$

We further simplified Eq. (3.9) with matrix notation, as given in Eq. (3.10):

$$
Y_N = \sqrt{P/M} \cdot H_{N \times M} X_M + Z_N \tag{3.10}
$$

where

$$
Y_N = \begin{bmatrix} y_1 \\ y_2 \\ \vdots \\ y_N \end{bmatrix}, \quad X_M = \begin{bmatrix} x_1 \\ x_2 \\ \vdots \\ x_M \end{bmatrix}, \quad Z_N = \begin{bmatrix} z_1 \\ z_2 \\ \vdots \\ z_N \end{bmatrix}, \quad H_{N \times M} = \begin{bmatrix} h_{11} & h_{12} & \cdots & h_{1M} \\ h_{21} & h_{22} & \cdots & h_{2M} \\ \vdots & \vdots & \ddots & \vdots \\ h_{N1} & h_{N2} & \cdots & h_{NM} \end{bmatrix}
$$

With OFDM, each subcarrier is described by Eq. (3.10).

A generalization of the Shannon capacity formula for M transmit antennas and N receive antennas is given by Eq. (3.11) (Foschini, 1996):

$$C\,(\text{bps/Hz}) = \log_2[\det(I_N + {}^\rho/_M \cdot HH^*)] \tag{3.11}$$

where H^* is the conjugate transpose of H.

In matrix notation, the MIMO/SDM system described in Eq. (3.10) has a similar mathematical structure to the SISO system given in Eq. (3.1). Conceptually, the equalizer for the MIMO/SDM system may also be designed with a similar structure as the SISO equalizer in Eq. (3.3). In matrix form, the received signal is divided by the channel to extract the estimate of the transmitted data, given in Eq. (3.12):

$$\hat{X} = \left(\sqrt{{}^\rho/_M} \cdot H\right)^{-1} Y$$
$$= X + \left(\sqrt{{}^\rho/_M} \cdot H\right)^{-1} Z \tag{3.12}$$

The equalizer is given in Eq. (3.12) as a matrix inversion for simplicity, but is actually only valid where $M = N$. The representation of the data estimate is generalized for $M = N$ and $M \neq N$ as follows:

$$\hat{X} = \frac{H^*}{\sqrt{{}^\rho/_M} \cdot H^*H} Y$$
$$= X + \frac{H^*}{\sqrt{{}^\rho/_M} \cdot H^*H} Z$$

It almost seems too good to be true that with such simplicity one may increase the data rate of a system by merely adding additional transmit and receive antennas. Through the use of an example it is shown that there are bounds on this increase. Consider the SISO system of Eq. (3.1), where now the antenna port of the transmitting device is directly connected to the antenna port of the receiving device. Such is a typical setup for conductive testing. Mathematically, h is equal to 1, and $y = \sqrt{\rho} \cdot x + z$. With enough transmit power, the system is very reliable. Now consider a 2×2 MIMO/SDM system with each antenna port of the transmitter device connected to each antenna port of the receiver device. The channel H for such system is equal to $\begin{bmatrix} 1 & 1 \\ 1 & 1 \end{bmatrix}$. To extract the transmitted data from the receive channel, we must perform an inverse of the channel matrix. In this case, the matrix is singular and cannot be inverted. The receiver fails, even in high SNR. As we see, the channel matrix must be well conditioned for the MIMO system to achieve large increases in data rate from spatial multiplexing.

3.4   MIMO environment

If the terms in the channel matrix are randomly distributed and uncorrelated with each other, statistically the channel matrix is well conditioned and invertible the vast majority of the time. Fortunately, a fading multipath channel creates such an environment.

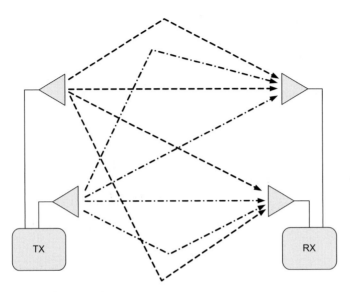

Figure 3.5 Multipath fading environment.

In an indoor environment, rays bounce off floors, ceilings, walls, furniture, etc. when propagating between the transmitter and the receiver (Rappaport, 1996). The paths of the rays are different as a function of the location of the transmit and receive antennas, as illustrated in Figure 3.5. How the paths propagate determines the amount of fading due to cancellation, delay, and correlation.

With a channel matrix having terms which are complex Gaussian distributed, uncorrelated with each other, and the same channel matrix across all subcarriers, a quantitative benefit of MIMO is realized. In a wideband OFDM system such as 802.11n, such a model represents a flat Rayleigh fading across the band. This is not very realistic, but allows for a simple analytical comparison between SISO and MIMO capacity. A more realistic channel model for 802.11n is provided in Section 3.5.

In Figure 3.6 a 2×2 MIMO system is compared with a SISO system and demonstrates the large gain that is obtained with MIMO systems. The complementary cumulative distribution function (CCDF) is plotted versus capacity to show how often the capacity exceeds a certain level. Capacity of each system is simulated by Eq. (3.11) at an SNR equal to 20 dB. For reference the maximum data rate of 802.11a is 54 Mbps in a 20 MHz channel, resulting in a spectral efficiency of 2.7 bps/Hz. In a flat fading environment, with ideal conditions (i.e. no hardware impairments, synchronization loss, etc.) such a capacity is achieved 95% of the time. At a comparable percentage, a 2×2 MIMO system achieves three times the capacity. In reality, various impairments and realistic channel environments inhibit the ability to increase the capacity by that amount in all situations. But in subsequent chapters it is shown that a two stream 802.11n system has a maximum data rate which is 2.8 times that of 802.11a in a 20 MHz channel, so this three fold increase can be approached in some circumstances.

A significant impact on MIMO capacity is antenna correlation. As the correlation increases, the condition of the channel matrix degrades. To ensure that the channel is

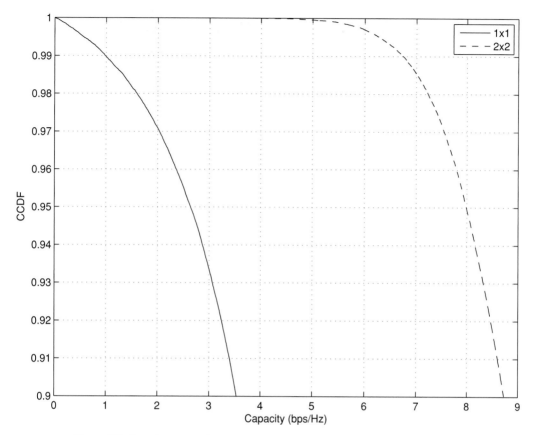

Figure 3.6 Capacity comparison between 2×2 MIMO and SISO with SNR $= 20\,$dB.

uncorrelated "enough," an antenna spacing of at least half a wavelength is generally used. In addition, the antennas should be designed with low mutual coupling between the elements to minimize channel correlation. A non-line-of-sight (NLOS) environment may also reduce correlation since highly correlated direct paths between the transmit and receive devices are not present. However, this typically means a reduction in SNR at the receiver.

Performance of a MIMO/OFDM system varies with many environmental conditions: LOS or NLOS based on obstructions, delay spread from delayed responses, antenna correlation, and Doppler from mobility. These aspects of the environment are addressed in Section 3.5 where the 802.11n channel model is presented.

3.5   802.11n propagation model

A set of channel models and path loss models were created during the development of the 802.11n standard (Erceg *et al.*, 2004). Due to the nature of typical WLAN deployments, the propagation models were developed based mostly on indoor measurements. A key

Table 3.1 Channel models

Model	RMS delay Spread (ns)	Environment	Example
A	0	N/A	N/A
B	15	Residential	Intra-room, room-to-room
C	30	Residential/small office	Conference room, classroom
D	50	Typical office	Sea of cubes, large conference room
E	100	Large office	Multi-story office, campus small hotspot
F	150	Large space (indoors/outdoors)	Large hotspot, industrial, city square

component of the channel models which affects MIMO system performance is the correlation matrix.

The channel models are composed of impulse responses for a range of indoor environments. Measurement data from both the 2.4 GHz and 5 GHz frequency bands were incorporated to develop a set of impulse responses and antenna correlation matrices applicable to both bands. The channel model is a function of frequency, as well as Doppler.

3.5.1   Impulse response

There are six channel models, each with a different impulse response corresponding to different indoor environments (Stephens *et al.*, 2004). An overview of the models is given in Table 3.1.

Channel model A is flat fading Rayleigh, and is not applicable to a wideband system like 802.11. This channel model was only included for its utility in analytic modeling. The most commonly used channel models are B, D, and E. These three models were largely used to compare proposals during the development of the 802.11n standard as outlined in Stephens (2004). These were also used for physical layer (PHY) packet error rate (PER) performance. In addition, the three simulation scenarios selected for MAC network simulations in Stephens (2004), residential, large enterprise, and hot spot, also use channel models B, D, and E, respectively, as described in Stephens *et al.* (2004).

The designs of the impulse responses are based on the cluster model initially developed by Saleh and Valenzuela (1987). In Figure 3.7, the vertical bars indicate a delayed response. These are located on 10 ns intervals. The lines indicate the extent of each cluster. Notice that the clusters overlap, and the delayed responses during the overlap are a composite of the power from the overlapping clusters. In addition, the power of the delayed responses of each cluster decays linearly on a log-scale. The power delay profile, separated into clusters, is given in Appendix 3.1 for each channel model. To compute the power for each tap at each delay, the powers of the taps in overlapping clusters are summed at each delay.

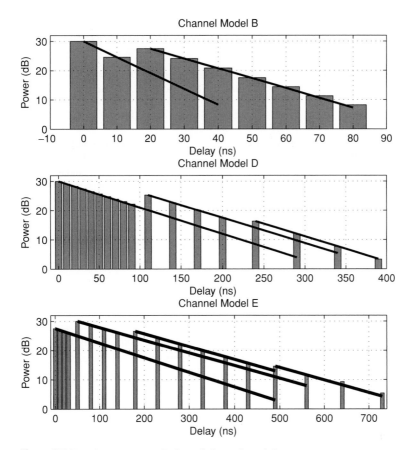

Figure 3.7 Impulse response of selected channel models.

Each tap of the impulse response, h, is Ricean distributed and comprises a fixed and random component. The fixed component is due to a constant LOS path between the transmitter and receiver. The NLOS component is random and Rayleigh distributed:

$$h = \sqrt{P} \left(\sqrt{\frac{K}{K+1}} e^{j\phi} + \sqrt{\frac{1}{K+1}} X \right) \tag{3.13}$$

where X is a complex Gaussian random variable with zero mean and unit variance, ϕ is derived from the angle of arrival/departure of the LOS component, K is the Ricean K-factor, and P is the power of the tap. To compute the total power for each tap at each delay, the powers of the taps in overlapping clusters, given in Appendix 3.1, are summed at each delay.

The K-factor values for each channel model are given in Table 3.2. The K-factor only applies to the first tap of the impulse response; all other taps have a K-factor of 0. For a NLOS channel, the K-factor for all taps is 0. The distance between the transmitter and receiver determines whether the channel should be modeled as LOS or NLOS.

Table 3.2 K-factor

Channel	LOS		NLOS	
	K-factor Tap index 1	K-factor All other taps	K-factor Tap index 1	K-factor All other taps
A	1	0	0	0
B	1	0	0	0
C	1	0	0	0
D	2	0	0	0
E	4	0	0	0
F	4	0	0	0

This is discussed further in Section 3.5.5. For the 802.11n channel model the angle of arrival/departure of the LOS component is fixed at $45°$.

3.5.2   Antenna correlation

The model for the impulse response outlined in Section 3.5.1 and detailed in Eq. (3.13) represents a SISO model. For a MIMO system, the taps are created for each element of the H matrix in Eq. (3.9). Therefore, h in Eq. (3.13) is modified for the ith receive antenna and jth transmit antenna pair, as follows:

$$h_{ij} = \sqrt{P} \left(\sqrt{\frac{K}{K+1}} e^{j\phi_{ij}} + \sqrt{\frac{1}{K+1}} X_{ij} \right) \tag{3.14}$$

There is an H matrix for each tap of the impulse response in a MIMO channel:

$$H = \sqrt{P} \left(\sqrt{\frac{K}{K+1}} \begin{bmatrix} e^{j\phi_{11}} & e^{j\phi_{12}} & \cdots & e^{j\phi_{1M}} \\ e^{j\phi_{21}} & e^{j\phi_{22}} & \cdots & e^{j\phi_{2M}} \\ \vdots & \vdots & \ddots & \vdots \\ e^{j\phi_{N1}} & e^{j\phi_{N2}} & \cdots & e^{j\phi_{NM}} \end{bmatrix} + \sqrt{\frac{1}{K+1}} \begin{bmatrix} X_{11} & X_{12} & \cdots & X_{1M} \\ X_{21} & X_{22} & \cdots & X_{2M} \\ \vdots & \vdots & \ddots & \vdots \\ X_{N1} & X_{N2} & \cdots & X_{NM} \end{bmatrix} \right) \tag{3.15}$$

Correlation is then applied to the random elements X_{ij} to incorporate antenna correlation into the channel model, as follows:

$$[X] = [R_{Rx}]^{1/2} [\widehat{X}] \left([R_{Tx}]^{1/2} \right)^{T} \tag{3.16}$$

where R_{Rx} and R_{Tx} are the receive and transmit correlation matrices, respectively. $[R]^{1/2}$ is defined as a matrix square root, where $[R]^{1/2} \cdot [R]^{1/2} = R$. $\widehat{X}$ is an independent, complex Gaussian random variable with zero mean and unit variance. The correlation matrices, R_{Rx} and R_{Tx}, are defined as follows:

$$
R_{\text{Tx}} = \begin{bmatrix}
1 & \rho_{\text{Tx}12} & \rho_{\text{Tx}13} & \cdots & & \rho_{\text{Tx}1M} \\
\rho_{\text{Tx}21} & 1 & \rho_{\text{Tx}23} & \cdots & & \rho_{\text{Tx}2M} \\
\rho_{\text{Tx}31} & \rho_{\text{Tx}32} & 1 & \ddots & & \vdots \\
\vdots & \vdots & \ddots & \ddots & & \\
& & & & & \rho_{\text{Tx}(N-1)M} \\
\rho_{\text{Tx}N1} & \rho_{\text{Tx}N2} & \cdots & \rho_{\text{Tx}N(M-1)} & & 1
\end{bmatrix}
$$

$$
R_{\text{Rx}} = \begin{bmatrix}
1 & \rho_{\text{Rx}12} & \rho_{\text{Rx}13} & \cdots & & \rho_{\text{Rx}1M} \\
\rho_{\text{Rx}21} & 1 & \rho_{\text{Rx}23} & \cdots & & \rho_{\text{Rx}2M} \\
\rho_{\text{Rx}31} & \rho_{\text{Rx}32} & 1 & \ddots & & \vdots \\
\vdots & \vdots & \ddots & \ddots & & \\
& & & & & \rho_{\text{Rx}(N-1)M} \\
\rho_{\text{Rx}N1} & \rho_{\text{Rx}N2} & \cdots & \rho_{\text{Rx}N(M-1)} & & 1
\end{bmatrix}
\tag{3.17}
$$

where $\rho_{\text{Tx}ij}$ are the complex correlation coefficients between the ith and jth transmitting antennas, and $\rho_{\text{Rx}ij}$ are the complex correlation coefficients between the ith and jth receiving antennas.

3.5.2.1 Correlation coefficient

In the 802.11n channel model, a complex correlation coefficient is derived based on the power angular spectrum (PAS) formulation. This section follows the description given in Erceg *et al.* (2004), Salz and Winters (1994), and Schumacher *et al.* (2002).

The PAS for each tap is a function of the angular spread (AS) and angle of incidence (angle of arrival (AoA) or angle of departure (AoD), depending on Tx or Rx) of each cluster. The angular spread and angle of incidence for each cluster (the 802.11n model assumes that all taps in a cluster have the same angular spread and angle of incidence) is given for each channel model in Appendix 3.1. The shape of the PAS distribution commonly used for 802.11n is truncated Laplacian. The PAS distribution over the angle for each tap is given by

$$
\text{PAS}(\phi) = \frac{1}{A} \sum_{k=1}^{N_C} \frac{p_k}{\sigma_k} \exp\left[\frac{-\sqrt{2}\,|\phi - \psi_k|}{\sigma_k} \right]
\tag{3.18}
$$

where N_C is the number of clusters, and for each cluster k, p_k is the tap power, σ_k is the tap AS, and ψ_k is the tap angle of incidence. Since the PAS is a probability density function, it must fulfill the requirement that $\int_{-\pi}^{\pi} \text{PAS}(\phi)\,d\phi = 1$. Therefore A is equal to $\int_{-\pi}^{\pi} \sum_{k=1}^{N_C}(p_k/\sigma_k)\exp[\sqrt{2}|\phi - \psi_k|/\sigma_k]\,d\phi$. Figure 3.8 illustrates the distribution function for the third tap of channel model B for each cluster for Rx (using parameters from Table 3.5). The sum over the clusters at each angle results in $\text{PAS}(\phi)$.

For a uniform linear antenna array, the correlation of the fading between two antennas spaced D apart is described by Lee (1973). The correlation functions are given in Erceg *et al.* (2004), as follows:

$$
R_{XX}(D) = \int_{-\pi}^{\pi} \cos\left(\frac{2\pi D}{\lambda} \sin\phi \right) \text{PAS}(\phi)\,d\phi
\tag{3.19}
$$

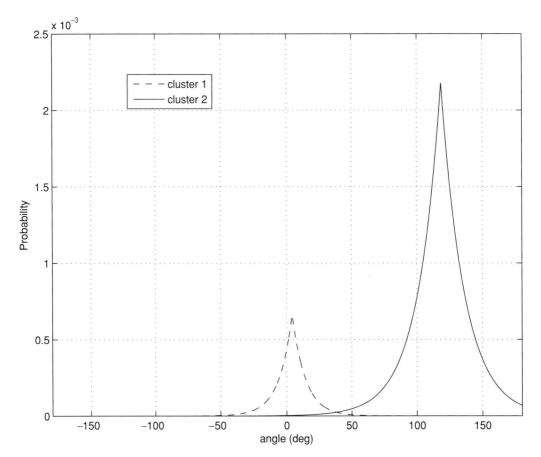

Figure 3.8 Distribution function for each cluster.

and

$$R_{XY}(D) = \int_{-\pi}^{\pi} \sin\left(\frac{2\pi D}{\lambda} \sin\phi\right) \text{PAS}(\phi) \, d\phi \qquad (3.20)$$

where R_{XX} is the correlation function between the real parts of the fading (or the imaginary parts), R_{XY} is the correlation function between the real and imaginary parts of the fading, and λ is the wavelength.

As an example, for the receiver and tap index 3 of channel model B (using parameters from Table 3.5), R_{XX} is equal to -0.52 and R_{XY} is equal to 0.38 with D equal to $\lambda/2$.

The complex correlation coefficients $\rho_{\text{Tx}ij}$ between the ith and jth transmitting antennas, and $\rho_{\text{Rx}ij}$ between the ith and jth receiving antennas in Eq. (3.17) is described by Eq. (3.21) (Erceg *et al.*, 2004):

$$\rho = R_{XX}(D) + jR_{XY}(D) \qquad (3.21)$$

To continue the third tap example, the antenna correlation matrix R_{Rx} is equal to

$$\begin{bmatrix} 1 & -0.52 + j \cdot 0.38 \\ -0.52 - j \cdot 0.38 & 1 \end{bmatrix}$$

When comparing proposals in the development of 802.11n, the antenna spacing was set to $\lambda/2$ in Stephens (2004).

3.5.3　Doppler model

Typically in an indoor WLAN model the transmitter and the receiver are stationary. However, people moving in between the transmitter and receiver could cause the channel to change. If the total transmission time of the packet or packet exchange is long enough to be affected by this motion, a Doppler model may be included in the fading characteristics of the impulse response. Due to the nature of the indoor environment, the Doppler spectrum is quite different from the typical models for mobile cellular channels. A bell shaped Doppler spectrum is used in the 802.11n channel model (Erceg $et\ al.$, 2004), as follows:

$$S(f) = \frac{1}{1 + A\left(\frac{f}{f_d}\right)^2} \qquad |f| \le f_{max} \qquad (3.22)$$

where f_d is the Doppler spread, f_{max} is the maximum frequency component of the Doppler power spectrum, and A is a constant used to set the value of $S(f)$.

In the 802.11n channel model the value of $S(f_d)$ is set to 0.1, therefore A is equal to 9. The Doppler spread f_d is defined as v_0/λ, where v_0 is the environmental speed. The value for v_0 in 802.11n is equal to 1.2 km/h. Therefore the values for Doppler spread in the 5 GHz band is approximately 6 Hz and in the 2.4 GHz band is approximately 3 Hz. f_{max} limits the range of frequencies, and in the 802.11n channel model, f_{max} is set to five times f_d.

The 802.11n channel model Doppler spectrum for 2.4 GHz and 5.25 GHz is illustrated in Figure 3.9. As a note, the relative bandwidth of the Doppler spectrum is independent of the absolute carrier frequency, which is seen in Eq. (3.22).

The autocorrelation function of the bell shaped spectrum is given by $R(\Delta t) = (\pi f_d/\sqrt{A}) \cdot \exp(-(2\pi f_d/\sqrt{A}) \cdot \Delta t)$ and the coherence time is given by $T_c = (\sqrt{A}/2\pi f_d) \cdot \ln(2)$.

3.5.3.1　Modified Doppler model for channel model F

Channel model F is used for large indoor or outdoor environments for large hotspot or industrial conditions, as in Table 3.1. The Doppler model for channel model F includes the scenario of reflections off of a moving object, such as a vehicle. Therefore for channel model F, an extra Doppler component was added to the third tap to account for

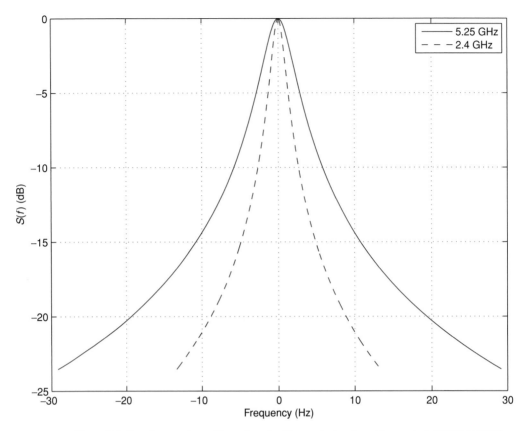

Figure 3.9 Doppler spectrum. Reproduced with permission from Erceg *et al.* (2004) © IEEE.

this higher speed component, as in Eq. (3.23):

$$S(f) = \frac{1}{1 + 9\left(\frac{f}{f_d}\right)^2} + \frac{1}{2}\frac{1}{1 + B\left(\frac{f - f_{veh}}{f_{veh}}\right)^2} \qquad |f| \le f_{max} \qquad (3.23)$$

where f_{veh} is defined as v_1/λ, where v_1 is the speed of the moving vehicle in the environment. The value for v_1 in the 802.11n channel model is equal to 40 km/h. The term B is used to set the bandwidth of the extra Doppler component. This extra Doppler component causes a spike at a frequency f_{veh}. The bandwidth of this spike to defined to be $0.02 \cdot f_{veh}$ when the spectrum is 10 dB below the peak of the spike. Solving for B, by setting $S(f_{veh}(1 \pm \frac{0.02}{2}))/S(f_{veh}) = 0.1$, results in B equal to 90 000. We define f_{max} to be $f_{veh} \cdot (1 + 5 \cdot (0.02/2)) + 5 \cdot f_d$.

The Doppler spectrum for the third tap of channel model F at a center frequency of 2.4 GHz and 5.25 GHz is illustrated in Figure 3.10. Note that the relative bandwidth and location of the spike in the Doppler spectrum is independent of the absolute carrier frequency, which is seen in Eq. (3.23).

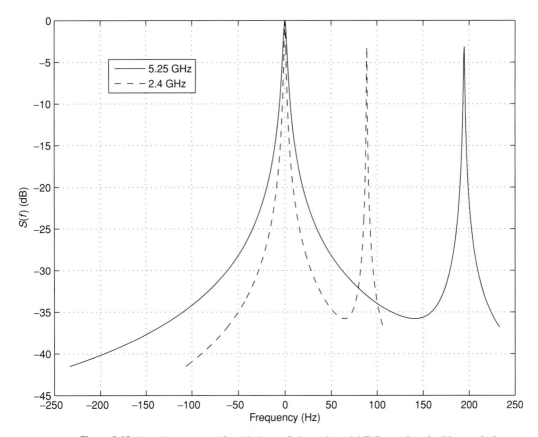

Figure 3.10 Doppler spectrum for third tap of channel model F. Reproduced with permission from Erceg (2004) © IEEE.

3.5.4   Physical layer impairments

Several physical layer (PHY) impairments were added to the channel model. These include Tx and Rx phase noise, power amplifier (PA) non-linearity, and carrier frequency and clock symbol offset. The noise figure is addressed in Section 3.5.5.

When comparing proposals in the development of 802.11n, the carrier frequency offset was set to -13.675 ppm at the receiver, with the clock symbol having the same relative offset.

3.5.4.1   Phase noise

Systems with high order modulation are sensitive to phase noise. Therefore, when comparing proposals for 802.11n, phase noise was added at both the transmitter and receiver in the PHY layer simulations. A single-pole, single-zero phase noise model was

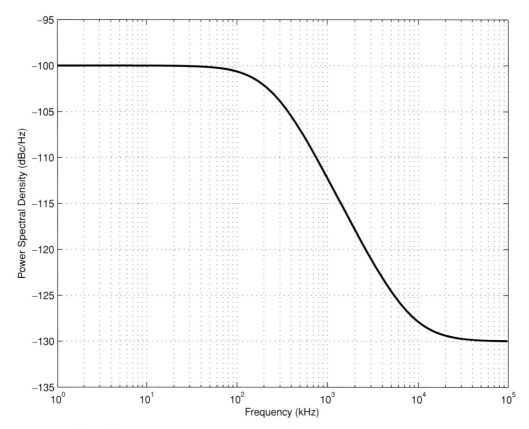

Figure 3.11 Phase noise power spectral density.

utilized with the following specifications (Stephens *et al.*, 2004):

$$\text{PSD}(f) = \text{PSD}(0)\frac{[1 + (f/f_z)^2]}{[1 + (f/f_p)^2]}$$

$$\text{PSD}(0) = -100\,\text{dBc/Hz} \qquad\qquad (3.24)$$

$$f_p = 250\,\text{kHz}$$

$$f_z = 7905.7\,\text{kHz}$$

where PSD is the phase noise power spectral density, f_p is the pole frequency, and f_z is the zero frequency. This model results in PSD(∞) equal to -130 dBc/Hz. The phase noise power spectral density is illustrated in Figure 3.11. Some people believe this to be a lower phase noise than could typically be achieved in practice.

3.5.4.2   Power amplifier non-linearity

A power amplifier (PA) of a certain size, class, and drive current is most efficiently used when driven to saturation for maximum output power. However, when operated at saturation, the PA exhibits non-linear behavior when amplifying the input signal. The distortion caused by the PA on the transmitted waveform causes spectral re-growth,

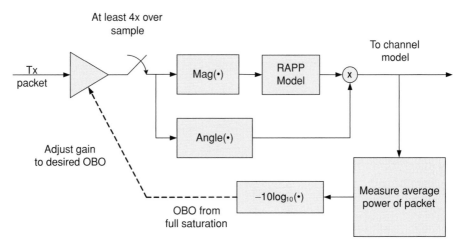

Figure 3.12 Power amplifier model.

impacting the transmitter's ability to meet the specified spectral mask. In addition, the distortion impacts the Tx error vector magnitude (EVM), increasing the packet error rate at the receiver.

The 802.11n MIMO/OFDM system is especially sensitive to PA non-linearity, since the transmitted waveform has high dynamic range and uses high order modulation. Therefore in order to properly model the PHY, especially with high order QAM modulation, a model for PA non-linearity was included in the PHY simulations when comparing proposals.

Figure 3.12 illustrates the selected model for the power amplifier adapted from Webster (2000) to be used in simulations during the development of 802.11n. The gain on the transmitted packet is adjusted to the desired output backoff (OBO). With the waveform over-sampled by at least a factor of four, a non-linear distortion is applied to the amplitude of the signal. The Rapp PA model was selected as the function to model this distortion, and is given in Eq. (3.25):

$$A_{\text{out}} = \frac{A_{\text{in}}}{\left(1 + A_{\text{in}}^{2p}\right)^{\frac{1}{2p}}} \tag{3.25}$$

For the simulation of different proposals for the 802.11n development, p is set to 3 (Stephens, 2004). The PA distortion between the input amplitude and output amplitude is illustrated in Figure 3.13.

For simulations, the recommended transmitted power, at full saturation, was 25 dBm. The total transmit power was limited to no more than 17 dBm. Therefore, with the recommended settings, the OBO from full saturation is 8 dB.

Due to the non-linear structure of the power amplifier model, one way to achieve the desired OBO of the transmitted waveform is to iteratively adjust the gain. This is depicted in Figure 3.12 by the dashed line between the measure of the OBO and the gain applied to the transmitted packet.

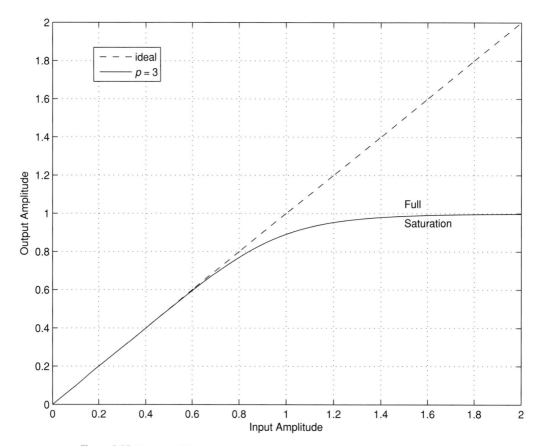

Figure 3.13 Rapp model.

3.5.5   Path loss

To determine the achievable range between the transmitter and receiver, a path loss model and the noise figure of the receiver are required. For 802.11n proposal comparisons, the input referenced total noise figure from the antenna to the output of the ADC was set to 10 dB in Stephens (2004). Real hardware implementations typically achieve a better noise figure.

The path loss model used for indoor propagation was defined in Erceg *et al.* (2004) and consists of the free space (FS) loss (slope of 2) up to a breakpoint distance, and a slope of 3.5 after the breakpoint distance. The breakpoint distances for the channel models are given in Table 3.3. Also included in the path loss model is the shadow fading loss due to large scale obstructions. The path loss model is given as follows:

$$
\begin{aligned}
L(d) &= L_{\text{FS}}(d) + \text{SF} & d \le d_{\text{BP}} \\
L(d) &= L_{\text{FS}}(d_{\text{BP}}) + 35\log_{10}\left(\frac{d}{d_{\text{BP}}}\right) + SF & d > d_{\text{BP}}
\end{aligned}
\tag{3.26}
$$

Table 3.3 Path loss model parameters

Channel model	Breakpoint distance d_{BP} (m)	Path loss slope		Shadow fading std. dev. (dB)		Channel conditions	
		Before d_{BP}	After d_{BP}	Before d_{BP}	After d_{BP}	Before d_{BP}	After d_{BP}
A	5	2	3.5	3	4	LOS	NLOS
B	5	2	3.5	3	4	LOS	NLOS
C	5	2	3.5	3	5	LOS	NLOS
D	10	2	3.5	3	5	LOS	NLOS
E	20	2	3.5	3	6	LOS	NLOS
F	30	2	3.5	3	6	LOS	NLOS

where d is the separation distance between the transmitter and receiver in meters, L_{FS} is the free space path loss in dB, d_{BP} is the breakpoint distance in meters, and SF is the shadow fading loss in dB. The definition for free space path loss is

$$L_{FS}(d) = 20 \log_{10}(d) + 20 \log_{10}(f) - 147.5 \tag{3.27}$$

where $L_{FS}(d)$ is in dB and f is the frequency in Hz.

The shadow fading loss is modeled by a log-normal distribution (Gaussian in dB) with zero mean, as follows:

$$p_{SF}(x) = \frac{1}{\sqrt{2\pi}\sigma_{SF}} \exp\left(-x^2/2\sigma_{SF}^2\right) \tag{3.28}$$

where σ_{SF} is the standard deviation of the shadow fading. The breakpoint distance is also the transition for the shadow fading standard deviation. Table 3.3 gives the shadow fading standard deviation before and after the breakpoint.

As discussed in Section 3.5.1 and in Table 3.2, the Ricean K-factor for the taps of the impulse response are a function of whether the channel should be modeled as LOS or NLOS. For all channel models, LOS conditions are used before the breakpoint and NLOS after the breakpoint, using the parameters outlined in Table 3.3.

3.6   Linear receiver design

Equations (3.3) and (3.12) demonstrate a linear equalization technique to extract data from the received signal commonly known as the zero-forcing (ZF) algorithm, which is described in detail in Proakis (1989). During a large flat fade (h approaches 0), or a deep notch in a frequency selective fading channel, noise is highly amplified in the estimate of the data using a ZF equalizer.

An alternative approach is to select equalizer weights (W) based on minimizing the mean-square-error (MSE). The estimate of x is given in Eq. (3.29), based on the received signal (Y) given in Eq. (3.10):

$$Y = \sqrt{\rho/M} \cdot H \cdot X + Z$$
$$\hat{X} = W \cdot Y \tag{3.29}$$
$$= W \cdot \sqrt{\rho/M} \cdot H \cdot X + W \cdot Z$$

The minimum mean-square-error (MMSE) estimate minimizes the mean-square value of the error vector $e = \hat{X} - X$. With the definition of $\hat{X}$ and Y in Eq. (3.29), the expression for MSE is given by

$$J_{M \times M} = \frac{\rho}{M} W H H^* W^* + W \Phi_Z W^* - 2\sqrt{\frac{\rho}{M}} \text{Re}(WH) + I \tag{3.30}$$

where $J = E[e \cdot e^*]$, H^* is the conjugate transpose of H, Φ_Z is the noise covariance matrix, and I is the identity matrix from the signal covariance. By minimizing the MSE expression with respect to W, we arrive at a solution for W:

$$W = \sqrt{\frac{\rho}{M}} H^* \left(\frac{\rho}{M} H H^* + \Phi_Z \right)^{-1} \tag{3.31}$$

The diagonal terms of the MSE in Eq. (3.30) are between 0 and 1 (Proakis, 1989) with MMSE weights. The output SNR for the ith data stream for MMSE is given by

$$\text{SNR}_i = \frac{1 - J_i}{J_i} \tag{3.32}$$

where J_i is the ith diagonal element of the MSE matrix given in Eq. (3.30).

On the other hand, with ZF the MSE is unbounded. With ZF (and the simplification of $M = N$ in Eq. (3.10)), the weights are equal to $\left(\sqrt{\rho/M} \cdot H \right)^{-1}$. By replacing W in Eq. (3.30) and setting $\Phi_Z = I$ (since we defined the noise term Z as Normal(0,1)), the MSE for ZF is $((\rho/M) \cdot H^* H)^{-1}$. The output SNR for ZF is the inverse of the diagonal terms of the MSE:

$$\text{SNR}_i = \frac{1}{\text{diag}_i \left(\left(\frac{\rho}{M} \cdot H^* H \right)^{-1} \right)} \tag{3.33}$$

In Figure 3.14, MMSE (dashed line) and ZF (solid line) are compared for three input SNRs, 0, 10, and 20 dB, with two transmit and two receive antennas. Each element of the channel matrix H is modeled as independent, identically distributed Rayleigh fading. At higher input SNR, the output SNR between MMSE and ZF is similar. This is seen by equivalently expressing the weights for ZF as $\sqrt{(\rho/M)} H^* ((\rho/M) H H^*)^{-1}$. Comparing this to the MMSE weights in Eq. (3.31), as the input SNR increases the contribution of Φ_Z goes to zero, the weights for ZF and MMSE converge. However, at lower input SNR levels, the improvement of MMSE over ZF exceeds 5 dB as illustrated in Figure 3.14 at the CCDF level of 90% or more.

The advantage of MMSE over ZF is further demonstrated by examining capacity. The formula for capacity based on output SNR is given by Eq. (3.34):

$$C = \sum_{i=1}^{M} \log_2 (1 + SNR_i) \tag{3.34}$$

where M is the number of data streams.

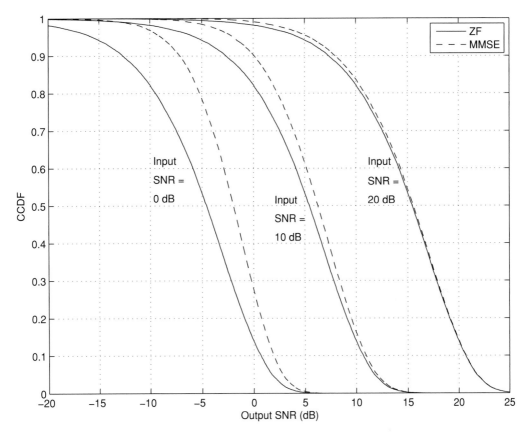

CCDF

Output SNR (dB)

Figure 3.14 Output SNR comparison between MMSE and ZF.

To simulate capacity, the output SNR data provided in Figure 3.14 is substituted into Eq. (3.34). The capacity results for ZF (solid line) and MMSE receiver (dashed line) are illustrated in Figure 3.15 for an input SNR of 0, 10, and 20 dB. The results are compared to the Shannon capacity (dotted line) given by Eq. (3.11). With an input SNR of 0 dB, MMSE capacity is comparable to Shannon capacity at a probability of 90%. With a ZF receiver, the capacity is substantially lower. As the SNR increases, the capacity of MMSE converges to the capacity of ZF. Furthermore, the capacity of MMSE diverges from the Shannon capacity limit.

3.7   Maximum likelihood estimation

An alternative to a linear receiver design is the use of maximum likelihood (ML) estimation. Referring to the communication system defined by Eq. (3.10), an ML estimator maximizes the probability that a signal X_i was transmitted given the received signal Y, as shown below (Proakis, 1989):

$$P(X_i \text{ was transmitted} \mid Y) = \frac{p(Y \mid X_i) \cdot P(X_i)}{p(Y)} \quad i = 1, \ldots, M \quad (3.35)$$

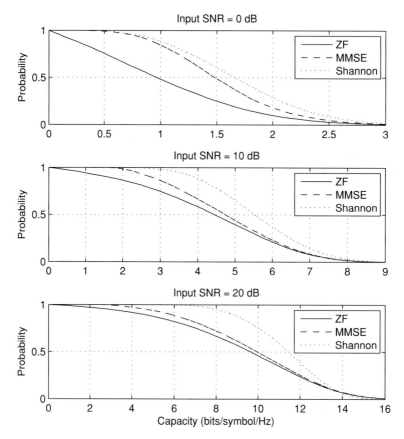

Figure 3.15 Comparison between MMSE and ZF capacity and Shannon capacity.

where X is a set of M equally likely transmitted signals. $p(Y)$ is identical for all signals, and for equally likely signals $P(X_i) = 1/M$ for all i, so neither term affects the decision. Therefore the maximum likelihood criterion becomes (Proakis, 1989)

$$\hat{X} = \arg\max_i p(Y \mid X_i) \quad i = 1, \ldots, M \qquad (3.36)$$

With an estimate of the channel at the receiver, $\tilde{H}$, this is expanded as follows (van Nee *et al.*, 2000):

$$\hat{X} = \arg\min_i \left\| Y - \tilde{H} \cdot X_i \right\| \qquad (3.37)$$

where the maximum likelihood estimate is a function of the minimum Euclidean distance.

An advantage of ML decoding is in the fact that it obtains a diversity order equal to the number of receive antennas (van Nee *et al.*, 2000). In comparison, a MMSE receiver only obtains a diversity order equal to the number of receive antennas minus the number of spatial streams plus one. For example, a three antenna MMSE receiver receives a two spatial stream transmission with diversity order of two. However, an ML decoder achieves a diversity order of two with only two receive antennas, thereby reducing the RF implementation complexity and cost. Or conversely, when comparing a two antenna MMSE receiver and a two antenna ML receiver, the ML receiver provides an additional diversity order. Being independent of the number of spatial streams is a very important property, since an ML receiver maintains its diversity order as the number of spatial streams grow with MIMO/SDM.

The main disadvantage of ML is that the estimator complexity grows exponentially with the constellation size and number of spatial streams. In this respect, ML is much more complex than MMSE. For example, with 64-QAM and two spatial streams, ML requires calculating and sorting $64^2 = 4096$ Euclidean norms. With three spatial streams this increases to $64^3 = 262\,144$.

Sub-optimal implementations of ML for complexity reduction is an on-going area of research. Two areas in particular include lattice reduction and spherical decoding. The basic approach with lattice reduction is to perform the initial detection with a basis change of the channel matrix. The basis change allows for a ZF-like receiver structure, but avoiding the issues of channel matrix singularities which degrade a ZF receiver. Assuming the streams are independent after the receiver greatly reduces the number of Euclidean norms which must be computed. Degradation to theoretical ML performance arises from the lattice reduction not resulting in completely independent streams. A lattice reduction technique for MIMO systems first appeared in Yao and Warnell (2002). An extension of the reduction technique is provided in Berenguer *et al.* (2004). A comprehensive survey of closest point searches in lattices is given in Argrell *et al.* (2002).

With spherical decoding, only Euclidian norms which fall within a specified sphere are considered in the search for X_i in Eq. (3.37). As the radius of the sphere is decreased, the complexity of the estimator is reduced. However, if the radius of the sphere is too small, the algorithm could fail to find any point inside the sphere (Hochwald and ten Brink, 2003). Further discussion (including performance curves) of list sphere decoding is given in Hochwald and ten Brink (2003).

References

Argrell, E., Eriksson, T., Vardy, A., and Zeger, K. (2002). Closest point search in lattices. *IEEE Transactions on Information Theory*, **48**(8), 2201–14.

Berenguer, I., Adeane, J., Wassell, I. J., and Wang, X. (2004). Lattice-reduction-aided receivers for MIMO-OFDM in spatial multiplexing systems. *15th Personal Indoor Multimedia and Radio Communications (PIMRC 04)*, Barcelona, September 2004.

Erceg, V., Schumacher, L., Kyritsi, P., *et al.* (2004). *TGn Channel Models*, IEEE 802.11-03/940r4.

Foschini, G. J. (1996). Layered space-time architecture for wireless communication in a fading environment when using multi-element antennas. *Bell Labs Technical Journal*, Autumn, 41–59.

Foschini, G. J. and Gans, M. J. (1998). On the limits of wireless communications in a fading environment when using multiple antennas. *Wireless Personal Communications*, **6**, 311–35.

Hochwald, B. M. and ten Brink, S. (2003). Achieving near-capacity on a multiple-antenna channel. *IEEE Transactions on Communications*, **51**(3), 389–99.

Jakes, W. C. (1974). *Microwave Mobile Communications*. New York: Wiley.

Lee, W. C. Y. (1973). Effects on correlation between two mobile radio base-station antennas. *IEEE Transactions on Communications*, **COM-21**, 1214–24.

Perahia, E. and Pottie, G. J. (1996). Adaptive antenna arrays and equalization for indoor digital radio. *International Conference on Communications*, June 23–27, Dallas, Tx.

Proakis, J. G. (1989). *Digital Communications*. New York: McGraw-Hill.

Rappaport, T. S. (1996). *Wireless Communications. Principles and Practice*. New Jersey: Prentice Hall.

Saleh, A. A. M. and Valenzuela, R. A. (1987). A statistical model for indoor multipath propagation. *IEEE Journal of Selected Areas in Communications*, **5**, 128–37.

Salz, J. and Winters, J. H. (1994). Effect of fading correlation on adaptive arrays in digital mobile radio. *IEEE Transactions on Vehicle Technology*, **43**, 1049–57.

Schumacher, L., Pedersen, K. I., and Mogensen, P. E. (2002). From antenna spacings to theoretical capacities – guidelines for simulating MIMO systems. *Proc. 13th International Symposium on Personal, Indoor, and Mobile Radio Communications*, **2**, 587–592.

Stephens, A. (2004), *IEEE 802.11 TGn Comparison Criteria*, IEEE 802.11-03/814r31.

Stephens, A., Bjerke, B., Jechoux, B., *et al.* (2004). *Usage Models*, IEEE 802.11-03/802r23.

van Nee, R., van Zelst, A., and Awater, G. (2000). Maximum likelihood decoding in a space division multiplexing system. *Proceedings of the Vehicular Technology Conference*, May 15–18, Tokyo, **1**, 6–10.

Webster, M. (2000). *Suggested PA Model for 802.11 HRB*, IEEE 802.11-00/294.

Yao, H. and Wornell, G. W. (2002). Lattice-reduction-aided detectors for MIMO communication systems. *Proceedings of IEEE Globecom 2002*, Taipei, Taiwan, 424–8.

Appendix 3.1: 802.11n channel models

Tables 3.4–3.11 give the parameters for the 802.11n channel models from Erceg *et al.* (2004). Empty entries in the tables indicate that there is no channel tap at the corresponding time for the particular cluster number.

Table 3.4 Channel model A (Erceg *et al.*, 2004)

			Cluster 1			
Tap index	Excess delay [ns]	Power [dB]	AoA [°]	AS Rx [°]	AoD [°]	AS Tx [°]
1	**0**	0	45	40	45	40

Table 3.5 Channel model B (Erceg *et al.*, 2004)

Tap index	Excess delay [ns]	Cluster 1					Cluster 2				
		Power [dB]	AoA [°]	AS Rx [°]	AoD [°]	AS Tx [°]	Power [dB]	AoA [°]	AS Rx [°]	AoD [°]	AS Tx [°]
1	0	0	4.3	14.4	225.1	14.4					
2	10	−5.4	4.3	14.4	225.1	14.4					
3	20	−10.8	4.3	14.4	225.1	14.4	−3.2	118.4	25.2	106.5	25.4
4	30	−16.2	4.3	14.4	225.1	14.4	−6.3	118.4	25.2	106.5	25.4
5	40	−21.7	4.3	14.4	225.1	14.4	−9.4	118.4	25.2	106.5	25.4
6	50						−12.5	118.4	25.2	106.5	25.4
7	60						−15.6	118.4	25.2	106.5	25.4
8	70						−18.7	118.4	25.2	106.5	25.4
9	80						−21.8	118.4	25.2	106.5	25.4

Table 3.6 Channel model C (Erceg *et al.*, 2004)

Tap index	Excess delay [ns]	Cluster 1					Cluster 2				
		Power [dB]	AoA [°]	AS Rx [°]	AoD [°]	AS Tx [°]	Power [dB]	AoA [°]	AS Rx [°]	AoD [°]	AS Tx [°]
1	0	0	290.3	24.6	13.5	24.7					
2	10	−2.1	290.3	24.6	13.5	24.7					
3	20	−4.3	290.3	24.6	13.5	24.7					
4	30	−6.5	290.3	24.6	13.5	24.7					
5	40	−8.6	290.3	24.6	13.5	24.7					
6	50	−10.8	290.3	24.6	13.5	24.7					
7	60	−13.0	290.3	24.6	13.5	24.7	−5.0	332.3	22.4	56.4	22.5
8	70	−15.2	290.3	24.6	13.5	24.7	−7.2	332.3	22.4	56.4	22.5
9	80	−17.3	290.3	24.6	13.5	24.7	−9.3	332.3	22.4	56.4	22.5
10	90	−19.5	290.3	24.6	13.5	24.7	−11.5	332.3	22.4	56.4	22.5
11	110						−13.7	332.3	22.4	56.4	22.5
12	140						−15.8	332.3	22.4	56.4	22.5
13	170						−18.0	332.3	22.4	56.4	22.5
14	200						−20.2	332.3	22.4	56.4	22.5

Table 3.7 Channel model D (Erceg *et al.*, 2004)

Tap index	Excess delay [ns]	Cluster 1					Cluster 2					Cluster 3				
		Power [dB]	AoA [°]	AS Rx [°]	AoD [°]	AS Tx [°]	Power [dB]	AoA [°]	AS Rx [°]	AoD [°]	AS Tx [°]	Power [dB]	AoA [°]	AS Rx [°]	AoD [°]	AS Tx [°]
1	0	0	158.9	27.7	332.1	27.4										
2	10	−0.9	158.9	27.7	332.1	27.4										
3	20	−1.7	158.9	27.7	332.1	27.4										
4	30	−2.6	158.9	27.7	332.1	27.4										
5	40	−3.5	158.9	27.7	332.1	27.4										
6	50	−4.3	158.9	27.7	332.1	27.4										
7	60	−5.2	158.9	27.7	332.1	27.4										
8	70	−6.1	158.9	27.7	332.1	27.4										
9	80	−6.9	158.9	27.7	332.1	27.4										
10	90	−7.8	158.9	27.7	332.1	27.4										
11	110	−9.0	158.9	27.7	332.1	27.4	−6.6	320.2	31.4	49.3	32.1					
12	140	−11.1	158.9	27.7	332.1	27.4	−9.5	320.2	31.4	49.3	32.1					
13	170	−13.7	158.9	27.7	332.1	27.4	−12.1	320.2	31.4	49.3	32.1					
14	200	−16.3	158.9	27.7	332.1	27.4	−14.7	320.2	31.4	49.3	32.1					
15	240	−19.3	158.9	27.7	332.1	27.4	−17.4	320.2	31.4	49.3	32.1	−18.8	276.1	37.4	275.9	36.8
16	290	−23.2	158.9	27.7	332.1	27.4	−21.9	320.2	31.4	49.3	32.1	−23.2	276.1	37.4	275.9	36.8
17	340						−25.5	320.2	31.4	49.3	32.1	−25.2	276.1	37.4	275.9	36.8
18	390											−26.7	276.1	37.4	275.9	36.8

Table 3.8 Channel model E, clusters 1 and 2 (Erceg *et al.*, 2004)

Tap index	Excess delay [ns]	Cluster 1					Cluster 2				
		Power [dB]	AoA [°]	AS Rx [°]	AoD [°]	AS Tx [°]	Power [dB]	AoA [°]	AS Rx [°]	AoD [°]	AS Tx [°]
1	0	−2.6	163.7	35.8	105.6	36.1					
2	10	−3.0	163.7	35.8	105.6	36.1					
3	20	−3.5	163.7	35.8	105.6	36.1					
4	30	−3.9	163.7	35.8	105.6	36.1					
5	50	−4.5	163.7	35.8	105.6	36.1	−1.8	251.8	41.6	293.1	42.5
6	80	−5.6	163.7	35.8	105.6	36.1	−3.2	251.8	41.6	293.1	42.5
7	110	−6.9	163.7	35.8	105.6	36.1	−4.5	251.8	41.6	293.1	42.5
8	140	−8.2	163.7	35.8	105.6	36.1	−5.8	251.8	41.6	293.1	42.5
9	180	−9.8	163.7	35.8	105.6	36.1	−7.1	251.8	41.6	293.1	42.5
10	230	−11.7	163.7	35.8	105.6	36.1	−9.9	251.8	41.6	293.1	42.5
11	280	−13.9	163.7	35.8	105.6	36.1	−10.3	251.8	41.6	293.1	42.5
12	330	−16.1	163.7	35.8	105.6	36.1	−14.3	251.8	41.6	293.1	42.5
13	380	−18.3	163.7	35.8	105.6	36.1	−14.7	251.8	41.6	293.1	42.5
14	430	−20.5	163.7	35.8	105.6	36.1	−18.7	251.8	41.6	293.1	42.5
15	490	−22.9	163.7	35.8	105.6	36.1	−19.9	251.8	41.6	293.1	42.5
16	560						−22.4	251.8	41.6	293.1	42.5

Table 3.9 Channel model E, clusters 3 and 4 (Erceg *et al.*, 2004)

Tap index	Excess delay [ns]	Cluster 3					Cluster 4				
		Power [dB]	AoA [°]	AS Rx [°]	AoD [°]	AS Tx [°]	Power [dB]	AoA [°]	AS Rx [°]	AoD [°]	AS Tx [°]
9	180	−7.9	80.0	37.4	61.9	38.0					
10	230	−9.6	80.0	37.4	61.9	38.0					
11	280	−14.2	80.0	37.4	61.9	38.0					
12	330	−13.8	80.0	37.4	61.9	38.0					
13	380	−18.6	80.0	37.4	61.9	38.0					
14	430	−18.1	80.0	37.4	61.9	38.0					
15	490	−22.8	80.0	37.4	61.9	38.0	−20.6	182.0	40.3	275.7	38.7
16	560						−20.5	182.0	40.3	275.7	38.7
17	640						−20.7	182.0	40.3	275.7	38.7
18	730						−24.6	182.0	40.3	275.7	38.7

Table 3.10 Channel model F, clusters 1, 2, and 3 (Erceg et al., 2004)

Tap index	Excess delay [ns]	Cluster 1					Cluster 2					Cluster 3				
		Power [dB]	AoA [°]	AS Rx [°]	AoD [°]	AS Tx [°]	Power [dB]	AoA [°]	AS Rx [°]	AoD [°]	AS Tx [°]	Power [dB]	AoA [°]	AS Rx [°]	AoD [°]	AS Tx [°]
1	0	−3.3	315.1	48.0	56.2	41.6										
2	10	−3.6	315.1	48.0	56.2	41.6										
3	20	−3.9	315.1	48.0	56.2	41.6										
4	30	−4.2	315.1	48.0	56.2	41.6										
5	50	−4.6	315.1	48.0	56.2	41.6	−1.8	180.4	55.0	183.7	55.2					
6	80	−5.3	315.1	48.0	56.2	41.6	−2.8	180.4	55.0	183.7	55.2					
7	110	−6.2	315.1	48.0	56.2	41.6	−3.5	180.4	55.0	183.7	55.2					
8	140	−7.1	315.1	48.0	56.2	41.6	−4.4	180.4	55.0	183.7	55.2					
9	180	−8.2	315.1	48.0	56.2	41.6	−5.3	180.4	55.0	183.7	55.2	−5.7	74.7	42.0	153.0	47.4
10	230	−9.5	315.1	48.0	56.2	41.6	−7.4	180.4	55.0	183.7	55.2	−6.7	74.7	42.0	153.0	47.4
11	280	−11.0	315.1	48.0	56.2	41.6	−7.0	180.4	55.0	183.7	55.2	−10.4	74.7	42.0	153.0	47.4
12	330	−12.5	315.1	48.0	56.2	41.6	−10.3	180.4	55.0	183.7	55.2	−9.6	74.7	42.0	153.0	47.4
13	400	−14.3	315.1	48.0	56.2	41.6	−10.4	180.4	55.0	183.7	55.2	−14.1	74.7	42.0	153.0	47.4
14	490	−16.7	315.1	48.0	56.2	41.6	−13.8	180.4	55.0	183.7	55.2	−12.7	74.7	42.0	153.0	47.4
15	600	−19.9	315.1	48.0	56.2	41.6	−15.7	180.4	55.0	183.7	55.2	−18.5	74.7	42.0	153.0	47.4
16	730						−19.9	180.4	55.0	183.7	55.2					

Table 3.11 Channel model F, clusters 4, 5, and 6 (Erceg *et al.*, 2004)

Tap index	Excess delay [ns]	Cluster 4					Cluster 5					Cluster 6				
		Power [dB]	AoA [°]	AS Rx [°]	AoD [°]	AS Tx [°]	Power [dB]	AoA [°]	AS Rx [°]	AoD [°]	AS Tx [°]	Power [dB]	AoA [°]	AS Rx [°]	AoD [°]	AS Tx [°]
13	400	−8.8	251.5	28.6	112.5	27.2										
14	490	−13.3	251.5	28.6	112.5	27.2										
15	600	−18.7	251.5	28.6	112.5	27.2										
16	730						−12.9	68.5	30.7	291.0	33.0					
17	880						−14.2	68.5	30.7	291.0	33.0	−16.3	246.2	38.2	62.3	38.0
18	1050											−21.2	246.2	38.2	62.3	38.0

4 PHY interoperability with 11a/g legacy OFDM devices

One of the functional requirements in the development of the 802.11n standard was that some modes of operation must be backward compatible with 802.11a (and 802.11g if 2.4 GHz was supported) as described in Stephens (2005). Furthermore, the 802.11n standard development group also decided that interoperability should occur at the physical layer. This led to the definition of a mandatory mixed format (MF) preamble in 802.11n. In this chapter, we first review the 802.11a packet structure, transmit procedures, and receive procedures to fully understand the issues in creating a preamble that is interoperable between 802.11a and 802.11n devices. For further details regarding 802.11a beyond this review, refer to clause 17 in IEEE (2007a). Following this overview, the mixed format preamble, which is part of 802.11n, is discussed.

4.1 11a packet structure review

The 802.11a packet structure is illustrated in Figure 4.1.

The Short Training field (STF) is used for start-of-packet detection and automatic gain control (AGC) setting. In addition, the STF is also used for initial frequency offset estimation and initial time synchronization. This is followed by the Long Training field (LTF), which is used for channel estimation and for more accurate frequency offset estimation and time synchronization. Following the LTF is the Signal field (SIG), which contains the rate and length information for the packet. Example rates are BPSK, rate $\frac{1}{2}$ encoding and 64-QAM, rate $\frac{3}{4}$ encoding. Following this is the Data field. The first 16 bits of the Data field contain the Service field. An example of an 802.11a transmit waveform is given in Figure 4.2.

4.1.1 Short Training field

The STF is 8 µs in length. In the time domain, the STF contains ten repetitions of a 0.8 µs symbol. The STF is defined based on the frequency domain sequence given in Eq. (4.1) (IEEE, 2007a):

$$
\begin{aligned}
S_{-26,26} = \sqrt{13/6} \{ & 0, 0, 1+j, 0, 0, 0, -1-j, 0, 0, 0, 1+j, 0, 0, 0, -1-j, \\
& 0, 0, 0, -1-j, 0, 0, 0, 1+j, 0, 0, 0, 0, 0, 0, 0, -1-j, \\
& 0, 0, 0, -1-j, 0, 0, 0, 1+j, 0, 0, 0, 1+j, 0, 0, 0, 1+j, \\
& 0, 0, 0, 1+j, 0, 0 \}
\end{aligned}
\tag{4.1}
$$

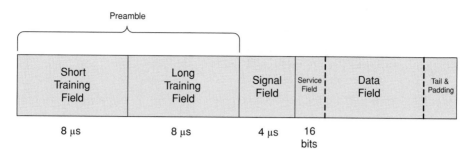

Figure 4.1 802.11a packet structure.

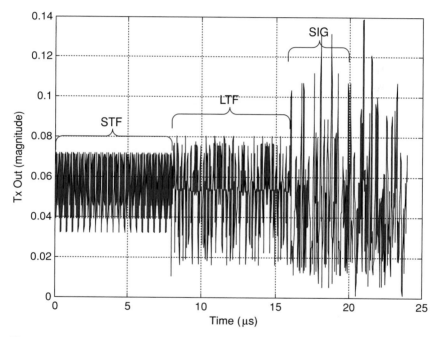

Figure 4.2 802.11a transmit waveform.

The sequence uses 12 of the 52 subcarriers. A 64-point IFFT creates a 3.2 μs time domain sequence with a pattern which repeats four times (resulting in 0.8 μs periodicity). This is illustrated in Figure 4.3, with a sampling rate of 20 Msamples/s. This sequence may then be repeated two and half times to create ten short symbol repetitions.

The sequence was chosen to have good correlation properties and a low peak-to-average power so that its properties are preserved even after clipping or compression by an overloaded analog front end. The cross-correlation between the STF and one short training symbol is illustrated in Figure 4.4. There is a separation of over 10 dB between a correlation peak and a correlation sidelobe.

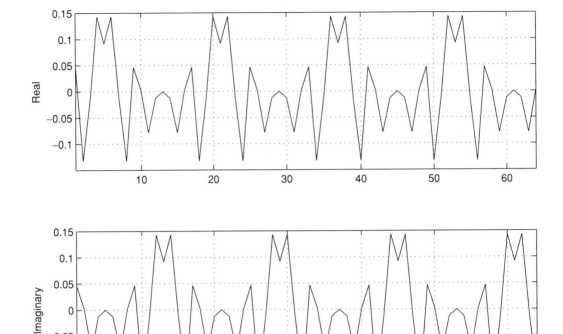

Figure 4.3 Construction of short training symbols.

The correlation peaks may be used to derive an initial time estimate in a time domain based timing acquisition method. It is important to note that the cross-correlation properties of the STF are degraded by delayed responses in the channel. In Figure 4.5, we give an example of the cross-correlation with a channel modeled by two taps. The curve represented by the solid line depicts the channel with the second tap delayed by 50 ns. The curve represented by the dashed line depicts the channel with the second tap delayed by 400 ns. With 50 ns, the cross-correlation peak is a bit wider than the peak illustrated in Figure 4.4. However, with 400 ns, extra correlation peaks arise which may degrade timing acquisition. This issue had a major impact on the design of the MF preamble in 802.11n and is further discussed in Section 4.2.1.

The repetitive nature of the STF can also be used by correlating a 0.8 μs short symbol with the previous symbol (commonly termed auto-correlation since the received signal is correlated with a delayed version of itself). This technique may be used for packet detection, whereby the correlation value exceeding a threshold indicates a packet detect. Additionally, this correlation value may be used to set the AGC. This approach proved less sensitive to longer delay spreads as compared to the cross-correlation; however, it is potentially less effective in noisier conditions.

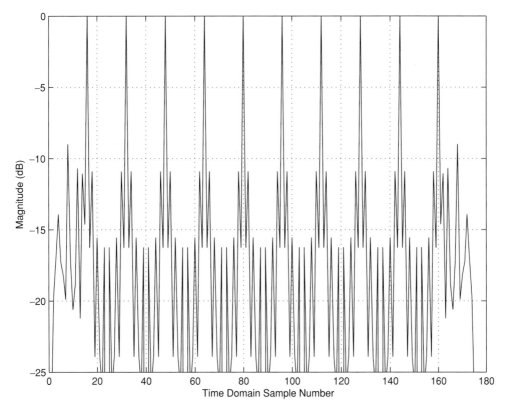

Figure 4.4 Short Training field correlation.

The STF is also used for initial frequency offset estimation. Due to the symbol repetition, the difference in phase between two samples in the STF separated by 0.8 µs (or 16 samples if sampling at 20 MHz) allows an estimate of the frequency offset. This operation can resolve a frequency offset between the transmitting device and receiving device of up to ±625 kHz (±1/0.8 µs/2).

4.1.2   Long Training field

The LTF is also 8 µs in length, and is composed of two 3.2 µs long training symbols prepended by a 1.6 µs cyclic prefix. The cyclic prefix is comprised of the second half of the long training symbol. The long training symbol is a 64-point IFFT of the frequency domain sequence given in Eq. (4.2) (IEEE, 2007a). Numbering the subcarrier locations ranging from $-32, -31, \ldots, -1, 0, 1, \ldots, 31$, the populated subcarriers are located from $-26, -25, \ldots, -2, -1, 1, 2, \ldots, 25, 26$. All the populated subcarriers have the value of $+1$ or -1. The subcarrier at DC, L_0, is not populated.

$$
\begin{aligned}
L_{-26,26} = \{ & 1, 1, -1, -1, 1, 1, -1, 1, -1, 1, 1, 1, 1, 1, 1, -1, -1, 1, 1, -1, 1, \\
& -1, 1, 1, 1, 1, 1, 0, 1, -1, -1, 1, 1, -1, 1, -1, 1, -1, -1, -1, -1, \\
& -1, 1, 1, -1, -1, 1, -1, 1, -1, 1, 1, 1, 1, 1 \}
\end{aligned} \tag{4.2}
$$

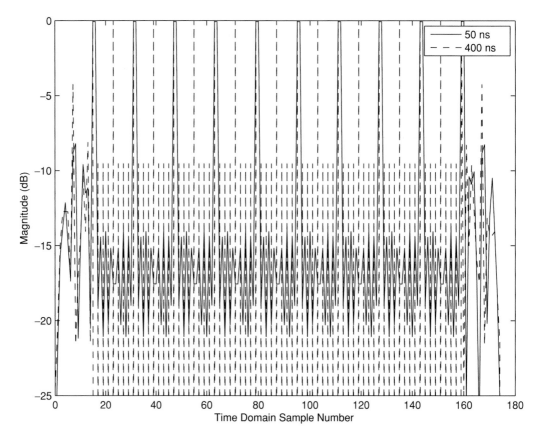

Figure 4.5 STF cross-correlation with a delayed response.

The cross-correlation between the entire preamble and the long training symbol is illustrated in Figure 4.6, with a sampling rate of 20 Msamples/s. The correlation peaks may be used to derive a more accurate time estimate in a time domain based timing acquisition method. Alternatively, the 1.6 μs cyclic prefix of the LTF may be used for correlation, with either cross-correlation or auto-correlation. Additionally, the transition between the STF and LTF may be used for more accurate time acquisition, e.g. by determining when the STF auto-correlation subsides.

Due to the symbol repetition, the difference in phase between two samples in the LTF separated by 3.2 μs (or 64 samples if sampling at 20 MHz) allows a more accurate estimate of the carrier frequency offset. This operation can resolve a frequency offset of up to ± 156.25 kHz ($\pm 1/3.2$ μs/2), so it should be used in conjunction with the STF frequency offset estimate.

The other main purpose of the LTF is for channel estimation. The receiver extracts the two long training symbols from the LTF. An FFT is performed on the symbols and the training subcarriers are extracted. The subcarriers from the first long training symbol are averaged with the subcarriers from the second symbol to reduce the effect of noise by 3 dB. Subsequently for each subcarrier k we have the basic communication system

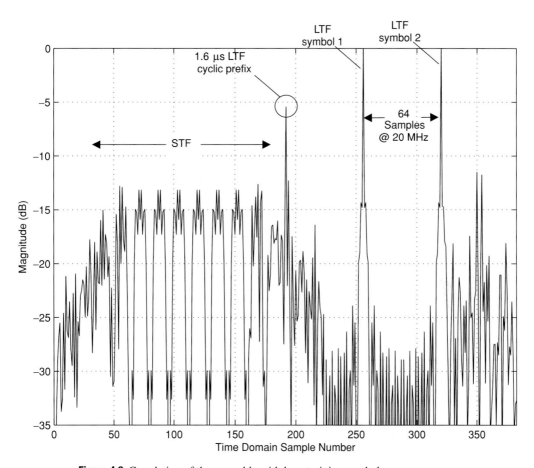

Figure 4.6 Correlation of the preamble with long training symbol.

equation, $y_k = h_k \cdot L_k + z_k$. The values for L_k are given in Eq. (4.2). The known training symbol information is divided out of the received signal leaving the channel estimate for each subcarrier k, $\hat{h}_k = y_k / L_k$.

Since the channel estimate is used to equalize the subsequent OFDM symbols, noise on the channel estimate propagates through the packet during data detection. To reduce the noise on the channel estimate, subcarrier smoothing may be employed. A simple approach is to perform a weighted average of the channel estimate at subcarrier k with its adjacent neighbors, as shown in Eq. (4.3):

$$\tilde{h}_k = \frac{a \cdot \hat{h}_{k-1} + b \cdot \hat{h}_k + a \cdot \hat{h}_{k+1}}{2 \cdot a + b} \tag{4.3}$$

The nature of the channel must be taken into consideration with subcarrier smoothing. In a low delay spread, flat fading channel, channel taps on adjacent subcarriers are highly correlated and subcarrier smoothing provides a significant noise reduction benefit. However, in a highly frequency selective fading channel, adjacent channel taps may not

Table 4.1 Rate information

Rate bit field	Data rate (Mbps)	Modulation	Code rate
1101	6	BPSK	$1/2$
1111	9	BPSK	$3/4$
0101	12	QPSK	$1/2$
0111	18	QPSK	$3/4$
1001	24	16-QAM	$1/2$
1011	36	16-QAM	$3/4$
0001	48	64-QAM	$2/3$
0011	54	64-QAM	$3/4$

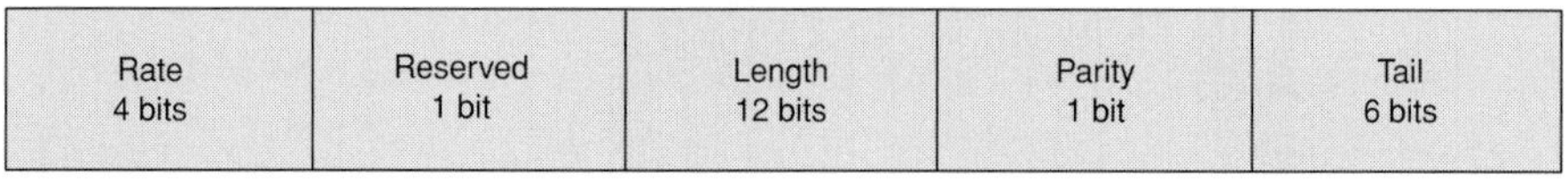

Rate 4 bits	Reserved 1 bit	Length 12 bits	Parity 1 bit	Tail 6 bits

Figure 4.7 Signal field.

be correlated and more sophisticated smoothing algorithms must be applied otherwise improper averaging degrades the channel estimate.

4.1.3   Signal field

The Signal field (SIG) consists of 24 information bits, illustrated in Figure 4.7. The SIG itself is transmitted using BPSK modulation and rate $1/2$ binary convolution code (BCC) (followed by interleaving) to maximize the probability of reception. The BCC is described in Section 4.1.5. Not only is it important for the intended receiver to correctly decode the SIG, but also nearby stations need to correctly decode the SIG in order to properly defer the channel access.

The SIG consists of a single 4 μs symbol (a 3.2 μs OFDM symbol prepended by a 0.8 μs cyclic prefix). The waveform uses 52 subcarriers, as in the LTF. The 48 coded bits are BPSK modulated on to 48 subcarriers. Four additional subcarriers are used as pilots for phase and frequency tracking and training. Numbering the subcarrier locations ranging from $-32, -31, \ldots, -1, 0, 1, \ldots, 31$, the populated subcarriers are located at $-26, -25, \ldots, -2, -1, 1, 2, \ldots, 25, 26$. The pilot subcarriers are $-21, -7, 7, 21$. The remaining 48 subcarriers are populated with the coded SIG bits.

The SIG contains packet information to configure the receiver: rate (modulation and coding) and length (amount of data being transmitted in octets). The details of the rate field are given in Table 4.1. The reserved bit was intended for future use. The parity bit gives the even parity over the first 17 bits. The tail bits are set to zero and used to flush the encoder and decoder, since the SIG is separately encoded from the data field.

The parity and reserve bits played a large role in the development of the 11n mixed format (MF) preamble structure. A single parity bit in the SIG proved to be problematic in lower SNR conditions. In very noisy conditions, the probability of a false positive with a single parity bit approaches 50%. This happens especially when no signal is present at the receiver. A false positive incorrectly indicates a valid SIG, which leads the receiver to use the length and rate values to defer transmission. In a noisy environment or when no packet is present, these fields are random bits, and the device cannot receive or transmit a valid packet for potentially a long period of time. This significantly reduces throughput.

This problem is partially mitigated by first measuring the received signal level and only processing signals which exceed a threshold. However, some manufacturers additionally used the reserve bit as extra parity. To maintain backward compatibility with devices in the field, it was decided not to use the reserve bit in the SIG to indicate a new 802.11n MF packet. This then required an auto-detection between 802.11a and 802.11n MF packets, which is discussed further in Section 4.2.2.1.

4.1.4   Data field

The Data field consists of the Service field, data bits, tail bits, and, if necessary, pad bits. The first 16 bits of the Data field contain the Service field. The first 7 bits of the Service field are the scrambler initialization bits. They are used to synchronize the descrambler and are set to zero to enable estimation of the initial state of the scrambler in the receiver. The remaining 9 bits are reserved and also set to zero.

Following the Service field are the data bits, which in turn are followed by six tail bits which are set to zero. Pad bits (also zeros) are added to the end of the packet so the resulting length fills up the last data symbol.

A length-127 frame synchronous scrambler is used in 802.11a which uses the generator polynomial $G(D) = D^7 + D^4 + 1$. The 127-bit sequence generated repeatedly by the scrambler is (leftmost used first), 00001110 11110010 11001001 00000010 00100110 00101110 10110110 00001100 11010100 11100111 10110100 00101010 11111010 01010001 10111000 1111111, if the all ones initial state is used (IEEE, 2007a). When transmitting, the initial state of the scrambler is actually set to a pseudo random non-zero state for each packet. Scrambling with a different sequence provides peak-to-average protection via retransmission. After the scrambling operation on the Data field, the six tail bits are replaced by six unscrambled zero bits. Pad bits are stripped off at the receiver.

The Data field consists of a stream of symbols, each data symbol 4 µs in length. A data symbol is comprised of a 3.2 µs OFDM symbol prepended by a 0.8 µs cyclic prefix. The waveform uses 52 subcarriers, as in the LTF and SIG. The data bits are encoded and modulated on to 48 subcarriers, as per the rate information in the SIG. Four additional subcarriers are used as pilots for phase and frequency tracking and training. Numbering the subcarrier locations $-32, -31, \ldots, -1, 0, 1, \ldots, 31$, the populated subcarriers are located at $-26, -25, \ldots, -2, -1, 1, 2, \ldots, 25, 26$. The pilot subcarriers are located on subcarriers $-21, -7, 7, 21$ with values as follows:

$$P_k = \begin{cases} 1 & k = -21, -7, 7 \\ -1 & k = 21 \\ 0 & \text{otherwise} \end{cases} \tag{4.4}$$

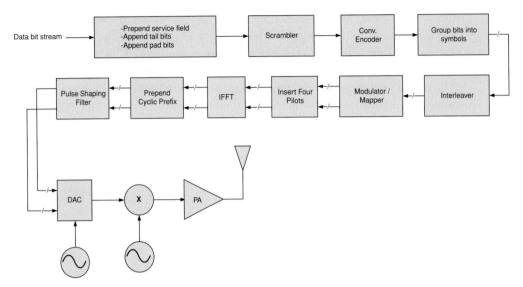

Figure 4.8 Transmitter block diagram for Data field.

The polarity of the pilot subcarriers is controlled by the sequence, p_n, which is a cyclic extension of the following 127 element sequence (IEEE, 2007a):

$$
\begin{aligned}
p_{0\ldots126v} = \{&1, 1, 1, 1, -1, -1, -1, 1, -1, -1, -1, -1, 1, 1, -1, 1, -1, -1, 1, \\
&1, -1, 1, 1, -1, 1, 1, 1, 1, 1, 1, -1, 1, 1, 1, 1, -1, 1, 1, -1, -1, 1, 1, 1, 1, \\
&-1, 1, -1, -1, -1, 1, -1, 1, -1, -1, 1, -1, -1, 1, 1, 1, 1, 1, 1, -1, \\
&-1, 1, 1, -1, -1, 1, -1, 1, -1, 1, 1, -1, -1, -1, 1, 1, -1, -1, -1, \\
&-1, 1, -1, -1, 1, -1, 1, 1, 1, 1, -1, 1, -1, 1, -1, 1, -1, -1, -1, \\
&-1, -1, 1, -1, 1, 1, -1, 1, -1, 1, 1, 1, -1, -1, 1, -1, -1, -1, 1, \\
&1, 1, -1, -1, -1, -1, -1, -1, -1\}
\end{aligned}
$$

(4.5)

The sequence p_n can be generated by the scrambler. The remaining 48 subcarriers are populated with the coded data bits.

4.1.5   Packet encoding process

The packet encoding process begins by creating the STF and LTF as described in Sections 4.1.1 and 4.1.2. The SIG is produced based on the selected rate and length, following the description in Section 4.1.3. The encoding of the Data field follows the block diagram in Figure 4.8.

The data bits are prepended by the Service field and appended by six tail zero bits. If necessary, additional pad zero bits are also appended to the Data field. The extended sequence is then scrambled. The initial state of the scrambler is set to a non-zero random value. At the receiver, the first seven bits of the Service field give the initial state of the scrambler. The same scrambler is used at the receiver to descramble the received data. After the scrambler, the six scrambled tail bits are replaced by unscrambled zero bits. This causes the convolution encoder, which follows, to return to the zero state.

The scrambled data is encoded with a rate $\frac{1}{2}$ binary convolutional encoder (BCC). The encoder has a constraint length of 7 and a generator polynomial of $\{133, 171\}$ octal. If necessary, the sequence is punctured to the selected code rate if a higher rate is chosen. Further detail is provided in IEEE (2007a, Clause 17.3.5.5) and van Nee and Prasad (2000).

The coded data is grouped into a number of bits per symbol and each group is block interleaved with two permutations. The first permutation maps adjacent coded bits onto non-adjacent subcarriers. In this manner, adjacent coded bits are not impacted by the same frequency selective fade. The second permutation ensures that adjacent coded bits are distributed between less and more significant bits of the modulation constellation. This ensures that long strings of bits are not mapped to the lower reliability, less significant constellation bits.

Following the interleaver, the groups of bits are modulated according to the selected rate and mapped onto the corresponding subcarriers. There are four modulation formats possible: BPSK, QPSK, 16-QAM, and 64-QAM. At this stage, the modulated data are complex numbers. For each symbol, the pilot subcarriers are inserted into their corresponding subcarriers. Subcarrier locations for data and pilots are given in Section 4.1.4.

Each symbol group of subcarriers is transformed into the time domain by an IFFT. A cyclic prefix is prepended to the symbol. A pulse shaping function may be applied to each symbol to smooth transitions between symbols. Smoothing transitions between symbols reduces the spectral sidelobes and may be necessary to meet the spectral mask. A pulse shaping function is not specified in the standard; however, the following example is given in IEEE (2007a) for 802.11a:

$$
w_T(t) = \begin{cases} \sin^2\left(\dfrac{\pi}{2}\left(0.5 + \dfrac{t}{T_{TR}}\right)\right) & -\dfrac{T_{TR}}{2} < t < \dfrac{T_{TR}}{2} \\ 1 & \dfrac{T_{TR}}{2} \leq t < T - \dfrac{T_{TR}}{2} \\ \sin^2\left(\dfrac{\pi}{2}\left(0.5 - \dfrac{t-T}{T_{TR}}\right)\right) & T - \dfrac{T_{TR}}{2} \leq t < T + \dfrac{T_{TR}}{2} \end{cases} \tag{4.6}
$$

where T_{TR} is the transition time of the windowing function, which creates an overlap between adjacent symbols to smooth the transition. With a sampling rate of 20 Msample/s, symbol duration of 4 μs, and $T_{TR} = 100$ ns, the windowing function is given by Eq. (4.7) (IEEE, 2007a):

$$
w_T(n) = \begin{cases} 1 & 1 < n < 79 \\ 0.5 & 0, 80 \\ 0 & \text{otherwise} \end{cases} \tag{4.7}
$$

An alternative approach is to use a 40 MHz IFFT with additional windowing.

Lastly, each of the OFDM symbols is appended one after another. The digital baseband signal is converted to analog with a DAC, which is then followed by an up-converter to RF and a power amplifier.

Further details on the encoding process are given in IEEE (2007a, Clause 17) (otherwise known as 802.11a), van Nee and Prasad (2000) and Halford (2001). An example of data bits encoded into an 802.11a packet is given in IEEE (2007a, Annex G).

4.1.6   Receive procedure

Typically, the receiver waits for a signal to exceed a threshold before initiating the packet reception procedure. The minimum requirement in 802.11a is to detect the start of a valid transmission received at a power level of at least -82 dBm within 4 µs with a probability greater than 90%. Due to competitive pressures, most products exceed this specification by 5–10 dB.

The short training symbols are processed to set the AGC, initial frequency correction, and initial timing acquisition. This is followed by processing the long training symbols for fine frequency correction and fine timing adjustment. The LTF is also used to generate the channel estimate.

The next 4 µs symbol is the SIG. The receiver follows the same symbol processing as the data, except for the de-scrambler. The bits in the SIG are not scrambled. Once the SIG bits are extracted, the receiver evaluates the parity bit for errors in the SIG. If the parity indicates an error, the receiver returns to an idle state. Upon no errors being detected by the parity bit, the receiver is configured based on rate and length information. The rate determines the modulation type and coding rate. The length determines the duration of the data stream in octets.

Following the SIG is the Data field, for which a receiver block diagram is illustrated in Figure 4.9. For each 4 µs symbol, the 0.8 µs GI is removed and an FFT is performed. Using the channel estimate from the LTF, the subcarrier data is equalized as described in Section 3.6. The four pilot subcarriers are used for phase rotation correction, and may be used for further frequency offset and symbol timing correction.

Trellis coded modulation (TCM), which combines modulation and coding, was generally accepted as the best approach for improved performance. However, with the commercial availability and popularity of Viterbi decoders, researchers began investigating schemes for maximizing performance with separate modulation and coding entities to enable reuse of the Viterbi decoder. In Zehavi (1992) it was shown that in a fading channel the code diversity is improved by bit-wise interleaving at the encoder output. The analysis in Claire *et al.* (1998) goes further and demonstrates that on some channels there is a downside to combining demodulation and decoding. For this reason, the 802.11a system is based on bit interleaved coded modulation (BICM). Therefore, the subcarrier data is then demapped by finding bit probabilities from the closest constellation points to the equalized signal. This is followed by the deinterleaver, which is the inverse of the interleaver function. A comprehensive presentation of BICM is given in Caire *et al.* (1998), and further information on low-complexity demapping of BICM is given in van Nee and Prasad (2000) and Akay and Ayanoglu, 2004a).

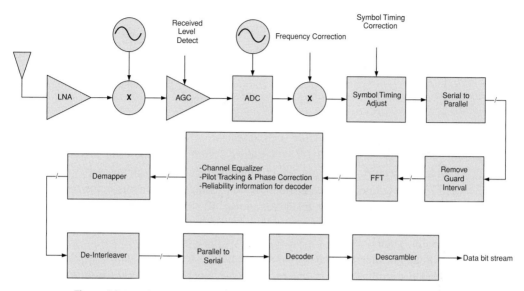

Figure 4.9 Receiver block diagram for Data field.

After the parallel to serial conversion (see Figure 4.9), the data is treated as a stream rather than symbol blocks. In a conventional receiver with ZF or MMSE, the output of the demapper and deinterleaver is passed directly to the decoder, typically a Viterbi decoder. If the code rate is higher than $\frac{1}{2}$, zero metrics are inserted into the decoder in place of the puncture bits.

However, with a frequency selective channel, decoding performance substantially improves by using the channel estimates to weight the decoder metrics (Akay and Ayanoglu, 2004b). The bits associated with subcarriers that are faded are de-weighted (less reliable). This forces the portions of the decoder trellis for which the SNR is low due to the channel fade to have lower branch metrics and have a smaller contribution to the path metric. The subcarriers with high SNR are larger branch metrics and are emphasized in the path metric. Based on the maximum likelihood estimator, we can mathematically demonstrate the benefit of applying channel estimates to weight the decoder metrics.

The maximum likelihood estimation was discussed in Section 3.7 and given in Eq. (3.37). For soft-decision decoding in an AWGN channel with fading, the probability density function of the demodulator output is given by Eq. (4.8) (Proakis, 1989):

$$p(y \mid x_i) = \frac{1}{\sqrt{2\pi}} \exp\left[\frac{(y - h \cdot x_i)^2}{2\sigma^2}\right] \tag{4.8}$$

Extrapolating from this, the maximum likelihood estimate of the transmitted data is given by Eq. (4.9) (Akay and Ayanoglu, 2004b):

$$\hat{X} = \min_i \left|\frac{Y - H \cdot X_i}{\sigma}\right|^2 \tag{4.9}$$

where Y is the received signal, σ^2 is the variance of the noise, H and $\tilde{H}$ are the channel tap and channel estimate, respectively, and X_i is the set of possible transmitted data. By factoring, we arrive at the expression for the ML estimator given in Eq. (4.10) (Akay and Ayanoglu, 2004b):

$$\hat{X} = \min_{i} \frac{|\tilde{H}|^2}{\sigma^2} \left| \frac{Y}{\tilde{H}} - X_i \right|^2 \tag{4.10}$$

The $Y/\tilde{H}$ term in Eq. (4.10) is the expression for a ZF receiver. The demapping operation is given by the $|Y/\tilde{H} - X_i|$ term. Therefore, to approach ML performance with a ZF receiver, we simply need to include the extra weighting step, given by the $|\tilde{H}|^2/\sigma^2$ term. Since with BPSK there is no coupling between bits in the constellation, whose information is lost in a low-complexity BICM demapping operation, Eq. (4.10) provides ML with a ZF receiver.

As illustrated in Figure 4.9, the final step in the receive procedure is descrambling the Data field. The first seven bits of the Service field give the initial state of the descrambler. The same scrambler is used to descramble the data stream. Bits received after the data length indicated by the SIG are discarded.

Further details on the 802.11a packet reception process are given in IEEE (2007a, Clause 17) (otherwise known as 802.11a), van Nee and Prasad (2000), and Halford (2001).

4.2　Mixed format high throughput packet structure

New high throughput (HT) packet formats are used to transmit multiple spatial streams. The mixed format (MF) HT preamble begins with legacy (802.11a) fields to ensure backward compatibility with 802.11a and OFDM 802.11g. These are followed by high throughput training fields. An overview of the fields is given in Figure 4.10. The MF preamble may be used for either 20 MHz or 40 MHz channel width. However, this section focuses on the general functionality of the MF preamble, and only describes the mandatory 20 MHz waveform. The optional 40 MHz waveform is covered in Section 5.1. In addition, other optional features such as sounding packets, short guard interval, STBC, and LDPC and their impact on the waveform design are addressed in Sections 12.6, 5.5, 6.3, and 6.4, respectively.

A plot of a full 802.11n MF waveform with two spatial streams is given in Figure 4.11. The legacy part of the preamble comprises the first 20 μs. The HT part of the preamble comprises the next 20 μs, with two spatial streams. This is then followed by the HT data field.

The following sections describe the fields of the MF preamble in detail.

4.2.1　Non-HT portion of the MF preamble

The first part of the MF preamble consists of non-HT (legacy) training fields identical to 802.11a, as illustrated in Figure 4.10. It contains a Non-HT Short Training field (L-STF),

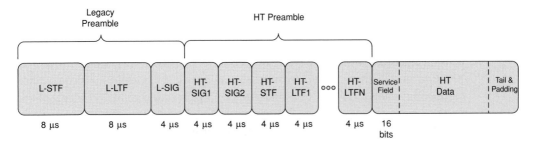

Figure 4.10 Mixed format preamble.

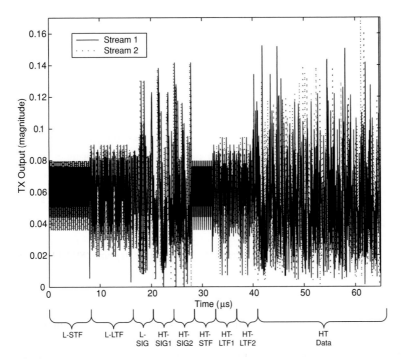

Figure 4.11 802.11n MF with two spatial streams.

non-HT Long Training field (L-LTF), and Non-HT Signal field (L-SIG) such that legacy devices can detect the preamble. However, as discussed below, an 802.11n HT device also needs to use these fields for training and mandatory demodulation of the L-SIG.

Referring to Section 4.1.1, the L-STF is used for start-of-packet detection, AGC setting, initial frequency offset estimation, and initial time synchronization by legacy devices. The L-LTF is used for further fine frequency offset estimation and time synchronization. The L-LTF is also used to generate channel estimates for receiving and equalizing the L-SIG. With proper decoding of the L-SIG, legacy 802.11a/g devices extract the rate and length information and are able to determine the proper time to remain off air. This is discussed in more detail in Section 10.6.9.

Since HT devices perform similar acquisition functions as legacy devices using the L-STF and L-LTF, the issue arises of how to transmit the legacy part of the MF preamble that is a single spatial stream with 802.11n devices that in most cases have multiple transmit antennas. To understand the issue, consider a single antenna legacy device with 17 dBm output transmit power. To maintain similar cost and complexity, a comparable two transmit antenna HT device has two power amplifiers each with 14 dBm output power for the same total output power. If the legacy part of the MF preamble is transmitted from a single antenna of the HT device, the maximum power and range during legacy training is reduced by 3 dB. Moreover, with four antennas, the power on a single transmit RF chain and antenna drops by 6 dB.

To alleviate this transmit power disparity, it is necessary to transmit the legacy training from all antennas of an HT device. However, since the legacy preamble is single stream, transmitting the same signal from each antenna may cause unintentional and undesirable beamforming effects. Since the L-STF is used for AGC setting, large variations in the power of the L-STF also results in saturation or quantization errors.

4.2.1.1 Cyclic shifts

During development of the 802.11n standard, several approaches for how to transmit the legacy training from all antennas were investigated. One approach that was considered was transmitting a subset of subcarriers on each antenna. However, with four transmit antennas, the L-STF, which consists of only twelve active subcarriers, would consist of only three active subcarriers per antenna. This is a very small number of subcarriers per antenna and this approach also suffers from legacy compatibility issues.

Hence, to decorrelate the signals transmitted from different antennas with a single spatial stream and alleviate undesirable beamforming effects, different cyclic shifts (T_{CS}) are applied to each of the signals on the different antennas. Consider an OFDM symbol $s(t)$ over a symbol period T. Replace $s(t)$ with $s(t - T_{CS})$ when $0 \leq t < T + T_{CS}$, and with $s(t - T_{CS} - T)$ when $T + T_{CS} \leq t \leq T$. The cyclic shifted signal is defined as

$$s_{CS}(t; T_{CS})\,|_{T_{CS} < 0} = \begin{cases} s(t - T_{CS}) & 0 \leq t < T + T_{CS} \\ s(t - T_{CS} - T) & T + T_{CS} \leq t \leq T \end{cases} \qquad (4.11)$$

The cyclic shift is applied to each OFDM symbol in the packet separately. Equivalently, in the frequency domain, where $S(f)$ is the Fourier transform of $s(t)$, the cyclic shifted symbol can be described as

$$S_{CS}(f) = S(f)e^{-j2\pi f T_{CS}} \qquad (4.12)$$

A two antenna example is given in Figure 4.12, with a -200 ns cyclic shift applied to the signal on the second antenna.

The L-STF is used for AGC setting. Thus, in order to properly adjust the gain setting in the receiver, the power of the STF needs to match the power of the Data field. The ratio between the power of the L-STF and the power of the Data field as measured by

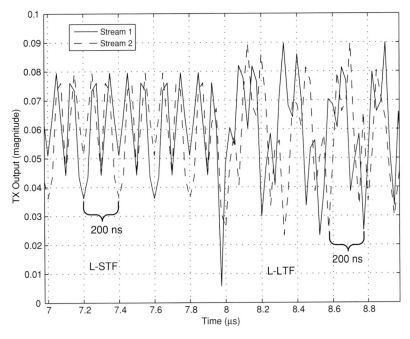

Figure 4.12 Example of cyclic shift applied to second antenna.

the receiver with the propagation modeled by channel model D is given in Figure 4.13. The top graph demonstrates two transmit antennas, and the bottom graph demonstrates four transmit antennas. Several combinations of cyclic shifts are given. For both graphs, the legend gives the cyclic shift on the signals transmitted on each antenna.

With both two and four transmit antennas, having no cyclic shifts on any of the signals results in large variation between the power of the L-STF and Data field, as illustrated by the dash-dotted lines. When the power of the L-STF is smaller than the Data field, the gain is set too high causing saturation or clipping of the data. When the power of the L-STF is larger than the Data field, the gain is set too low causing quantization of the data. With two antennas, the power variation between L-STF and Data field diminishes as the cyclic shift exceeds –200 ns. However with four antennas, cyclic shifts of up to –600 ns are required to minimize variation.

4.2.1.2   Legacy compatibility
Legacy OFDM devices receive the HT packet with MF preamble and process the L-STF, L-LTF, and L-SIG. As discussed in Section 4.1.1 and illustrated in Figure 4.5, the performance of legacy OFDM devices that employ a cross-correlation receiver is degraded by delayed responses. The MF preamble with cyclic shifts applied to the signals on the additional antennas appears to the legacy OFDM devices as a signal convolved with a channel with a delayed response. The larger the cyclic shift, the larger the delay spread, and the more the legacy receiver is degraded.

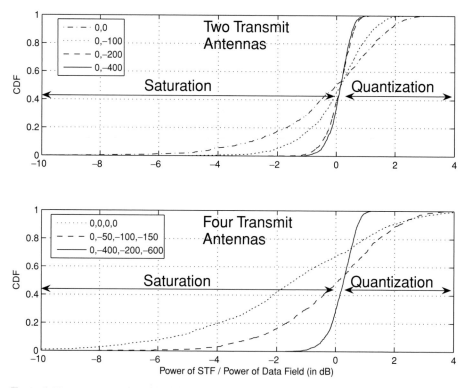

Figure 4.13 Power of the L-STF relative to power of the Data field for channel model D.

During development of the standard, measurements were made by Aoki and Takeda (2005) with 802.11a devices with cross-correlation receivers to test the sensitivity to cyclic shifts. Figure 4.14 illustrates the results of their experiment. Since the legacy device is only able to process up to the L-SIG in the MF preamble, only the SIG error rate was measured. The line marked with diamonds (♦) is the performance of the receiver when receiving an 802.11a transmission, to which we are comparing. The line marked with squares (■) demonstrates the legacy performance when receiving a two antenna transmission with 50 ns cyclic shift on the second antenna. The line marked with triangles (▲) demonstrates the legacy performance when receiving a two antenna transmission with 400 ns cyclic shift on the second antenna. As seen from the results, with 50 ns cyclic shift the SIG error rate is comparable to 802.11a performance. However, with 400 ns cyclic shift the SIG error rate degrades substantially. A compromise was reached to define a maximum magnitude of 200 ns cyclic shift on the non-HT part of the MF preamble.

Table 4.2 gives the values of the cyclic shifts that are applied to the L-STF, L-LTF, L-SIG, and the HT-SIG of the MF preamble.

With more than four transmit antenna chains, specific values for the cyclic shifts on the additional transmit chains are not given. However, the standard states that the shifts on the additional antennas must be between −200 ns and 0 ns.

Table 4.2 Cyclic shift for non-HT portion of the packet (IEEE, 2007b)

Number of Tx chains	Cyclic shift for Tx chain 1 (ns)	Cyclic shift for Tx chain 2 (ns)	Cyclic shift for Tx chain 3 (ns)	Cyclic shift for Tx chain 4 (ns)
1	0	–	–	–
2	0	−200	–	–
3	0	−100	−200	–
4	0	−50	−100	−150

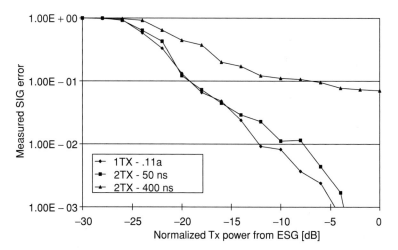

Figure 4.14 Preamble performance with cross-correlation 802.11a receiver in channel model D. Reproduced with permission from Aoki and Takeda (2005) © IEEE.

4.2.1.3 Non-HT Short Training field

In 802.11a, the scaling of the Short Training field was set to make the power of the symbol equal to that of the Long Training field. In 802.11n, the scaling of the symbol is set to make the total power of the time domain signal (summed over all transmit chains) equal to one. Thus, the L-STF frequency domain sequence S_k for the 20 MHz channel is identical to that of 802.11a, given by Eq. (4.1), except as above, the scaling of the sequence for L-STF is $\sqrt{1/2}$ rather than $\sqrt{13/6}$. The time domain L-STF waveform for the 20 MHz channel bandwidth for each transmit chain i_{TX} is defined by the equation below:

$$r_{\text{L-STF}}^{(i_{TX})}(t) = \frac{1}{\sqrt{N_{TX} \cdot 12}} \sum_{k=-26}^{26} S_k \exp\left(j\,2\pi\,k\,\Delta_F\left(t - T_{CS}^{i_{TX}}\right)\right) \tag{4.13}$$

where $\Delta_F = 312.5\,\text{kHz}$ (20 MHz/64); $T_{CS}^{i_{TX}}$ represents the cyclic shift for transmit chain i_{TX} with values given by Table 4.2; and N_{TX} is the number of transmit chains.

As described in Section 4.1.1, the L-STF may be created by performing a 64-point IFFT over the frequency domain sequence resulting in a 3.2 μs time domain sequence, which is repeated two and a half times to create the 8 μs L-STF waveform. A windowing function may be applied to the L-STF waveform to smooth transitions between it and the L-LTF (refer to Section 4.1.5, and Eqs. (4.6) and (4.7)).

4.2.1.4   Non-HT Long Training field

The L-LTF frequency domain sequence for the 20 MHz channel is identical to that of 802.11a, given by Eq. (4.2). The time domain L-LTF waveform for the 20 MHz channel bandwidth for each transmit chain i_{TX} is defined by the equation below:

$$r_{L-LTF}^{(i_{TX})}(t) = \frac{1}{\sqrt{N_{TX} \cdot 52}} \sum_{k=-26}^{26} L_k \exp\left(j2\pi \, k\Delta_F \left(t - T_{GI2} - T_{CS}^{i_{TX}}\right)\right) \quad (4.14)$$

where $\Delta_F = 312.5$ kHz (20 MHz/64); $T_{CS}^{i_{TX}}$ represents the cyclic shift for transmit chain i_{TX} with values given by Table 4.2; T_{GI2} is the 1.6 µs guard interval; and N_{TX} is the number of transmit chains.

As described in Section 4.1.2, the resulting long training symbol created by performing a 64-point IFFT over the frequency domain sequence is a 3.2 µs time domain sequence. The 8 µs L-LTF waveform may be created by repeating the long training symbol twice and prepended by a 1.6 µs cyclic prefix. A windowing function may be applied to the L-LTF waveform to smooth transitions between it and the L-STF and L-SIG (refer to Section 4.1.5, and Eqs. (4.6) and (4.7)).

4.2.1.5   Non-HT Signal field

The construction of the Non-HT Signal field is identical to the Signal field in 802.11a and described in Section 4.1.3. Although some fields have been co-opted to protect 802.11n from 802.11a devices, e.g. the rate field is always set to 6 Mbps and the length field is set so that 802.11a devices correctly defer transmission for at least the entire duration of the HT packet.

Furthermore, the waveform derived from the 48 complex data d_k, $k = 0, \ldots, 47$, may need to be transmitted over multiple antennas. Legacy cyclic shifts are applied to each signal on each transmit antenna chain i_{TX} as described in Section 4.2.1.1. The L-SIG waveform for each transmit antenna chain i_{TX} is given below:

$$r_{L-SIG}^{(i_{TX})}(t) = \frac{1}{\sqrt{N_{TX} \cdot 52}} \sum_{k=-26}^{26} (D_k + p_0 P_k) \exp\left(j2\pi k\Delta_F (t - T_{GI} - T_{CS}^{i_{TX}})\right) \quad (4.15)$$

where D_k and $M^r(k)$, given by

$$D_k = \begin{cases} 0 & k = 0, \pm 7, \pm 21 \\ d_{M^r(k)} & \text{otherwise} \end{cases}$$

$$M^r(k) = \begin{cases} k + 26 & -26 \leq k \leq -22 \\ k + 25 & -20 \leq k \leq -8 \\ k + 24 & -6 \leq k \leq -1 \\ k + 23 & 1 \leq k \leq 6 \\ k + 22 & 8 \leq k \leq 20 \\ k + 21 & 22 \leq k \leq 26 \end{cases}$$

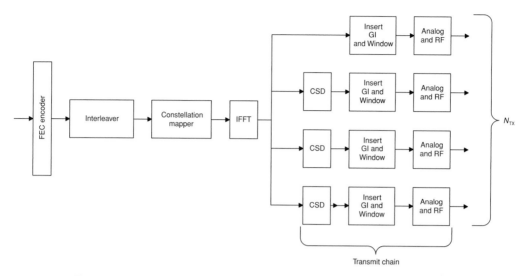

Figure 4.15 Transmitter block diagram for L-SIG and HT-SIG of the HT mixed format packet. Reproduced with permission from IEEE (2007b) © IEEE.

are the mapping of complex data values to subcarrier locations while leaving space for pilots, DC, and guard band; $T_{CS}^{i_{TX}}$ represents the non-HT cyclic shift for transmit chain i_{TX} with values given by Table 4.2; T_{GI} is the 0.8 μs guard interval; P_k is the same as legacy OFDM and is given by Eq. (4.4); and p_0 is the same as legacy OFDM and is given by Eq. (4.5).

As described in Section 4.1.3, the resulting waveform is a 3.2 μs time domain OFDM symbol. A 0.8 μs cyclic prefix is prepended, resulting in a 4.0 μs symbol. A windowing function may be applied to the L-SIG waveform to smooth transitions between it and the L-LTF and HT-SIG (refer to Section 4.1.5, and Eqs. (4.6) and (4.7)). The transmitter block diagram for L-SIG is illustrated in Figure 4.15.

4.2.2   HT portion of the MF preamble

As illustrated in Figure 4.10, the HT portion of the MF preamble consists of the High Throughput Signal field (HT-SIG), High Throughput Short Training field (HT-STF), and High Throughput Long Training field (HT-LTF). The two OFDM symbols of the HT-SIG, HT-SIG$_1$ and HT-SIG$_2$, contain new signaling information and are also used for auto-detection between HT MF and legacy OFDM packets. The HT-STF is used to redo the AGC setting. The HT-LTFs are used for MIMO channel estimation. The HT Data field follows the HT-LTFs.

4.2.2.1   High Throughput Signal field

The HT-SIG contains 48 information bits, divided into two symbols, HT-SIG$_1$ and HT-SIG$_2$, each containing 24 information bits. We require 32 bits to signal the new HT

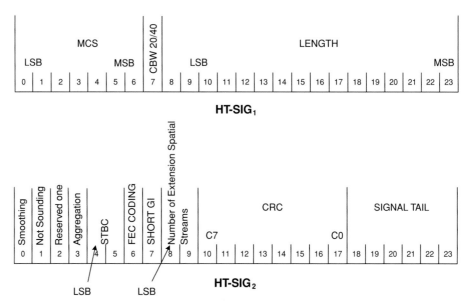

Figure 4.16 Format of HT-SIG$_1$ and HT-SIG$_2$. Reproduced with permission from IEEE (2007b) © IEEE.

information and modes. In addition, the HT-SIG includes an 8-bit CRC for strong error detection, as opposed to the single parity in the L-SIG. As in the L-SIG, the last 6 bits are tail bits (set to zero) to terminate the trellis of the convolution code since the HT-SIG is also encoded with a rate $\frac{1}{2}$ BCC. The HT-SIG is illustrated in Figure 4.16.

In HT-SIG$_1$, the first 7 bits are used to signal which of the modulation coding schemes (MCS) 0 to 76 is used for transmission of the Data field. The MCSs consist of BPSK, QPSK, 16-QAM, and 64-QAM, identical to 802.11a. The coding rates include 1/2, 2/3, and 3/4 the same as 802.11a, and a new rate 5/6. MCS 0 to 7 consist of single stream, MCS 8 to 15 consist of two streams, MCS 16 to 23 consist of three streams, and MCS 24 to 31 consist of four streams. MCS 0 to 31 have equal modulation on each stream. MCS 32 only applies to 40 MHz operation and is discussed in Chapter 5. MCS 33 to 76 have unequal modulation on each stream and are discussed in Chapter 12. The MCS parameters and data rates for 20 MHz, equal modulation, and 800 ns guard interval are defined in Tables 4.7–4.10. Values 77 to 127 are reserved.

The CBW 20/40 bit is used to indicate whether the transmission has 20 MHz or 40 MHz bandwidth. This information is useful to allow a 20 MHz receiver to determine that it is not capable of receiving the transmission and to terminate reception early, reducing power consumption. Sixteen bits are used for the Length field to allow for long packets in the range of 0 to 65 535 bytes. Longer packets than 802.11a are needed to improve efficiency at higher data rates.

In HT-SIG$_2$, the Smoothing bit, Not Sounding bit, and Number of Extension Spatial Streams bits are all related to Tx beamforming. With Tx beamforming or spatial expansion, the resulting channel between the transmitter and receiver could exceed 800 ns.

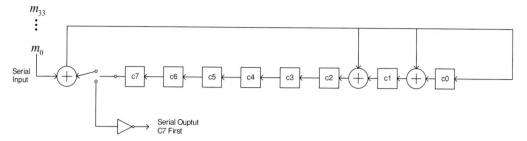

m_{33}

⋮

m_0

Serial Input

c7 c6 c5 c4 c3 c2 c1 c0

Serial Ouptut
C7 First

Figure 4.17 HT-SIG CRC calculation. Reproduced with permission from IEEE (2007b) © IEEE.

High delay spread results in the decorrelation of adjacent subcarriers. In addition, with certain methods of Tx beamforming, a phase discontinuity could exist between adjacent subcarriers. Both of these conditions degrade the channel estimate when subcarrier smoothing is used (refer to Section 4.1.2 for a discussion on subcarrier smoothing). Under these conditions, the transmitter should set the Smoothing bit to 0 to inform the receiver that only per-carrier (unsmoothed) channel estimation should be employed.

When the Not Sounding bit is set to 0, the transmission is a sounding packet. A sounding packet is used to gather channel state information for Tx beamforming and link adaptation. To sound extra spatial dimensions, beyond that of the Data field, the Number of Extension Spatial Streams field is set to greater than zero. Further detail on sounding packets is found in Section 12.6.

The aggregate bit indicates whether the payload contains a single MPDU (aggregate bit set to 0) or an aggregate of MPDUs (A-MPDU) (aggregate bit set to 1). An A-MPDU contains delimiters between the individual MPDUs. The PHY reports to the MAC that the packet contains an A-MPDU immediately after decoding the HT-SIG. The MAC requires this information in order to extract the individual MPDUs; refer to Section 8.2.2.

The two STBC bits indicate the dimensionality of the STBC operation. When the bits have a value of 0, there is no STBC. The value of 3 is reserved. Other values are discussed in detail in Section 6.3.

The remaining three informational bits include a bit for FEC Coding that is set to 0 to indicate that the data is BCC encoded, or set to 1 to indicate LDPC encoding. The other bit is the Short GI bit, which is set to 1 to indicate that a 400 ns short guard interval is used, rather than the standard 800 ns length guard interval. These features are presented in Chapter 6. The last bit is a reserved bit that is always set to 1.

The CRC protects bits 0 to 33 of the HT-SIG (bits 0 to 23 of HT-SIG$_1$ and bits 0 to 9 of HT-SIG$_2$) (IEEE, 2007b). The generator polynomial is $G(D) = D^8 + D^2 + D + 1$. Figure 4.17 shows the operation of the CRC. First, the shift register is reset to all ones. The HT-SIG bits are then passed sequentially through the exclusive-or operation (XOR) at the input. When the last bit has entered, the output is generated by shifting the bits out of the shift register, c7 first, through an inverter.

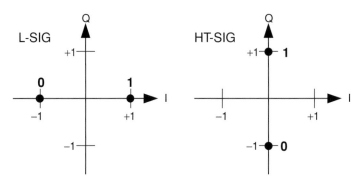

Figure 4.18 Data subcarrier constellations of L-SIG and HT-SIG. Reproduced with permission from IEEE (2007b) © IEEE.

As an example, if bits $\{m0 \ldots m33\}$ are given by

$$\{1111000100100110000000001110000000\}$$

the resulting CRC bits $\{c7 \ldots c0\}$ are $\{10101000\}$.

The 48 HT-SIG bits are BCC encoded with rate $\frac{1}{2}$ resulting in 96 bits. These are split into two symbols. Each symbol is interleaved, BPSK mapped, and has pilot subcarriers inserted as described in Section 4.1.3. The complex numbers generated by these steps are denoted as d_k, $k = 0, \ldots, 47$. The resulting time domain waveform for the two HT-SIG OFDM symbols is given in Eq. (4.16). Cyclic shifts are applied to the additional transmit antennas as with the L-SIG. To complete the waveform, a cyclic prefix is prepended to each OFDM symbol.

$$r_{HT-SIG}^{i_{TX}}(t) = \frac{1}{\sqrt{N_{TX} \cdot 52}} \sum_{n=0}^{1} \sum_{k=-26}^{26} \left(j \, D_{k,n} + p_{n+1} P_k \right)$$

$$\times \exp\left(j \, 2\pi \, k \Delta_F \left(t - n T_{SYM} - T_{GI} - T_{CS}^{i_{TX}} \right) \right) \qquad (4.16)$$

where D_k and $M^r(k)$ are the mapping of complex data values to subcarrier locations, as for L-SIG defined by Eq. (4.15); $T_{SYM} = 4 \, \mu s$ symbol time composed of 3.2 μs OFDM symbol time and 0.8 μs guard interval; $T_{CS}^{i_{TX}}$ is the non-HT cyclic shift for transmit chain i_{TX} with values given by Table 4.2; $T_{GI} = 0.8 \, \mu s$ guard interval; P_k denotes the pilot subcarriers defined to be the same as legacy OFDM and given by Eq. (4.4); and p_{n+1} is the pilot polarity defined to be the same as legacy OFDM and given by Eq. (4.5) – since the first value of the sequence is used for L-SIG, HT-SIG uses the next two values.

A windowing function may be applied to each of the HT-SIG symbols to smooth transitions between it and adjacent symbols (refer to Section 4.1.5, and Eqs. (4.6) and (4.7)). The transmitter block diagram for the HT-SIG symbols is illustrated in Figure 4.15.

Note that in Eq. (4.16) the data D_k are rotated by 90 degrees (multiplication by j). Since both the L-SIG and HT-SIG are BPSK modulated, the HT-SIG is rotated by 90 degrees relative to L-SIG, as illustrated in Figure 4.18.

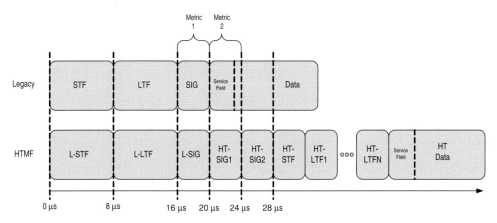

Figure 4.19 Timing of the preamble fields in legacy and HT MF.

The rotation allows for robust auto-detection between L-SIG and HT-SIG$_1$ to differentiate between legacy OFDM packets and HT MF packets.

An example auto-detection algorithm is as follows. For each symbol that is processed, compute a metric based on the power of the real part of the data subcarriers averaged over the 48 data subcarriers minus the power of the imaginary part of the data subcarriers averaged over the 48 data subcarriers. Figure 4.19 illustrates the relative timing of the preamble fields between legacy and HT MF. During the time spanning 16 μs to 20 μs, the symbol corresponds to the SIG in a legacy OFDM packet or the L-SIG in the HT MF packet. For both packet formats, the metric is a large positive number. During the time spanning 20 μs to 24 μs, the symbol corresponds to the first data symbol in a legacy OFDM packet or HT-SIG$_1$ in the HT MF packet. For the first data symbol in a legacy OFDM packet, the metric ranges from large positive with BPSK modulation to 0 for QAM modulation. (Moreover, in the case of QAM modulation the packet cannot be HT, yet due to pipelining this knowledge is not yet available.) For the HT-SIG$_1$ of the HT MF packet, the metric is a large negative number. Finally, the metric from the time spanning 16 μs to 20 μs is compared with the metric from the time spanning 20 μs to 24 μs. With an appropriate threshold, we can differentiate between a legacy OFDM packet and an HT MF packet.

4.2.2.2 High Throughput Short Training field

As was discussed in Section 4.2.1.1, large cyclic shifts are required to decorrelate the signals when the same spatial stream is sent on the different transmit antenna chains. This was illustrated in Figure 4.13, where, with four transmit antennas, cyclic shifts with a magnitude of 600 ns are required for the STF to have comparable power to the Data field. However, for backward compatibility with legacy OFDM devices, a cyclic shift with only a maximum magnitude of 200 ns is needed for L-STF as discussed in Section 4.2.1.2. Therefore, an additional field is added after HT-SIG to enable accurate power estimates for AGC (automatic gain control) setting required for MIMO. The HT-STF allows for a more accurate AGC setting than L-STF and reduces the dynamic range

Table 4.3 Cyclic shift for values for HT-STF, HT-LTF, and Data field (IEEE, 2007b)

Number of spatial streams	Cyclic shift for spatial stream 1 (ns)	Cyclic shift for spatial stream 2 (ns)	Cyclic shift for spatial stream 3 (ns)	Cyclic shift for spatial stream 4 (ns)
1	0	–	–	–
2	0	−400	–	–
3	0	−400	−200	–
4	0	−400	−200	−600

requirement on the ADC. The values of the cyclic shifts for different numbers of spatial streams are given in Table 4.3.

The time domain HT-STF waveform for 20 MHz channel bandwidth for each spatial stream i_{SS} is defined by Eq. (4.17). The subcarrier sequence S_k for the HT-STF is identical to that of the L-STF, defined in Eq. (4.13). For simplicity, we assume that the number of spatial streams is equal to the number of transmit antennas (direct mapping). Spatial mapping, which maps fewer spatial streams to more transmit antennas, is discussed in Section 4.2.3.8.

$$r_{HT-STF}^{i_{SS}}(t) = \frac{1}{\sqrt{N_{SS} \cdot 12}} \sum_{k=-26}^{26} S_k \exp\left(j\, 2\pi k \Delta_F\left(t - T_{CS}^{i_{SS}}\right)\right) \tag{4.17}$$

where $\Delta_F = 312.5$ kHz (20 MHz/64); $T_{CS}^{i_{SS}}$ represents the cyclic shift for spatial stream i_{SS} with values given by Table 4.3; and N_{SS} is the number of spatial streams.

The HT-STF symbol is 4.0 μs in length. A windowing function may be applied to the HT-STF symbol to smooth transitions between it and adjacent symbols (refer to Section 4.1.5, and Eqs. (4.6) and (4.7)).

4.2.2.3   High Throughput Long Training field

A new HT Long Training field is provided in the HT MF preamble for MIMO channel estimation. A 4 μs HT long training symbol is transmitted for each spatial stream indicated by the selected MCS, except in the case of three spatial streams where four long training symbols are transmitted. An orthogonal mapping matrix is used to generate the HT-LTF. In order to permit non-complex mapping matrix for three spatial streams, the extra long training symbol was added. The same cyclic shifts as used for HT-STF, defined in Table 4.3, are applied to the long training symbols to accurately match the power variation of the Data field.

The HT-LTF is illustrated in Figure 4.20. The HT long training frequency sequence is multiplied by a value (+1 or −1) from the orthogonal mapping matrix. Each long training symbol is cyclic shifted with values given in Table 4.3.

The HT long training symbol for 20 MHz operation is defined based on the frequency domain sequence given in Eq. (4.18) (IEEE, 2007b). This sequence is an extension of the legacy long training symbol (defined by Eq. (4.2)), where two additional subcarriers,

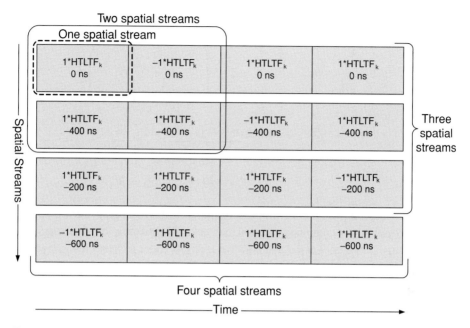

Figure 4.20 Construction of the HT-LTF.

−28 and −27, are filled with +1 and two additional subcarriers, 27 and 28, are filled with −1:

HTLTF$_{-28:28}$

$$
= \begin{bmatrix} 1, 1, 1, 1, -1, -1, 1, 1, -1, 1, -1, 1, 1, 1, 1, 1, 1, -1, -1, 1, 1, -1, 1, -1, \\ 1, 1, 1, 1, 0, 1, -1, -1, 1, 1, -1, 1, -1, 1, -1, -1, -1, -1, -1, 1, 1, -1, \\ -1, 1, -1, 1, -1, 1, 1, 1, 1, -1, -1 \end{bmatrix}
$$

(4.18)

The time domain HT-LTF waveform for 20 MHz channel bandwidth for each spatial stream i_{SS} and for each long training symbol n, $1 \leq n \leq N_{LTF}$, is defined by Eq. (4.19). For simplicity, we assume that the number of spatial streams is equal to the number of transmit antennas (direct mapping). Spatial mapping, which maps fewer spatial streams to more transmit antennas, is discussed in Section 4.2.3.8.

$$
r_{HT-LTF}^{n,i_{SS}}(t) = \sqrt{\frac{1}{N_{SS} \cdot 56}} \cdot \sum_{k=-28}^{28} [P_{HTLTF}]_{i_{SS},n} \, \text{HTLTF}_k
$$
$$
\times \exp\left(j \, 2\pi \, k \Delta_F \left(t - T_{GI} - T_{CS}^{i_{SS}}\right)\right)
$$

(4.19)

where T_{GI} is the 0.8 μs guard interval; $\Delta_F = 312.5$ kHz (20 MHz/64); $T_{CS}^{i_{SS}}$ represents the cyclic shift for spatial stream i_{SS} with values given by Table 4.3; and N_{SS} is the number of spatial streams.

The orthogonal mapping matrix is given in Eq. (4.20) (IEEE, 2007b):

$$
P_{\text{HTLTF}} = \begin{pmatrix} 1 & -1 & 1 & 1 \\ 1 & 1 & -1 & 1 \\ 1 & 1 & 1 & -1 \\ -1 & 1 & 1 & 1 \end{pmatrix}
\tag{4.20}
$$

The number of long training symbols is given in Eq. (4.21):

$$
N_{\text{LTF}} = \begin{cases} N_{\text{SS}} & \text{if } N_{\text{SS}} = 1, 2, 4 \\ 4 & \text{if } N_{\text{SS}} = 3 \end{cases}
\tag{4.21}
$$

To complete the waveform, a 0.8 μs cyclic prefix is prepended to each 3.2 μs OFDM symbol. A windowing function may be applied to each long training symbol to smooth transitions between it and adjacent symbols (refer to Section 4.1.5, and Eqs. (4.6) and (4.7)).

4.2.3   Data field

Many of the steps in the encoding process of the Data field are the same as 802.11a, a description of which will not be repeated. For an explanation of the 802.11a functions, refer to Section 4.1.5. The transmitter block diagram for the HT Data field is given in Figure 4.21.

The following sections describe the encoding process when using the mandatory binary convolutional code (BCC). Refer to Section 6.4 for the encoding steps when using the optional low density parity check (LDPC) code. In addition, refer to Section 6.3 for the waveform design when using the optional space-time block code (STBC).

4.2.3.1   Bit string

As in 802.11a, the Service field consisting of 16 zero bits is prepended to the data bits. With BCC, six (all zeros) tail bits are appended to the data bits. If necessary, the bit string is extended with pad bits (all zeros) such that the resulting length can be divided equally into an integer multiple of OFDM symbols. The encoding process for the optional LDPC code is described in Section 6.4.

The number of OFDM symbols in the Data field is as follows:

$$
N_{\text{SYM}} = \left\lceil \frac{8 \cdot \text{length} + 16 + 6}{N_{\text{DBPS}}} \right\rceil
\tag{4.22}
$$

where length is the value of the Length field in octets in the HT-SIG field illustrated in Figure 4.16; and N_{DBPS} is the number of data bits per OFDM symbol determined by the selected MCS.

The symbol $\lceil x \rceil$ denotes the smallest integer greater than or equal to x, corresponding to the *ceiling* function. The number of pad bits is therefore $N_{\text{SYM}} \cdot N_{\text{DBPS}} - 8 \cdot \text{length} - 16 - 6$.

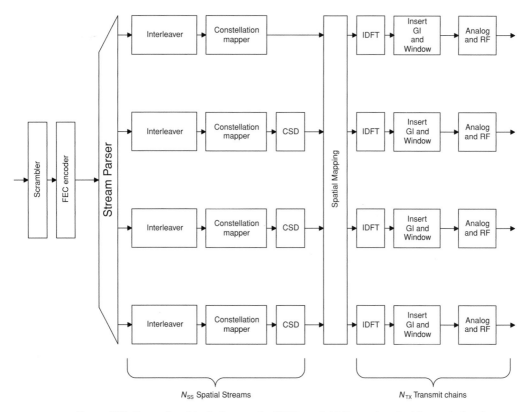

Figure 4.21 Transmitter block diagram for HT Data field. Reproduced with permission from IEEE (2007b) © IEEE.

4.2.3.2 Scrambling and encoding

The resulting data bit string is scrambled with the same scrambler as in 802.11a, described in Section 4.1.4. The six scrambled zero bits that served as tail bits following the data are replaced with the same number of non-scrambled zero bits. This returns the convolutional encoder to the zero state, and is described further in IEEE (2007a, Clause 17.3.5.2).

The scrambled bit string is encoded with the rate $\frac{1}{2}$ BCC. After encoding, the bit string is punctured to the selected data rate. Both encoding and puncturing are the same as in 802.11a and are described in Section 4.1.5.

4.2.3.3 Stream parsing

After encoding and puncturing, the bit string is re-arranged into a new set of bit strings equal to the number of spatial streams (N_{SS}). To simplify the description, this section only describes the stream parsing function for equal MCSs. For unequal MCSs, refer to Section 12.4. The output of the stream parser is N_{SS} bit strings, each of length N_{CBPSS} (coded bits per spatial stream).

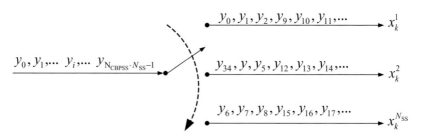

Figure 4.22 Stream parser for single encoder and equal MCS.

A block of s bits is alternately assigned to different spatial streams in a round robin fashion. Figure 4.22 illustrates an example of 64-QAM ($s = 3$) and three spatial streams.

The size of a block of s bits is determined from a single axis (real or imaginary) of a constellation point, defined in the equation below:

$$s = \max \left\{ 1, \frac{N_{BPSCS}}{2} \right\}$$

(4.23)

where N_{BPSCS} is the number of coded bits per single subcarrier for each spatial stream, or equivalently the modulation order. The output of the stream parser, $x_k^{i_{SS}}$, is a function of the input, y_i, and is defined as follows:

$$x_k^{i_{SS}} = y_i$$

$$i = (i_{SS} - 1) \cdot s + s \cdot N_{SS} \cdot \left\lfloor \frac{k}{s} \right\rfloor + k \bmod s$$

(4.24)

where $1 \leq i_{SS} \leq N_{SS}$ and $k = 0, 1, \ldots, N_{CBPSS} - 1$.

The symbol $\lfloor x \rfloor$ denotes the largest integer less than or equal to x, corresponding to the *floor* function. $k \bmod s$ is defined as the remainder from the division of the integer k by the integer s.

4.2.3.4 Interleaving

After the stream parser, each of the N_{SS} bit strings is interleaved using three permutations if BCC encoding was used. The first two interleaver permutations are based on the interleaver for 802.11a. The first permutation ensures that adjacent coded bits are mapped onto non-adjacent subcarriers (IEEE, 2007a). The second permutation ensures that adjacent coded bits are mapped alternately onto less and more significant bits of the constellation, avoiding long runs of low reliability bits (IEEE, 2007a). If more than one spatial stream exists, the third operation is applied, which performs a frequency rotation to the additional spatial streams.

In this section, we limit the description of the interleaver to 20 MHz operation and equal MCS to simplify the explanation. Interleaver design for 40 MHz is addressed in Section 5.1.4 and unequal MCS is addressed in Section 12.4.

The output of the first interleaver permutation, w_k^{iss}, is a function of the output of the stream parser, x_i^{iss}, and is defined as follows:

$$w_k^{iss} = x_i^{iss}$$

$$i = 4 \cdot N_{\text{BPSCS}} \cdot (k \bmod 13) + \left\lfloor \frac{k}{13} \right\rfloor \tag{4.25}$$

where $1 \leq iss \leq N_{SS}$, $k = 0, 1, \ldots, N_{\text{CBPSS}} - 1$.

Comparing the first permutation to that of 802.11a, the modulus and floor function use a factor of 16 because 802.11a consists of 48 data subcarriers, which is divisible by 16. In 802.11n 20 MHz operation, there are 52 data subcarriers, which is divisible by 13.

The output of the second interleaver permutation, y_k^{iss}, is a function of the output of the first interleaver permutation, w_i^{iss}, and is defined as follows:

$$y_k^{iss} = w_j^{iss}$$

$$j = s \cdot \left\lfloor \frac{k}{s} \right\rfloor + \left(k + N_{\text{CBPSS}} - \left\lfloor \frac{13 \cdot k}{N_{\text{CBPSS}}} \right\rfloor \right) \bmod s \tag{4.26}$$

where $1 \leq iss \leq N_{SS}$, $k = 0, 1, \ldots, N_{\text{CBPSS}} - 1$, and s is the same block size as with the stream parser, defined by Eq. (4.23).

If there are more than one spatial stream, the third permutation is performed. The output of the third interleaver permutation, z_k^{iss}, is a function of the output of the second interleaver permutation, y_k^{iss}, and is defined as follows:

$$z_k^{iss} = y_r^{iss}$$

$$r = \left(k - \left(((iss - 1) \cdot 2) \bmod 3 + 3 \cdot \left\lfloor \frac{iss - 1}{3} \right\rfloor \right) \cdot 11 \cdot N_{\text{BPSCS}} \right) \bmod N_{\text{CBPSS}} \tag{4.27}$$

where $1 \leq iss \leq N_{SS}$, and $k = 0, 1, \ldots, N_{\text{CBPSS}} - 1$.

4.2.3.5 Modulation mapping

The subcarriers in 802.11n are BPSK, QPSK, 16-QAM, or 64-QAM modulated based on the selected MCS in the same method as in 802.11a (refer to IEEE (2007a, Clause 17.3.5.7) or van Nee and Prasad (2000) for further details). The bit streams at the output of the interleaver are accordingly mapped to complex constellation points, with the notation shown in Eq. (4.28) for the data streams. Each 20 MHz HT OFDM symbol consists of 52 complex data values for each stream indexed by k.

$$d_{k,l,n}, \quad k = 0, 1, \ldots, 51, \ l = 1, \ldots, N_{SS}, \ n = 0, 1, \ldots, N_{\text{SYM}} - 1 \tag{4.28}$$

where N_{SYM} is the number of OFDM symbols defined in Eq. (4.22), and N_{SS} is the number of spatial streams.

Table 4.4 Pilot subcarrier values for 20 MHz transmission
(IEEE, 2007b)

N_{SS}	i_{SS}	$\Psi_{i_{SS},0}^{(N_{SS})}$	$\Psi_{i_{SS},1}^{(N_{SS})}$	$\Psi_{i_{SS},2}^{(N_{SS})}$	$\Psi_{i_{SS},3}^{(N_{SS})}$
1	1	1	1	1	−1
2	1	1	1	−1	−1
2	2	1	−1	−1	1
3	1	1	1	−1	−1
3	3	1	−1	1	−1
3	3	−1	1	1	−1
4	1	1	1	1	−1
4	2	1	1	−1	1
4	3	1	−1	1	1
4	4	−1	1	1	1

4.2.3.6   Pilot subcarriers

In 802.11n 20 MHz operation, the pilot subcarriers are located in the same subcarriers as in 802.11a, −21, −7, 7, and 21, as given in Eq. (4.29) (IEEE, 2007b):

$$P_{(i_{SS},n)}^{-28,28} = \{0, 0, \ 0, 0, 0, 0, 0, \ \Psi_{i_{SS},n\oplus4}^{(N_{SS})}, 0, 0, 0, 0, 0, 0, 0, 0, 0, 0, 0, 0, 0,$$

$$\Psi_{i_{SS},(n+1)\oplus4}^{(N_{SS})}, 0, 0, 0, 0, 0, 0, 0, 0, 0, 0, 0, 0, 0, 0, \Psi_{i_{SS},(n+2)\oplus4}^{(N_{SS})}, \qquad (4.29)$$

$$0, 0, 0, 0, 0, 0, 0, 0, 0, 0, 0, 0, 0, \Psi_{i_{SS},(n+3)\oplus4}^{(N_{SS})}, 0, 0, 0, 0, 0, 0, 0\}$$

where $n \oplus a$ indicates symbol number modulo integer a and the patterns $\Psi_{i_{SS},n}^{(N_{SS})}$ are defined in Table 4.4.

For each spatial stream there is a different pilot subcarrier pattern. In addition, for each number of spatial streams, the set of pilot subcarrier patterns is different. And finally, the pattern is cyclically shifted from symbol to symbol due to the modulo indexing operation of $\Psi_{i_{SS},n\oplus4}^{(N_{SS})}$ in Eq. (4.29).

4.2.3.7   Transmission in 20 MHz HT format

The 52 data subcarriers and the four pilot subcarriers for each spatial stream are combined and an IFFT is performed to generate the OFDM data symbol. For simplicity, we assume the number of spatial streams is equal to the number of transmit antennas (direct mapping). Spatial mapping, which maps fewer spatial streams to more transmit antennas, is discussed further in Section 4.2.3.8. The resulting time domain waveform for the OFDM data symbols is given by Eq. (4.30). To complete the waveform, a cyclic prefix is prepended to each OFDM symbol:

$$r_{HT-DATA}^{n,i_{SS}}(t) = \sqrt{\frac{1}{N_{SS} \cdot 56}} \cdot \sum_{n=0}^{N_{SYM}-1} \sum_{k=-28}^{28} \left(D_{k,i_{SS},n} + p_{n+3} P_{(i_{SS},n)}^k \right)$$

$$\times \exp\left(j2\pi k \Delta_F \left(t - nT_{SYM} - T_{GI} - T_{CS}^{i_{SS}} \right) \right) \qquad (4.30)$$

where $D_{k,l,n}$ and $M^r(k)$, is defined as

$$D_{k,l,n} = \begin{cases} 0 & k = 0, \pm 7, \pm 21 \\ d_{M^r(k),l,n} & \text{otherwise} \end{cases}$$

$$M^r(k) = \begin{cases} k + 28 & -28 \le k \le -22 \\ k + 27 & -20 \le k \le -8 \\ k + 26 & -6 \le k \le -1 \\ k + 25 & 1 \le k \le 6 \\ k + 24 & 6 \le k \le 20 \\ k + 23 & 22 \le k \le 28 \end{cases}$$

are the mapping of complex data values to subcarrier locations; T_{GI} is the 0.8 μs guard interval; T_{SYM} is the 4 μs symbol time, composed of 3.2 μs OFDM symbol time and 0.8 μs guard interval; $\Delta_F = 312.5$ kHz (20 MHz/64); T_{CS}^{iss} represents the cyclic shift for spatial stream i_{SS} with values given by Table 4.3; N_{SS} is the number of spatial streams; N_{SYM} is the number of OFDM symbols defined in Eq. (4.22); p_n is the pilot polarity defined the same as legacy OFDM and given by Eq. (4.5) since the first three values of the sequence are used for L-SIG and HT-SIG, indexing into the pilot polarity sequence is offset by three; and $P_{(i_{SS},n)}^k$ denotes the pilot subcarriers, defined by Eq. (4.29).

A windowing function may be applied to each data symbol to smooth transitions between it and adjacent symbols (refer to Section 4.1.5, and Eqs. (4.6) and (4.7)).

4.2.3.8 Spatial expansion

Consider a device with four antennas wanting to transmit two spatial streams. One reason for the usefulness of this type of configuration is if the current channel environment does not support additional spatial streams. To maximize the output power, each transmit chain needs to be utilized for transmitting the signal. The question arises of how to map fewer spatial streams to more transmit antenna chains without causing unintentional beamforming and making the best use of the available power amplifiers and transmit chains.

Continuing the two spatial stream example, we have complex data for an individual subcarrier and OFDM symbol,

$$\begin{bmatrix} d_1 \\ d_2 \end{bmatrix}$$

After applying the cyclic shift defined for two spatial streams, the complex data is given by

$$\begin{bmatrix} d_1 \\ d_2 \exp(-j\,2\pi\,k\,\Delta_F\,(-400\,\text{ns})) \end{bmatrix}$$

As a first step we can repeat the two streams over the additional antennas, as follows:

$$\frac{1}{\sqrt{2}} \begin{bmatrix} d_1 \\ d_2 \exp(-j\,2\pi\,k\,\Delta_F\,(-400\,\text{ns})) \\ d_1 \\ d_2 \exp(-j\,2\pi\,k\,\Delta_F\,(-400\,\text{ns})) \end{bmatrix}$$

However, if such a signal was transmitted, simultaneous transmission of the same signals has undesired beamforming effects. To decorrelate the signals we multiply the signals by a cyclic shift. An example is given below:

$$
\frac{1}{\sqrt{2}}
\begin{bmatrix}
d_1 \\
d_2 \exp\left(-j\, 2\pi\, k\Delta_F\,(-400\,\text{ns})\right) \\
d_1 \\
d_2 \exp\left(-j\, 2\pi\, k\Delta_F\,(-400\,\text{ns})\right)
\end{bmatrix}
\begin{matrix}
\times \\
\times \\
\times \\
\times
\end{matrix}
\begin{bmatrix}
\exp\left(-j\, 2\pi\, k\Delta_F\,(0\,\text{ns})\right) \\
\exp\left(-j\, 2\pi\, k\Delta_F\,(0\,\text{ns})\right) \\
\exp\left(-j\, 2\pi\, k\Delta_F\,(-200\,\text{ns})\right) \\
\exp\left(-j\, 2\pi\, k\Delta_F\,(-200\,\text{ns})\right)
\end{bmatrix}
$$

$$
= \frac{1}{\sqrt{2}}
\begin{bmatrix}
d_1 \exp\left(-j\, 2\pi\, k\Delta_F\,(0\,\text{ns})\right) \\
d_2 \exp\left(-j\, 2\pi\, k\Delta_F\,(-400\,\text{ns})\right) \\
d_1 \exp\left(-j\, 2\pi\, k\Delta_F\,(-200\,\text{ns})\right) \\
d_2 \exp\left(-j\, 2\pi\, k\Delta_F\,(-600\,\text{ns})\right)
\end{bmatrix}
$$

With this selection of cyclic shifts we only attain a 200 ns separation between copies of the same signal. One might be tempted to use a set of cyclic shifts consisting of [0 ns; 0 ns; −400 ns; +400 ns] which would result in an overall cyclic shift on the data of [0 ns; −400 ns; 0 ns; −400 ns] separating the same signals by 400 ns. However, in order for the HT-STF and the HT-LTF to match the data, the same spatial mapping matrix applied to the data must be applied to those fields. In the first case, the spatial mapping applied to the short training symbol would result in

$$
\frac{1}{\sqrt{2}}
\begin{bmatrix}
S\cdot \exp\left(-j\, 2\pi\, k\Delta_F\,(0\,\text{ns})\right) \\
S\cdot \exp\left(-j\, 2\pi\, k\Delta_F\,(-400\,\text{ns})\right) \\
S\cdot \exp\left(-j\, 2\pi\, k\Delta_F\,(-200\,\text{ns})\right) \\
S\cdot \exp\left(-j\, 2\pi\, k\Delta_F\,(-600\,\text{ns})\right)
\end{bmatrix}
$$

which matches the cyclic shifts in Table 4.3. In the second case, the spatial mapping applied to the short training symbol would result in

$$
\frac{1}{\sqrt{2}}
\begin{bmatrix}
S\cdot \exp\left(-j\, 2\pi\, k\Delta_F\,(0\,\text{ns})\right) \\
S\cdot \exp\left(-j\, 2\pi\, k\Delta_F\,(-400\,\text{ns})\right) \\
S\cdot \exp\left(-j\, 2\pi\, k\Delta_F\,(0\,\text{ns})\right) \\
S\cdot \exp\left(-j\, 2\pi\, k\Delta_F\,(-400\,\text{ns})\right)
\end{bmatrix}
$$

Since the same short training symbol is used over all streams, and subsequently all antennas, the resulting cyclic shift must be different for each transmit antenna chain. To summarize, the selection of the spatial mapping matrix must consider the properties of the HT-DATA as well as the HT-STF and HT-LTF.

The 802.11n standard does not specify spatial mapping matrices, though a few examples are given. Table 4.5 gives mapping matrices which result in cyclic shifts on the HT-STF and HT-LTF for each transmit antenna chain that match those in Table 4.3 for the equivalent number of spatial streams. In matrix notation, the resulting signal is $Q_k X_k$, where Q_k is the spatial expansion matrix of dimension $N_{TX} \times N_{SS}$ and X_k is the signal vector of dimension $N_{SS} \times 1$.

The waveform equations for HT-STF, HT-LTF, and HT-DATA defined by Eqs. (4.17), (4.19), and (4.30) are modified to incorporate the spatial mapping the matrix Q which

Table 4.5 Spatial expansion matrices

N_{TX}	N_{SS}	Q_k
2	1	$\dfrac{1}{\sqrt{2}}\begin{bmatrix} 1 \\ \exp\left(-j\,2\pi\,k\Delta_F\,(-400\,\text{ns})\right) \end{bmatrix}$
3	1	$\dfrac{1}{\sqrt{3}}\begin{bmatrix} 1 \\ \exp\left(-j\,2\pi\,k\Delta_F\,(-400\,\text{ns})\right) \\ \exp\left(-j\,2\pi\,k\Delta_F\,(-200\,\text{ns})\right) \end{bmatrix}$
4	1	$\dfrac{1}{2}\begin{bmatrix} 1 \\ \exp\left(-j\,2\pi\,k\Delta_F\,(-400\,\text{ns})\right) \\ \exp\left(-j\,2\pi\,k\Delta_F\,(-200\,\text{ns})\right) \\ \exp\left(-j\,2\pi\,k\Delta_F\,(-600\,\text{ns})\right) \end{bmatrix}$
3	2	$\sqrt{\dfrac{2}{3}}\begin{bmatrix} 1 & 0 \\ 0 & 1 \\ \exp\left(-j\,2\pi\,k\Delta_F\,(-200\,\text{ns})\right) & 0 \end{bmatrix}$
4	2	$\dfrac{1}{\sqrt{2}}\begin{bmatrix} 1 & 0 \\ 0 & 1 \\ \exp\left(-j\,2\pi\,k\Delta_F\,(-200\,\text{ns})\right) & 0 \\ 0 & \exp\left(-j\,2\pi\,k\Delta_F\,(-200\,\text{ns})\right) \end{bmatrix}$
4	3	$\dfrac{\sqrt{3}}{2}\begin{bmatrix} 1 & 0 & 0 \\ 0 & 1 & 0 \\ 0 & 0 & 1 \\ \exp\left(-j\,2\pi\,k\Delta_F\,(-600\,\text{ns})\right) & 0 & 0 \end{bmatrix}$

maps the N_{SS} spatial streams to the i_{TX} transmit antenna chain, as follows:

$$r_{Field}^{(i_{TX})}(t) = \frac{1}{\sqrt{N_{SS} \cdot N_{Field}^{Tone}}} \sum_{k=-N_{SR}}^{N_{SR}} \sum_{i_{SS}=1}^{N_{SS}} [Q_k]_{i_{TX},i_{SS}}\, X_k^{(i_{SS})}$$
$$\times \exp\left(j\,2\pi\,k\Delta_F\left(t - T_{GI}^{Field} - T_{CS}^{i_{SS}}\right)\right) \tag{4.31}$$

where T_{GI}^{Field} is the 0.8 μs guard interval for HT-LTF and HT-DATA, 0 for HT-STF; $N_{Field}^{Tone} = 56$ for HT-LTF and HT-DATA, 12 for HT-STF; $N_{SR} = 28$ for HT-LTF and HT-DATA, 26 for HT-STF; N_{SS} is the number of spatial streams; $T_{CS}^{i_{SS}}$ represents the cyclic shift for spatial stream i_{SS} with values given by Table 4.3; and $X_k^{(i_{SS})} = S_k$ for HT-STF, $[P_{HTLTF}]_{i_{SS},n}$ HTLTF$_k$ for HT-LTF, and $(D_{k,i_{SS},n} + p_{n+3}P_{(i_{SS},n)}^k)$ for HT-DATA.

As a final note, transmit beamforming is also used to map spatial streams to transmit antenna chains and is addressed in Chapter 12.

4.2.4 HT MF receive procedure

This section describes the receive procedure for the basic modes of operation for 20 MHz. Special conditions due to optional features are addressed in their corresponding sections.

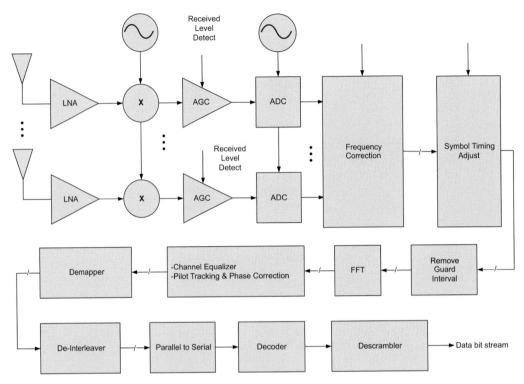

Figure 4.23 Receiver block diagram for MIMO.

Such features include GF preamble, sounding packet, short guard interval, LDPC, and STBC described in Sections 5.4, 12.6, 5.5, 6.4, and 6.3, respectively.

4.2.4.1   RF front end

A basic block diagram for a MIMO receiver is illustrated in Figure 4.23. A MIMO receiver design has multiple receive antenna chains each consisting of an antenna, LNA, down converter, AGC, and ADC. To maximize the dynamic range on each antenna branch, typically each AGC in each receive chain is individually adjusted.

Since the MF packet begins with the legacy preamble, the use of the L-STF for AGC, initial frequency correction, and initial time acquisition is similar to that of 802.11a, described in Section 4.1.6. However, 802.11n devices typically have multiple receive antennas, therefore correlation functions for start-of-packet detection and coarse timing should be summed over all the antennas to maximize detection performance. The coarse time adjustment is applied to all receive signals. Similarly, if in the transmitter all the transmit chains use the same oscillator, and in the receiver all the receive chains use the same oscillator, the frequency estimation described in Section 4.1.1 may also be summed over all receive antenna chains and applied to all received signals. Since the 802.11n standard mandates that the transmit center frequency and the symbol clock frequency for all transmit antennas is to be derived from the same reference oscillator, the requirement of common oscillator source is not onerous. Further description of time synchronization and frequency estimation for MIMO systems is given in van Zelst and Schenk (2004).

4.2.4.2   Legacy part of the preamble

Processing of the L-LTF and L-SIG follows next as described in Section 4.1.6. However, with the multiple receive antennas of a MIMO receiver, more robust processing may be performed on the single stream transmission of the L-SIG than with a single antenna legacy receiver. Reception of a single stream transmission by multiple receive antennas is generally described in Section 3.2. Furthermore, the benefits of maximal ratio combining are described in Section 6.1.

With a legacy packet, a check of the SIG parity bit to determine the validity of the packet would occur. However, if the packet is an HT packet, it is not necessary to fail the packet upon an invalid L-SIG parity bit. Therefore, the check of the L-SIG parity bit only needs to occur if it has been determined based on auto-detection that the packet is a legacy packet. The auto-detection between legacy and HT MF packets occurs during the next symbol following the L-SIG, which is the HT-SIG for an HT MF packet as described in Section 4.2.2.1.

4.2.4.3   High Throughput Signal field (HT-SIG)

Except for the 90 degree phase rotation, the HT-SIG field is modulated and encoded similarly to the L-SIG and the receive processing steps are the same as that of the L-SIG. With regard to decoding, the HT-SIG data are encompassed by two OFDM symbols and are treated as a single block when decoded. Since the frequency correction and time synchronization were computed based on L-STF and L-LTF, it advisable that phase rotation correction computed during the L-SIG and HT-SIG be used for further frequency offset and symbol timing correction before processing the HT-LTF. This minimizes the error in the channel estimates computed from the HT-LTF.

If signal loss occurs at any point during processing during the legacy part of the preamble and the HT-SIG, the receiver terminates reception. However, the device must not access the medium (defer) for as long as the received signal level remains above $-62\,\mathrm{dBm}$ (for 20 MHz operation). This ensures that even if the receiver is not able to process the signal, the fundamental nature of carrier sense multiple access is still maintained. This basic operation is termed clear channel assessment (CCA).

In the next step of the receive procedure, the HT-SIG CRC is checked for validity. With a valid determination of the HT-SIG CRC, the information bits in the HT-SIG (described in Section 4.2.2.1) are used to configure the receiver for reception of the packet. At this point even following signal loss, the device refrains from accessing the medium until the transmit time (TXTIME) computed based on the HT-SIG values has expired:

$$\mathrm{TXTIME} = T_{\mathrm{L-STF}} + T_{\mathrm{L-LTF}} + T_{\mathrm{L-SIG}} + N_{\mathrm{LTF}} \cdot T_{\mathrm{HT-LTF}} + T_{\mathrm{HT-SIG}} + T_{\mathrm{SYM}} \cdot N_{\mathrm{SYM}}$$

$$(4.32)$$

where $T_{\mathrm{L-STF}}$, $T_{\mathrm{L-LTF}}$, $T_{\mathrm{HT-SIG}} = 8\,\mu s$ duration; $T_{\mathrm{L-SIG}}$, $T_{\mathrm{HT-LTF}} = 4\,\mu s$ duration; $T_{\mathrm{SYM}} = 4\,\mu s$ duration for 800 ns guard interval operation; N_{SYM} is the number of OFDM symbols in the Data field defined by Eq. (4.22); and N_{LTF} is the number of long training symbols, determined from the MCS – the number of long training symbols is based on the number of spatial streams indicated by the MCS (excluding a sounding packet).

The computation of TXTIME changes for GF preamble, short guard interval, sounding packet, LDPC, and STBC. The modifications for TXTIME are described in Sections 5.4.4, 5.5, 12.6, 6.4, and 6.3, respectively.

TXTIME must be computed even if the indicated rate or mode is not supported. An exception to the TXTIME computation is if the reserved bit is decoded as 0. If the HT-SIG CRC is valid, then a reserved bit value of 0 indicates an invalid condition or a false positive of the CRC computation. Furthermore, if the MCS field is in the range 77–127 or the STBC field has a value of 3, TXTIME can not be computed. In these conditions, the device must not access the medium for as long as the received signal level remains above -62 dBm (for 20 MHz operation), rather than based on the TXTIME computation.

One further note on proper CCA, even though the GF preamble is optional, an HT device not capable of receiving such preamble type must still properly defer. Rules regarding CCA of a GF packet received by an HT non-GF capable device are discussed in Section 5.4.4.

4.2.4.4 High Throughput Training fields and MIMO channel estimation

Following the HT-SIG fields is the HT-STF that is used to readjust the AGC, as described in Section 4.2.2.2. The subsequent symbols comprise the HT-LTF used for MIMO channel estimation. To explain the steps to compute the channel estimates for the MIMO channel, we begin with a two stream, two transmit and two receive antenna example. The receiver processes each 3.2 µs long training symbol by removing the cyclic prefix from the HT-LTF and performs an FFT to extract the training subcarriers. For the first symbol in time, indicated by t_1, the frequency domain representation for each training subcarrier k at the receiver is represented by Eq. (4.33). Note that z is AWGN. Refer to Figure 3.4 in Section 3.3 for a general illustration of the MIMO system.

$$
\begin{aligned}
y_{1,t_1}^k &= h_{11}^k \cdot \text{HTLTF} \cdot e^{-j 2\pi k \Delta_F T_{CS}^1} + h_{12}^k \cdot \text{HTLTF} \cdot e^{-j 2\pi k \Delta_F T_{CS}^2} + z_{1,t_1}^k \\
y_{2,t_1}^k &= h_{21}^k \cdot \text{HTLTF} \cdot e^{-j 2\pi k \Delta_F T_{CS}^1} + h_{22}^k \cdot \text{HTLTF} \cdot e^{-j 2\pi k \Delta_F T_{CS}^2} + z_{2,t_1}^k
\end{aligned}
\tag{4.33}
$$

Then at the second symbol in time, t_2,

$$
\begin{aligned}
y_{1,t_2}^k &= h_{11}^k \cdot \text{HTLTF} \cdot (-1) \cdot e^{-j 2\pi k \Delta_F T_{CS}^1} + h_{12}^k \cdot \text{HTLTF} \cdot e^{-j 2\pi k \Delta_F T_{CS}^2} + z_{1,t_2}^k \\
y_{2,t_2}^k &= h_{21}^k \cdot \text{HTLTF} \cdot (-1) \cdot e^{-j 2\pi k \Delta_F T_{CS}^1} + h_{22}^k \cdot \text{HTLTF} \cdot e^{-j 2\pi k \Delta_F T_{CS}^2} + z_{2,t_2}^k
\end{aligned}
\tag{4.34}
$$

we see the effect of the orthogonal mapping matrix in Eq. (4.20). The "-1" term is from the first row and second column of the matrix. To simplify the expressions, we combine the channel tap h with the cyclic shift, $\tilde{h}_{i,m} = h_{i,m} e^{-j 2\pi k \Delta_F T_{CS}^m}$. As a note, the cyclic shifts applied to the long training symbols are also applied to the data symbols. Redefining the channel taps to incorporate the cyclic shift has the added advantage of automatically removing the effect of the cyclic shift when equalizing the data with the channel estimate, since the channel estimate is the combination of the channel tap and the cyclic shift. The simplification for Eqs. (4.33) and (4.34) is given in the equation below:

$$
\begin{aligned}
y_{1,t_1}^k &= \tilde{h}_{11}^k \cdot \text{HTLTF} + \tilde{h}_{12}^k \cdot \text{HTLTF} + z_{1,t_1}^k \\
y_{2,t_1}^k &= \tilde{h}_{21}^k \cdot \text{HTLTF} + \tilde{h}_{22}^k \cdot \text{HTLTF} + z_{2,t_1}^k \\
y_{1,t_2}^k &= \tilde{h}_{11}^k \cdot \text{HTLTF} \cdot (-1) + \tilde{h}_{12}^k \cdot \text{HTLTF} + z_{1,t_2}^k \\
y_{2,t_2}^k &= \tilde{h}_{21}^k \cdot \text{HTLTF} \cdot (-1) + \tilde{h}_{22}^k \cdot \text{HTLTF} + z_{2,t_2}^k
\end{aligned}
\tag{4.35}
$$

Once the two symbols are processed and the subcarriers are extracted, the subcarriers from symbol one are combined with the subcarriers from symbol two to derive the channel estimates. First the symbols from antenna one are combined, as follows. The subcarriers from the first symbol are subtracted by the subcarriers from the second symbol to derive the channel estimate for $\tilde{h}_{11}^k$. This cancels out the $\tilde{h}_{12}^k$ term. Then the subcarriers from the first symbol are added with the subcarriers from the second symbol to derive the channel estimate for $\tilde{h}_{12}^k$. This cancels out the $\tilde{h}_{11}^k$ term. This is expressed in the equation below:

$$
\begin{aligned}
\hat{h}_{11}^k &= \frac{y_{1,t_1}^k - y_{1,t_2}^k}{2 \cdot \text{HTLTF}} \\
\hat{h}_{12}^k &= \frac{y_{1,t_1}^k + y_{1,t_2}^k}{2 \cdot \text{HTLTF}}
\end{aligned}
\tag{4.36}
$$

Using the subcarriers from the second antenna, we similarly derive the channel estimate for $\tilde{h}_{21}^k$ and $\tilde{h}_{22}^k$, as follows:

$$
\begin{aligned}
\hat{h}_{21}^k &= \frac{y_{2,t_1}^k - y_{2,t_2}^k}{2 \cdot \text{HTLTF}} \\
\hat{h}_{22}^k &= \frac{y_{2,t_1}^k + y_{2,t_2}^k}{2 \cdot \text{HTLTF}}
\end{aligned}
\tag{4.37}
$$

Combining the subcarriers from the two symbols to extract the channel estimate is possible due to the orthogonal mapping matrix.

The derivation of the channel estimates can be generalized to an arbitrary number of spatial streams and receive antennas. First we define the received signal vector Y_{t_n} with the number of elements equal to the number of receive antennas (N_{RX}), and the Long Training Field index n, $1 \leq n \leq N_{\text{LTF}}$. The received signal vector from each symbol is combined into a single matrix, $[Y_{t_1}|Y_{t_2}|\cdots|Y_{t_{N_{\text{LTF}}}}]$ having the dimensions $N_{\text{RX}} \times N_{\text{LTF}}$, similarly with the noise vector Z; given in the equation below:

$$
\begin{aligned}
&\left[Y_{t_1}^k|Y_{t_2}^k|\cdots|Y_{t_{N_{\text{LTF}}}}^k\right] \\
&= \begin{bmatrix}
\tilde{h}_{11}^k & \tilde{h}_{12}^k & \cdots & \tilde{h}_{1N_{\text{SS}}}^k \\
\tilde{h}_{21}^k & \tilde{h}_{22}^k & \cdots & \tilde{h}_{2N_{\text{SS}}}^k \\
\vdots & \vdots & \ddots & \vdots \\
\tilde{h}_{N_{\text{RX}}1}^k & \tilde{h}_{N_{\text{RX}}2}^k & \cdots & \tilde{h}_{N_{\text{RX}}N_{\text{SS}}}^k
\end{bmatrix} \cdot \tilde{P} \cdot \text{HTLTF}_k + \left[Z_{t_1}^k|Z_{t_2}^k|\cdots|Z_{t_{N_{\text{LTF}}}}^k\right]
\end{aligned}
\tag{4.38}
$$

where $\tilde{P}$ is an $N_{SS} \times N_{LTF}$ subset of the orthogonal mapping matrix P_{HTLTF}. For example, with three spatial streams and four long training symbols,

$$\tilde{P} = \begin{pmatrix} 1 & -1 & 1 & 1 \\ 1 & 1 & -1 & 1 \\ 1 & 1 & 1 & -1 \end{pmatrix}$$

To compute the channel estimate matrix we multiply the received signal vector matrix by the transpose of $\tilde{P}$ and normalize, as follows:

$$\begin{bmatrix} \hat{h}_{11}^k & \hat{h}_{12}^k & \cdots & \hat{h}_{1N_{SS}}^k \\ \hat{h}_{21}^k & \hat{h}_{22}^k & \cdots & \hat{h}_{2N_{SS}}^k \\ \vdots & \vdots & \ddots & \vdots \\ \hat{h}_{N_{RX}1}^k & \hat{h}_{N_{RX}2}^k & \cdots & \hat{h}_{N_{RX}N_{SS}}^k \end{bmatrix} = \begin{bmatrix} Y_{t_1}^k | Y_{t_2}^k | \cdots | Y_{t_{N_{LTF}}}^k \end{bmatrix} \cdot \tilde{P}^T \cdot \frac{1}{N_{LTF} \cdot HTLTF_k}$$

$$(4.39)$$

4.2.4.5   Data field

Following the HT-LTF is the Data field, which is processed as illustrated in Figure 4.23. For each 4 µs symbol, the 0.8 µs GI is removed and an FFT is performed. The data may be equalized by a linear equalizer as described in Section 3.6, using the channel estimates from the HT-LTF (Eq. (4.39)).

In processing the pilot subcarriers to track the phase rotation, we again highlight the advantage of the receiver using a common oscillator for all the receive chains. A particular pilot subcarrier in the transmitted OFDM symbol in the frequency domain is described by the equation below:

$$\begin{bmatrix} y_1 \\ y_2 \\ \vdots \\ y_N \end{bmatrix} = \begin{bmatrix} h_{11} & h_{12} & \cdots & h_{1M} \\ h_{21} & h_{22} & \cdots & h_{2M} \\ \vdots & \vdots & \ddots & \vdots \\ h_{N1} & h_{N2} & \cdots & h_{NM} \end{bmatrix} \begin{bmatrix} p_1 \\ p_2 \\ \vdots \\ p_M \end{bmatrix} + \begin{bmatrix} z_1 \\ z_2 \\ \vdots \\ z_N \end{bmatrix}$$

or, in vector notation,

$$Y = HP + Z$$

If a common oscillator is employed, the received signal is described by

$$R = e^{j\phi}Y = HPe^{j\phi} + Ze^{j\phi}$$

Extracting the pilot subcarriers and equalizing with the channel estimate results in

$$H^{-1}R = Pe^{j\phi} + H^{-1}Ze^{j\phi}$$

for a ZF receiver. We may further enhance the estimation of the phase rotation by averaging over the spatial streams. As described in Akay and Ayanoglu (2004b), a weighted average using the channel estimates improves performance by de-weighting the subcarriers which are faded.

When a common oscillator is not employed, the received signal is described by the equation below:

$$
\begin{bmatrix} r_1 \\ r_2 \\ \vdots \\ r_N \end{bmatrix} = \begin{bmatrix} e^{j\phi_1} & 0 & \cdots & 0 \\ 0 & e^{j\phi_2} & \cdots & 0 \\ \vdots & \vdots & \ddots & \vdots \\ 0 & 0 & \cdots & e^{j\phi_N} \end{bmatrix} \begin{bmatrix} y_1 \\ y_2 \\ \vdots \\ y_N \end{bmatrix}
$$

or, in vector notation,

$$
R = \Phi Y = \Phi H P + \Phi Z.
$$

In this case, it is difficult to separate the phase rotation from the channel. Such implementations result in inferior pilot tracking performance.

4.2.4.6   Demapping, deinterleaving, decoding, and descrambling

The demapping operation on each spatial stream is the same for an 802.11a receiver. The deinterleaver uses the following operations to perform the inverse of the interleaving described in Section 4.2.3.4.

The index of the bit in the received block (per spatial stream i_{ss}) is denoted by k. The first permutation reverses the third (frequency rotation) permutation of the interleaver (defined in Eq. (4.27)) if there is more than one spatial stream:

$$
y_r^{i_{ss}} = z_k^{i_{ss}}
$$

$$
r = \left(k + \left(((i_{ss} - 1) \cdot 2) \bmod 3 + 3 \cdot \left\lfloor \frac{i_{ss} - 1}{3} \right\rfloor \right) \cdot 11 \cdot N_{BPSCS} \right) \bmod N_{CBPSS}
$$

(4.40)

where $1 \leq i_{ss} \leq N_{SS}$, $k = 0, 1, \ldots, N_{CBPSS} - 1$. N_{BPSCS} is the number of coded bits per single subcarrier for each spatial stream, or equivalently the modulation order, and N_{CBPSS} is the coded bits per spatial stream.

The second permutation in the deinterleaver reverses the second permutation in the interleaver (defined in Eq. (4.26)) for each spatial stream:

$$
w_j^{i_{ss}} = y_k^{i_{ss}}
$$

$$
j = s \cdot \text{floor}(k/s) + \left(k + \left\lfloor \frac{13 \cdot k}{N_{CBPSS}} \right\rfloor \right) \bmod s
$$

(4.41)

where $1 \leq i_{ss} \leq N_{SS}$, $k = 0, 1, \ldots, N_{CBPSS} - 1$, and s is defined in Eq. (4.23).

The third permutation in the deinterleaver reverses the first permutation of the interleaver (defined in Eq. (4.25)) for each spatial stream:

$$
x_i^{i_{ss}} = w_k^{i_{ss}}
$$

$$
i = 13 \cdot k - (N_{CBPSS} - 1) \cdot \left\lfloor \frac{k}{4 \cdot N_{BPSCS}} \right\rfloor
$$

(4.42)

where $1 \leq i_{ss} \leq N_{SS}$, $k = 0, 1, \ldots, N_{CBPSS} - 1$.

As the final step in the deinterleaving process, the stream parsing operation (described in Section 4.2.3.3) is reversed by taking blocks of s bits from each spatial stream in a round robin fashion to create a single bit stream for the decoder.

The resulting convolutional coded data stream is decoded typically with a Viterbi decoder. If the MCS indicates a rate higher than $\frac{1}{2}$, depuncturing is performed at the input of each decoder. Decoder performance can be substantially improved by using the MIMO channel estimates to weight the decoder metrics. Decoding is similar to 802.11a and described further in Section 4.1.6.

The first seven bits of the Service field give the initial state of the descrambler. The same scrambler used in 802.11a is used to descramble the data stream. Bits received after the data length, indicated by the HT-SIG, are discarded.

References

Akay, E. and Ayanoglu, E. (2004a). Low complexity decoding of bit-interleaved coded modulation for M-ary QAM. *International Conference on Communcations*, June 20–24, Paris, France, 901–5.

Akay, E. and Ayanoglu, E. (2004b). High performance Viterbi decoder for OFDM systems. *Vehicular Technology Conference*, May 17–19, Milan, Italy, 323–7.

Aoki, T. and Takeda, D. (2005). *Backward Compatibility of CDD Preambles*, IEEE 802.11-05/0006r1.

Caire, G., Taricco, G., and Biglieri, E. (1998). Bit-interleaved coded modulation. *IEEE Transactions on Information Theory*, **44**(3), 927–46.

Halford, S. (2001). Implementing OFDM in wireless designs. *Communications Design Conference*, October 1, San Jose, CA.

IEEE (2007a). *IEEE Standard for Information Technology – Telecommunications and Information Exchange Between Systems – Local and Metropolitan Area Networks – Specific Requirements. Part 11: Wireless LAN Medium Access Control (MAC) and Physical Layer (PHY) Specifications.* IEEE Std 802.11™-2007 (Revision of IEEE Std 802.11-1999).

IEEE (2007b). IEEE P802.11n™/D3.00, *Draft Amendment to STANDARD for Information Technology – Telecommunications and Information Exchange Between Systems – Local and Metropolitan Networks – Specific Requirements – Part 11: Wireless LAN Medium Access Control (MAC) and Physical Layer (PHY). Amendment 4: Enhancements for Higher Throughput.*

Proakis, J. G. (1989). *Digital Communications*. New York: McGraw-Hill.

Stephens, A. (2005). *802.11 TGn Functional Requirements*, IEEE 802.11-03/813r13.

van Nee, R. and Prasad, R. (2000). *OFDM for Wireless Multimedia Communications*. Boston, MA: Artech House.

van Zelst, A. and Schenk, T. C. W. (2004). Implementation of a MIMO OFDM-based wireless LAN system. *IEEE Transactions on Signal Processing*, **52**(2), 483–94.

Zehavi, E. (1992). 8-PSK trellis codes for a Rayleigh channel. *IEEE Transactions on Communications*, **40**, 873–84.

Appendix 4.1: 20 MHz basic MCS tables

The following tables consist of the MCSs for 20 MHz "basic" rates. As defined herein, the basic rates include equal modulation and the mandatory 800 ns guard interval. 40 MHz

MCSs are described in Section 5.1.4, unequal modulation is described in Section 12.4, and short guard interval is described in Section 5.5.

For 20 MHz, there are 52 data subcarriers and four pilot subcarriers in an OFDM symbol in an individual spatial stream.

Table 4.6 describes the symbols used in the subsequent tables.

Table 4.6 Symbols used for MCS parameters (IEEE, 2007b)

Symbol	Explanation
R	Code rate
N_{BPSCS}	Number of coded bits per single carrier for each spatial stream (or modulation order)
N_{CBPS}	Number of coded bits per OFDM symbol
N_{DBPS}	Number of data bits per OFDM symbol

Table 4.7 defines the 20 MHz MCS parameters and data rates for single spatial stream transmission. The data rate is calculated by dividing N_{DBPS} by the symbol time of 4 μs (3.2 μs OFDM symbol plus 0.8 μs guard interval).

Table 4.7 20 MHz MCS parameters for single spatial stream (IEEE, 2007b)

MCS index	Modulation	R	N_{BPSCS}	N_{CBPS}	N_{DBPS}	Data rate (Mbps)
0	BPSK	$^1/_2$	1	52	26	6.5
1	QPSK	$^1/_2$	2	104	52	13.0
2	QPSK	$^3/_4$	2	104	78	19.5
3	16-QAM	$^1/_2$	4	208	104	26.0
4	16-QAM	$^3/_4$	4	208	156	39.0
5	64-QAM	$^2/_3$	6	312	208	52.0
6	64-QAM	$^3/_4$	6	312	234	58.5
7	64-QAM	$^5/_6$	6	312	260	65.0

Table 4.8 defines the 20 MHz MCS parameters and data rates for two spatial stream transmission.

Table 4.8 20 MHz MCS parameters for two spatial streams (IEEE, 2007b)

MCS index	Modulation	R	N_{BPSCS}	N_{CBPS}	N_{DBPS}	Data rate (Mbps)
8	BPSK	$^1/_2$	1	104	52	13.0
9	QPSK	$^1/_2$	2	208	104	26.0
10	QPSK	$^3/_4$	2	208	156	39.0
11	16-QAM	$^1/_2$	4	416	208	52.0
12	16-QAM	$^3/_4$	4	416	312	78.0
13	64-QAM	$^2/_3$	6	624	416	104.0
14	64-QAM	$^3/_4$	6	624	468	117.0
15	64-QAM	$^5/_6$	6	624	520	130.0

Table 4.9 defines the MCS parameters and data rates for three spatial stream transmission.

Table 4.9 20 MHz MCS parameters for three spatial streams (IEEE, 2007b)

MCS index	Modulation	R	N_{BPSCS}	N_{CBPS}	N_{DBPS}	Data rate (Mbps)
16	BPSK	$1/2$	1	156	78	19.5
17	QPSK	$1/2$	2	312	156	39.0
18	QPSK	$3/4$	2	312	234	58.5
19	16-QAM	$1/2$	4	624	312	78.0
20	16-QAM	$3/4$	4	624	468	117.0
21	64-QAM	$2/3$	6	936	624	156.0
22	64-QAM	$3/4$	6	936	702	175.5
23	64-QAM	$5/6$	6	936	780	195.0

Table 4.10 defines the MCS parameters and data rates for four spatial stream transmission.

Table 4.10 20 MHz MCS parameters for four spatial streams (IEEE, 2007b)

MCS index	Modulation	R	N_{BPSCS}	N_{CBPS}	N_{DBPS}	Data rate (Mbps)
24	BPSK	$1/2$	1	208	104	26.0
25	QPSK	$1/2$	2	416	208	52.0
26	QPSK	$3/4$	2	416	312	78.0
27	16-QAM	$1/2$	4	832	416	104.0
28	16-QAM	$3/4$	4	832	624	156.0
29	64-QAM	$2/3$	6	1248	832	208.0
30	64-QAM	$3/4$	6	1248	936	234.0
31	64-QAM	$5/6$	6	1248	1040	260.0

5 High throughput

802.11n is the high throughput amendment to the 802.11 standard. This section describes all the aspects of the physical layer which increase the data rate. MIMO/SDM is a key feature of 802.11n, which is discussed in Chapters 3 and 4. The other significant increase in data rate is derived from the new 40 MHz channel width. This section also discusses short guard interval, Greenfield preamble, and other modifications to the 20 MHz waveform.

5.1   40 MHz channel

In the last several years, regulatory domains have made much more spectrum available for unlicensed operation in the 5.47–5.725 GHz band for wireless local area networking. The addition of the new spectrum has more than doubled the number of available 20 MHz channels in the USA and Europe. Table 5.9 and Table 5.10 in Appendix 5.1 describe the current allocation in the USA and Europe, and the corresponding 802.11 channel number. Even with doubling the channel width to create 40 MHz channels, there are still more channels available for frequency re-use than in the early days of 802.11a. Furthermore, products currently in the market with proprietary 40 MHz modes have demonstrated that the cost for 40 MHz products is roughly the same as for 20 MHz products. Therefore, with free spectrum and relatively no increase in hardware cost, doubling the channel bandwidth is the simplest and most cost effective way to increase data rate.

There are some issues that arise with the adoption of the new 40 MHz channel. Coexistence and interoperability between 20 MHz and 40 MHz devices and modes of operation need to be addressed. In addition, though a minor point, frequency re-use in 40 MHz is higher than with 20 MHz. Larger high density network deployments may need to consider this in their network plan. Finally, although the initial intention of the 40 MHz channel was for 5 GHz because of the additional new spectrum, 40 MHz channels and operation are permitted in IEEE (2007) in the 2.4 GHz band. Due to the limited spectrum and overlapping channels, 40 MHz operation in 2.4 GHz requires special attention, and is addressed in Section 10.3.3.2.

In this section simulation results of packet error rate versus SNR are provided for 40 MHz to demonstrate performance. The simulations are performed in channel model D, NLOS, as described in Section 3.5. Physical layer impairments are included in

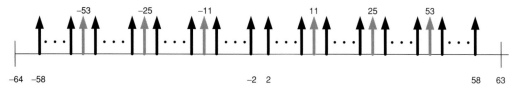

Figure 5.1 40 MHz subcarrier design.

the simulation, as described in Section 3.5.4. The equalizer is based on MMSE. The simulations include synchronization, channel estimation, and phase tracking.

5.1.1   40 MHz subcarrier design and spectral mask

The 40 MHz subcarrier design is illustrated in Figure 5.1. The frequency domain design for 40 MHz encompasses 128 subcarrier locations, with a subcarrier spacing the same as for 20 MHz, 40 MHz/128 = 312.5 kHz. For the guard band on the edges of the channel, there are six null subcarriers on the left side of the band, subcarriers −64, . . . , −59, and there are five null subcarriers on the right side of the band, 59, . . . , 63. In addition, there are three nulls around DC, −1, 0, 1 to ease the implementation of direct down conversion receivers. In comparison, the 20 MHz HT subcarrier design has four null subcarriers on the left side of the band, three null subcarriers on the right side, and only has one null subcarrier at DC. For 40 MHz, there is a total of 14 null subcarriers. There are 114 populated subcarriers in locations −58, . . . , −2, 2, . . . , 58. Of these, six are for pilot subcarriers in locations −53, −25, −11, 11, 25, 53. The wider bandwidth called for an additional two pilot subcarriers with respect to 20 MHz to accurately track the phase across the entire band. The remaining 108 subcarriers are dedicated to data. This is more than double the number of data subcarriers as 20 MHz which has 52 for HT and 48 for legacy, and is reflected in the 40 MHz data rates which are more than double those of 20 MHz. The MCS tables for 40 MHz, and the associated 40 MHz data rates are given in Tables 5.14–5.18 in Appendix 5.2. For example, MCS 15 for 20 MHz is 130 Mbps, and for 40 MHz is 270 Mbps.

With the new channel width and subcarrier design, a new 40 MHz spectral mask is necessary. The basic premise of the design of the 40 MHz spectral mask is that adjacent channel interference between neighboring 40 MHz devices is the same as the adjacent channel interference between neighboring 20 MHz HT devices. The new spectral mask is given in Figure 5.2. For reference, the new 20 MHz HT spectral mask is given in Figure 5.11.

When a 40 MHz device and a 20 MHz device are in neighboring channels, the 20 MHz device experiences more adjacent channel interference than when neighboring another 20 MHz device. Figure 5.3 illustrates a 20 MHz device with a neighboring 20 MHz device and a neighboring 40 MHz device, all at equal received power level. We see that the roll-off from the 40 MHz spectral mask is at levels of −20 dBr to −28 dBr within the 20 MHz pass band. This is interference which the 20 MHz device can not filter out. On the other hand, we see that the roll-off from the neighboring 20 MHz device

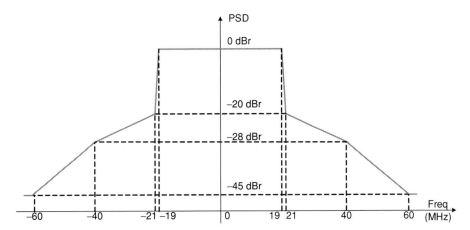

Figure 5.2 40 MHz spectral mask. Reproduced with permission from IEEE (2007) © IEEE.

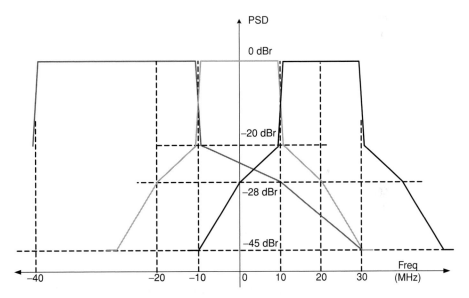

Figure 5.3 Adjacent channel interference between 20 MHz and 40 MHz.

is at levels of −20 dBr to −45 dBr within the 20 MHz pass band. This demonstrates the possibility of higher adjacent channel interference from the 40 MHz device than the 20 MHz device. Note, however, that regulatory restrictions remain, so near the edge of a regulatory band, the situation can be better.

The other component of adjacent channel interference is the relative power between the signal-of-interest and the interference. With the additional spectrum allocation in 5 GHz, most 5 GHz deployments are capable of placing spatially neighboring APs on non-neighboring channels. Therefore, APs on neighboring channels have a large spatial separation and the interference power is lower. This is a reason why the less stringent 40 MHz spectral mask may not have an effect on system performance.

The main reason why the 40 MHz spectral mask was not made tighter, was to minimize the increase in the cost and complexity and minimize the decrease in power efficiency of the transmit power amplifier and transmit filtering. The main source of out-of-band interference is due to spectral re-growth from the power amplifier. To meet a tighter mask would have meant further power backoff to run the power amplifier at a more linear operating point. Additional backoff reduces the output power, thereby reducing coverage. Conversely, a larger power amplifier could be used to maintain the same output power. However, this is more costly and consumes more DC power hence reducing battery life in power sensitive devices.

5.1.2   40 MHz channel design

Fundamental to the design of the 40 MHz channel is that only adjacent 20 MHz channels are combined to form a 40 MHz channel. Frequency band and associated channel numbering for 20 MHz is discussed Appendix 5.1. The 20 MHz channel numbers are given in Tables 5.9 and 5.10. A 40 MHz channel is specified by two fields: (Nprimary_ch, Secondary). The first field represents the channel number of the primary 20 MHz channel. This is the channel by which AP transmits all control and management frames in a mixed 20/40 MHz environment. In the same type of environment, all 20 MHz clients only associate to the primary channel since the beacon is only transmitted in the primary channel. The second field indicates whether the secondary 20 MHz channel is above or below the primary channel ($1 \rightarrow$ above, $-1 \rightarrow$ below). The secondary channel number is equal to (Nprimary_ch + Secondary*4). For example, a 40 MHz channel consisting of a Nprimary_ch = 36 and Secondary = 1 specifies the use of channel 36 for the primary channel and channel 40 for the secondary channel.

As can be seen from Tables 5.9 and 5.10 in Appendix 5.1, the 20 MHz channels in the 5 GHz band are non-overlapping. This makes coexistence and interoperability between 20 MHz and 40 MHz devices much simpler. Keeping with this same philosophy, the new 40 MHz channels are also non-overlapping, as given in Tables 5.11 and 5.12.

Unfortunately in the 2.4 GHz band, the channels are overlapping. To better adjust the selection of a 40 MHz channel to the environment, the new 40 MHz channel numbers are also overlapping. The channel numbers for 2.4 GHz are also given in Tables 5.11 and 5.12.

5.1.3   40 MHz mixed format preamble

In order to interoperate and coexist with legacy 802.11a/g OFDM devices the legacy portion of the 20 MHz MF preamble is replicated over the other 20 MHz portion of the 40 MHz band. This is illustrated in Figure 5.4. This allows a legacy OFDM device in either 20 MHz portion of the 40 MHz channel to process the L-STF, L-LTF, and decode the L-SIG to properly defer transmission. Therefore legacy OFDM devices in a 40 MHz BSS which operate on the primary channel coexist with the 40 MHz HT devices. Furthermore legacy OFDM devices in a neighboring BSS which overlaps the secondary channel of the 40 MHz BSS also coexist with the 40 MHz HT devices.

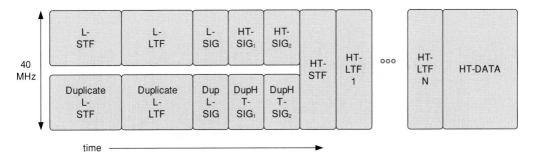

Figure 5.4 40 MHz mixed mode preamble.

Similarly, the HT-SIG fields are also replicated over the other 20 MHz portion of the 40 MHz channel. This allows 20 MHz HT devices in either portion of the 40 MHz channel to decode the HT-SIG fields and properly defer transmission. Therefore 20 MHz and 40 MHz HT devices in a 40 MHz BSS will coexist, and similarly with 20 MHz HT devices in a neighboring BSS on the secondary channel.

As noted in Section 4.2.2.1, when the CBW 20/40 bit in the HT-SIG is set to 0, a 20 MHz transmission is indicated. When the bit is set to 1, a 40 MHz transmission is indicated. This allows a 20 MHz device to terminate reception in the presence of a 40 MHz transmission. A 40 MHz device may try to auto-detect between 20 MHz and 40 MHz prior to decoding the HT-SIG. For example, at initial packet detection, a receive signal strength measure on either half of the band may be used to indicate that both portions contain energy, indicating 40 MHz. Since the subcarriers in the 40 MHz preamble are replicated, the upper and lower portions may be combined for extra frequency diversity and processing gain. This improves detection of L-SIG and HT-SIG. Care needs to be taken to not confuse a 20 MHz transmission in one portion of the 40 MHz channel and interference on the other 20 MHz portion of the 40 MHz channel with a 40 MHz transmission.

The 20 MHz OFDM symbol waveform equations for L-STF, L-LTF, L-SIG, and HT-SIG defined by Eqs. (4.13)–(4.16) are modified for 40 MHz transmission in two ways, see Eq. (5.1). The 20 MHz waveform is shifted lower by 32 subcarriers. A copy of the waveform is shifted higher by 32 subcarriers and phase rotated by 90 degrees (the j term preceding the second summation in Eq. (5.1)). The phase rotation on the upper subcarriers is added to provide a reduction in peak-to-average ratio of the 40 MHz waveform in some conditions. The general waveform equation for MF L-STF, L-LTF, L-SIG, and HT-SIG fields for 40 MHz transmission is given in the equation below.

$$r_{\text{Field}}^{(i_{\text{TX}})}(t) = \frac{1}{\sqrt{N_{\text{TX}} \cdot N_{\text{Field}}^{\text{Tone}}}} \left[\sum_{k=-26}^{26} X_k \exp\left(j \, 2\pi \, (k-32) \, \Delta_{\text{F}} \left(t - T_{\text{GI}}^{\text{Field}} - T_{\text{CS}}^{i_{\text{TX}}} \right) \right) \right.$$
$$\left. + j \sum_{k=-26}^{26} X_k \exp\left(j \, 2\pi \, (k+32) \, \Delta_{\text{F}} \left(t - T_{\text{GI}}^{\text{Field}} - T_{\text{CS}}^{i_{\text{TX}}} \right) \right) \right] \quad (5.1)$$

where N_{TX} is the number of transmit antenna chains; $N_{\text{Field}}^{\text{Tone}} = 24$ for L-STF, and 104 for L-LTF, L-SIG, and HT-SIG; $X_k = S_k$ in Eq. (4.13) for L-STF, L_k in Equation (4.14) for L-LTF, $(D_k + p_0 P_k)$ in Eq. (4.15) for L-SIG, and $(j D_{k,n} + p_{n+1} P_k)$ in Eq. (4.16) for

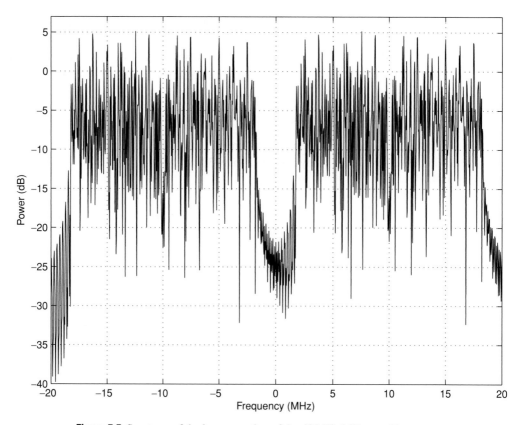

Figure 5.5 Spectrum of the legacy portion of the 40 MHz MF preamble.

HT-SIG; $T_{\text{GI}}^{\text{Field}}$ is the 1.6 μs guard interval for L-LTF, 0.8 μs for L-SIG and HT-SIG, 0 for L-STF; and $T_{\text{CS}}^{i_{\text{TX}}}$ is the cyclic shift for transmit antenna chain i_{TX} with values given by Table 4.2.

An illustration of the spectrum of the legacy portion of the 40 MHz MF preamble is given in Figure 5.5. As can be seen, the legacy portion of the preamble contains many null subcarriers in the area around DC.

Once the legacy portion of the preamble is complete, legacy compatibility no longer constrains the preamble design. The HT-STF, HT-LTF, and HT-DATA incorporate all the additional subcarriers available in 40 MHz based on the subcarrier design illustrated in Figure 5.1. Again, for peak-to-average ratio considerations, the upper portion of the subcarriers are phase rotated by 90 degrees. The 20 MHz waveform equations for HT-STF, HT-LTF, and HT-DATA defined by Eqs. (4.17), (4.19), and (4.30) are modified for 40 MHz, as follows:

$$
r_{\text{Field}}^{(i_{\text{SS}})}(t) = \frac{1}{\sqrt{N_{\text{SS}} \cdot N_{\text{Field}}^{\text{Tone}}}} \left[\sum_{k=-58}^{0} X_k \, \exp\left(j\, 2\pi \, k\Delta_{\text{F}} \left(t - T_{\text{GI}}^{\text{Field}} - T_{\text{CS}}^{i_{\text{SS}}} \right) \right) \right.
$$
$$
\left. + j \sum_{k=1}^{58} X_k \, \exp\left(j\, 2\pi \, k\Delta_{\text{F}} \left(t - T_{\text{GI}}^{\text{Field}} - T_{\text{CS}}^{i_{\text{SS}}} \right) \right) \right]
$$
(5.2)

where N_{SS} is the number of spatial streams; $N_{Field}^{Tone} = 24$ for HT-STF, and 114 for HT-LTF and HT-DATA; $X_k^{(iss)} = S_k$ for HT-STF, $[P_{HTLTF}]_{iss,n}$ HTLTF$_k$ for HT-LTF, and $(D_{k,iss,n} + p_{n+3}P_{(iss,n)}^k)$ for HT-DATA; T_{GI}^{Field}: 0.8 μs guard interval for HT-LTF and HT-DATA, 0 for HT-STF; and T_{CS}^{iss} is the cyclic shift for spatial stream i_{SS} with values given by Table 4.3.

The HT-STF subcarrier sequence for 40 MHz is derived from the 20 MHz L-STF sequence defined in Eq. (4.13), since its low peak-to-average ratio, high autocorrelation properties, and short repetition periods are still desirable. The 20 MHz sequence is shifted lower by 32 and a copy is shifted up by 32, as given in Eq. (5.3) (IEEE, 2007):

$$
\begin{aligned}
S_{-58,58} = \sqrt{1/2}\,\{0, 0, 1+j, 0, 0, 0, -1-j, 0, 0, 0, 1+j, 0, 0, 0, \\
-1-j, 0, 0, 0, -1-j, 0, 0, 0, 1+j, 0, 0, 0, 0, 0, 0, -1-j, 0, 0, 0, \\
-1-j, 0, 0, 0, 1+j, 0, 0, 0, 1+j, 0, 0, 0, 1+j, 0, 0, 0, 1 \\
+j, 0, 0, 0, 0, 0, 0, 0, 0, 0, 0, 0, 0, 0, 0, 0, 1+j, 0, 0, 0, -1-j, 0, 0, 0, 1 \\
+j, 0, 0, 0, -1-j, 0, 0, 0, -1-j, 0, 0, 0, 1+j, 0, 0, 0, 0, 0, 0, 0, \\
-1-j, 0, 0, 0, -1-j, 0, 0, 0, 1+j, 0, 0, 0, 1+j, 0, 0, 0, 1 \\
+j, 0, 0, 0, 1+j, 0, 0\}
\end{aligned}
\tag{5.3}
$$

The 40 MHz subcarrier sequence for HT-LTF is derived from the L-LTF sequence defined in Eq. (4.2), to similarly retain its peak-to-average ratio and autocorrelation properties. The sequence is constructed by first shifting the 20 MHz sequence lower by 32 and a copy is shifted up by 32. Next the equivalent 20 MHz DC subcarriers, -32 and 32, are both filled in with the value 1. The subcarriers near 40 MHz DC $[-5, -4, 3, -2, 2, 3, 4, 5]$ are filled with the values $[-1, -1, -1, 1, -1, 1, 1, -1]$, respectively. The subcarrier sequence is given in Eq. (5.4):

$$
\begin{aligned}
HTLTF_{-58,58} = \{1, 1, -1, -1, 1, 1, -1, 1, -1, 1, 1, 1, 1, 1, 1, -1, -1, 1, 1, -1, 1, \\
-1, 1, 1, 1, 1, 1, 1, -1, -1, 1, 1, -1, 1, -1, 1, -1, -1, -1, -1, \\
-1, 1, 1, -1, -1, 1, -1, 1, -1, 1, 1, 1, 1, -1, -1, -1, 1, 0, 0, 0, -1, \\
1, 1, -1, 1, 1, -1, -1, 1, 1, -1, 1, -1, 1, 1, 1, 1, 1, 1, -1, -1, 1, 1, \\
-1, 1, -1, 1, 1, 1, 1, 1, 1, -1, -1, 1, 1, -1, 1, -1, 1, -1, -1, -1, \\
-1, -1, 1, 1, -1, -1, 1, -1, 1, -1, 1, 1, 1, 1\}
\end{aligned}
\tag{5.4}
$$

As previously mentioned, there are six pilot subcarriers in 40 MHz. The pilot subcarriers are inserted at subcarrier locations $-53, -25, -11, 11, 25, 53$, as given in Eq. (5.5) (IEEE, 2007):

$$
\begin{aligned}
P_{i_{STS}}^{(-58..58,\,n)} = \{0, 0, 0, 0, 0, \Psi_{i_{STS},n\oplus6}^{(N_{STS})}, 0, 0, 0, 0, 0, 0, 0, 0, 0, 0, 0, 0, 0, 0, 0, 0, 0, 0, 0, 0, \\
0, 0, 0, 0, 0, 0, 0, 0, \Psi_{i_{STS},(n+1)\oplus6}^{(N_{STS})}, 0, 0, 0, 0, 0, 0, 0, 0, 0, 0, 0, 0, 0, \\
\Psi_{i_{STS},(n+2)\oplus6}^{(N_{STS})}, 0, 0, 0, 0, 0, 0, 0, 0, 0, 0, 0, 0, 0, 0, 0, 0, 0, 0, 0, 0, 0, 0, \\
\Psi_{i_{STS},(n+3)\oplus6}^{(N_{STS})}, 0, 0, 0, 0, 0, 0, 0, 0, 0, 0, 0, 0, \Psi_{i_{STS},(n+4)\oplus6}^{(N_{STS})}, 0, 0, 0, 0, 0, \\
0, 0, 0, 0, 0, 0, 0, 0, 0, 0, 0, 0, 0, 0, 0, 0, 0, 0, 0, 0, 0, 0, \Psi_{i_{STS},(n+5)\oplus6}^{(N_{STS})}, \\
0, 0, 0, 0, 0\}
\end{aligned}
\tag{5.5}
$$

Table 5.1 Pilot values for 40 MHz transmission (IEEE, 2007)

N_{SS}	i_{SS}	$\Psi_{i_{SS},0}^{(N_{SS})}$	$\Psi_{i_{SS},1}^{(N_{SS})}$	$\Psi_{i_{SS},2}^{(N_{SS})}$	$\Psi_{i_{SS},3}^{(N_{SS})}$	$\Psi_{i_{SS},4}^{(N_{SS})}$	$\Psi_{i_{SS},5}^{(N_{SS})}$
1	0	1	1	1	-1	-1	1
2	0	1	1	-1	-1	-1	-1
2	1	1	1	1	-1	1	1
3	0	1	1	-1	-1	-1	-1
3	1	1	1	1	-1	1	1
3	2	1	-1	1	-1	-1	1
4	0	1	1	-1	-1	-1	-1
4	1	1	1	1	-1	1	1
4	2	1	-1	1	-1	-1	1
4	3	-1	1	1	1	-1	1

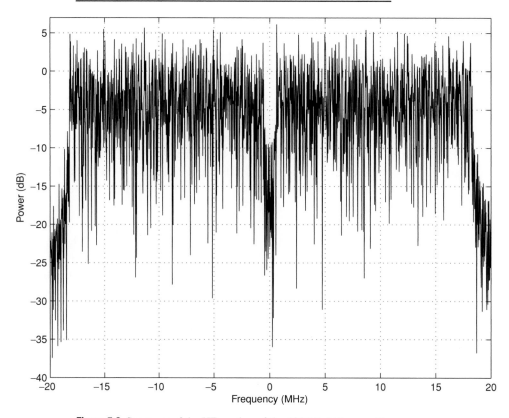

Figure 5.6 Spectrum of the HT portion of the 40 MHz MF preamble.

where $n \oplus a$ indicates symbol number modulo integer a and the patterns $\Psi_{i_{SS},n}^{(N_{SS})}$ are defined in Table 5.1.

An illustration of the spectrum of the HT portion of the 40 MHz MF preamble is given in Figure 5.6. As can be seen, in the HT portion of the preamble and data most of the null subcarriers in the area around DC are filled with data subcarriers for higher spectral efficiency.

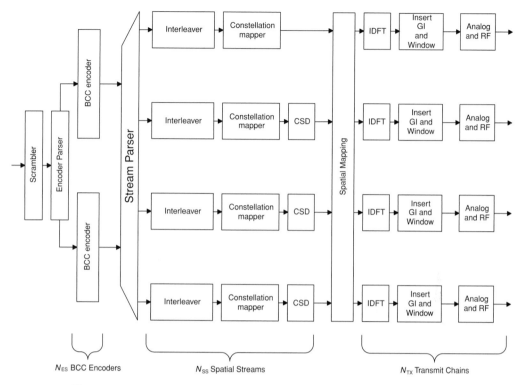

Figure 5.7 Transmitter block diagram for HT Data field with two encoders. Reproduced with permission from IEEE (2007) © IEEE.

To accommodate transmitting with more antennas than spatial streams, Eq. (5.2) may be modified to incorporate a spatial mapping matrix, which maps spatial streams to antennas, as described in Section 4.2.3.8.

5.1.4   40 MHz data encoding

The primary difference in data encoding between 20 MHz and 40 MHz is that all 20 MHz MCSs use a single encoder but some of the 40 MHz MCSs use two encoders. It was decided during 802.11n standard development that convolutional encoding and decoding of data rates above 300 Mbps with a single encoder and decoder carries an excessive implementation burden. Therefore, when the selected MCS has a data rate greater than 300 Mbps, computed based on the 800 ns guard interval, two encoders are used. With two encoders, a number of functions need to be modified. The transmitter block diagram for the HT Data field with two encoders is given in Figure 5.7.

The MCS parameters and data rates for 40 MHz, equal modulation, and 800 ns guard interval are defined in Tables 5.14–5.17 in Appendix 5.2, for one to four spatial streams, respectively. The tables indicate which require two encoders. For MCSs that only require one encoder, the data encoding is identical to 20 MHz and described in Section 4.2.3. The following sections describe the modifications necessary with two encoders.

5.1.4.1   Bit string with two encoders

With two BCC encoders, twelve tail bits, all zeros, are appended to the PSDU. The number of OFDM symbols in the Data field is as follows:

$$N_{\text{SYM}} = \left\lceil \frac{8 \cdot \text{length} + 16 + 12}{N_{\text{DBPS}}} \right\rceil \tag{5.6}$$

where length is the value of the Length field in octets in the HT-SIG field illustrated in Figure 4.16; and N_{DBPS} is the number of data bits per OFDM symbol determined by the selected MCS.

The symbol $\lceil x \rceil$ denotes the smallest integer greater than or equal to x, corresponding to the *ceiling* function. The number of pad bits is therefore $N_{\text{SYM}} \cdot N_{\text{DBPS}} - 8 \cdot \text{length} - 16 - 12$.

5.1.4.2   Scrambling, encoder parsing, and encoding with two encoders

The resulting data bit string is scrambled with the same scrambler as 802.11a, described in Section 4.1.4. When two BCC encoders are used, the scrambled bits are divided between two encoders by sending alternate bits to each encoder. The ith bit is sent to the jth encoder, where b is the data bit string, as follows in Eq. (5.7) (IEEE, 2007):

$$x_i^{(j)} = b_{2 \cdot i + j} \quad ; \quad j \in \{0, 1\}, \quad i \geq 0 \tag{5.7}$$

Equation (5.8) gives the tail bits which are replaced with zeros (IEEE, 2007):

$$x_i^{(j)} \quad : \quad j \in \{0, 1\} \quad ; \quad \frac{\text{length} \cdot 8 + 16}{2} \leq i \leq \frac{\text{length} \cdot 8 + 16}{2} + 5 \tag{5.8}$$

Each scrambled, parsed bit string is encoded with the rate $\frac{1}{2}$ BCC. After encoding, each encoded bit string is punctured to the selected code rate. Both encoding and puncturing are the same as 802.11a and described in Section 4.1.5.

In the receiver, the encoder parsing operation is reversed. In this, bits at the output of decoding are alternately selected to create the bit stream for the descrambler.

5.1.4.3   Stream parsing with two encoders

After encoding and puncturing, the two bit strings are re-arranged into a new set of bit strings equal to the number of spatial streams (N_{SS}). To simplify the description, this section only describes the stream parsing function for equal MCSs. For unequal MCSs, refer to Section 12.4. The output of the stream parser is N_{SS} bit strings, each of length N_{CBPSS} (coded bits per spatial stream).

With two encoders, a block of $s \cdot N_{\text{SS}}$ bits (s defined in Eq. (4.23) as $\max\{1, N_{\text{BPSCS}}/2\}$) from the output of each encoder is alternately used. The output of the two encoders is given by $y_i^{(j)}$, with bit i from encoder j. The new equation for the output of the stream parser with two encoders in Eq. (5.9) is modified from Eq. (4.24) as follows. In Eq. (4.24), y is a function of the bit stream index i only; however, in Eq. (5.9), y is also a function of the encoder index j. Furthermore, the sequence i is modified to account for the two encoders by modifying the component $\lfloor k/s \rfloor$ in

Eq. (4.24) to $\lfloor k/2 \cdot s \rfloor$ in Eq. (5.9):

$$x_k^{iss} = y_i^{(j)}$$
$$j = \lfloor \tfrac{k}{s} \rfloor \bmod 2 \qquad\qquad (5.9)$$
$$i = (i_{SS} - 1) \cdot s + s \cdot N_{SS} \cdot \lfloor \tfrac{k}{2 \cdot s} \rfloor + k \bmod s$$

where $1 \le i_{SS} \le N_{SS}$; $k = 0, 1, \ldots, N_{CBPSS} - 1$; $\lfloor v \rfloor$ is the largest integer less than or equal to v, corresponding to the *floor*; and $v \bmod w$ is the remainder from the division of the integer v by the integer w.

Once the bit strings are parsed into a new set of bit streams equal to the number of spatial streams, encoding of the data continues equivalently to the steps in Section 4.2.3.

In the receiver, stream parsing operation is reversed. Blocks of s bits from each spatial stream are taken in a round robin fashion and assigned to one decoder then blocks of s bits from each spatial stream are alternately assigned to the input of the other decoder.

5.1.5   MCS 32: High throughput duplicate format

Consider a 20 MHz device with 17 dBm output power. Typically, a 40 MHz device is designed with the same output power to maintain the same cost as a 20 MHz device. However, with the same output power, the 40 MHz device has 3 dB lower power spectral density. Therefore, the 40 MHz range is "3dB worse" than 20 MHz due to wider noise bandwidth.

To compensate for the reduced range in 40 MHz relative to 20 MHz, a special HT duplicate format, MCS 32, was created in 40 MHz. With MCS 32, the signal is duplicated on the upper and lower channel during the Data field. Combining the subcarriers in the upper and lower channel at the receiver improves the required SNR of the HT duplicate format resulting in a comparable maximum range for the lowest data rates of 20 MHz and 40 MHz. Figure 5.8 illustrates the packet error rate for 20 MHz MCS 0, 40 MHz MCS 0, and MCS 32.

The simulation uses a 2×2 antenna configuration for all three cases. 40 MHz, MCS 0 performs 1 dB better at a PER of 1% than 20 MHz, MCS 0 due to frequency diversity gain. With MCS 32, the required SNR for the lowest rate in 40 MHz is improved by 2 dB over MCS 0.

Table 5.2 demonstrates a link budget which compares the range for 20 MHz MCS 0, 40 MHz MCS 0, and 40 MHz MCS 32. For this link budget example, the transmitter power is 15 dBm. The transmit and receive antenna gains are 2 dBi, typical for omnidirectional antennas used on 802.11 devices. Both the transmitter and the receiver have two antennas. A typical receiver noise figure of 6 dB is used. The path loss is derived for channel model D based on Section 3.5.5 for the operating frequency of 5.2 GHz. The received signal strength indication (RSSI) is calculated as follows, with all parameters in dB:

$$\text{RSSI} = P_T + A_T - \text{PL} - \text{SF} + A_R$$

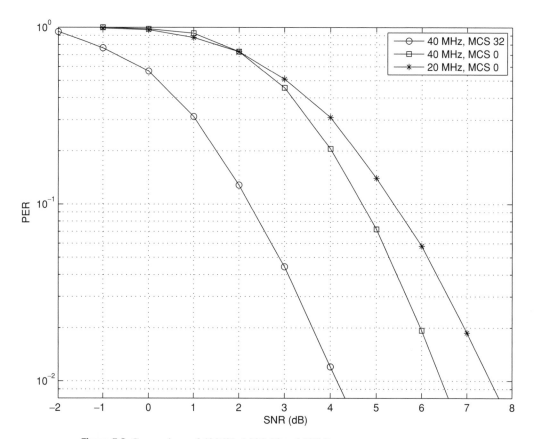

Figure 5.8 Comparison of 40 MHz MCS 32 to MCS 0.

where P_T is the transmit power; A_T, A_R are the transmit or receive antenna gain; PL is the path loss; and SF is shadow fading. With a noise figure of 6 dB, the thermal noise power is −95 dBm for 20 MHz and −92 dBm for 40 MHz. The distance between the transmitter and receiver is calculated to meet the required SNR for channel model D for the specific MCS at a packet error rate of 1%.

As can be seen from the link budget, the range for 40 MHz, MCS 32 and 20 MHz, MCS 0 is almost the same. With MCS 32, the range of 40 MHz is equivalent to the range of 20 MHz.

The data is transmitted on only one spatial stream with BPSK and rate ½ coding. The subcarrier design for the Data field matches that of legacy 802.11a 6 Mbps rate described in Section 4.1.4, and replicates the subcarriers in the lower and upper half of the channel thus maintaining the 6 Mbps data rate for MCS 32. The MCS parameters are given in Table 5.18. During the 802.11n standard development there was discussion of whether to use the 20 MHz HT data format for a higher HT duplicate format data rate. The rationale for using 52 subcarriers of legacy rather than 56 subcarriers of HT 20 MHz is due to the filtering requirements of 40 MHz devices.

Table 5.2 802.11n link budget

Parameter	Units	20 MHz MCS 0	40 MHz MCS 0	40 MHz MCS 32
Tx power	dBm	15	15	15
Tx antenna gain	dBi	2	2	2
Pathloss				
frequency	GHz	5.2	5.2	5.2
distance	**m**	98.2	86.6	100.8
breakpoint	m	10	10	10
shadow fading	dB	5	5	5
free space	dB	66.8	66.8	66.8
total loss	dB	106.5	104.6	106.9
Rx antenna gain	dBi	2	2	2
RSSI	dBm	− 87.5	− 85.6	− 87.9
Noise power				
NF	dB	6	6	6
BW	MHz	20	40	40
total	dBm	− 95.0	− 92.0	− 92.0
Required SNR	**dB**	**7.5**	**6.4**	**4.1**

The span of normal 40 MHz format is from subcarrier −58 to 58. The span of the HT duplicate format with replicating the 802.11a subcarrier format is also −58 to 58, requiring the same filtering. If the 20 MHz HT data format was used, however, the span would have been −60 to 60, exceeding that of the standard 40 MHz mode. The 802.11n standard development group decided to not have the HT duplicate format drive the filtering requirements of a 40 MHz device, and the 802.11a subcarrier format was selected as the basis for HT duplicate format.

There are some interesting points about the HT duplicate format waveform that should be highlighted. The legacy portion of the HT MF preamble, the HT-SIG, and the HT-LTF remain unchanged from the standard 40 MHz waveform. This means there are extra subcarriers in the HT-LTF that are not necessarily required for channel estimation. Furthermore, to maximize the SNR on the data subcarriers a scaling factor of 104 was selected. This has the implication of a mismatch in per-subcarrier scaling on the HT-LTF (which has a scaling factor of 114) relative to the Data field. This mismatch needs to be accounted for in the channel estimates or the equalizer. The waveform for the HT duplicate format Data field is given below:

$$
\begin{aligned}
r_{\text{HT-DUP}}^{(i_{\text{TX}})}(t) = \frac{1}{\sqrt{104}} \sum_{n=0}^{N_{\text{SYM}}-1} & \left[\sum_{k=-26}^{26} \left(D_{k,n} + p_{n+1} P_k \right) \exp\left(j 2\pi \left(k - 32 \right) \Delta_{\text{F}} \right) \right. \\
& \times \left(t - n \cdot T_{\text{SYM}} - T_{\text{GI}} \right) + j \sum_{k=-26}^{26} \left(D_{k,n} + p_{n+1} P_k \right) \exp\left(j 2\pi \left(k + 32 \right) \Delta_{\text{F}} \right) \\
& \left. \times \left(t - n \cdot T_{\text{SYM}} - T_{\text{GI}} \right) \right]
\end{aligned}
$$

$$(5.10)$$

where N_{SYM} is the number of OFDM symbols defined in Eq. (4.22); $\left(D_{k,n} + p_{n+1} P_k \right)$ are the data and pilots, defined in Eq. (4.15); $\Delta_{\text{F}} = 312.5 \, \text{kHz} \, (40 \, \text{MHz}/128)$; $T_{\text{SYM}} = 4 \, \mu\text{s}$

symbol time, composed of 3.2 μs OFDM symbol time and 0.8 μs guard interval; and $T_{GI} = 0.8$ μs guard interval.

To accommodate transmitting with more antennas than spatial streams, Eq. (5.10) may be modified to incorporate a spatial mapping matrix, which maps spatial streams to antennas, as described in Section 4.2.3.8.

5.1.6   20/40 MHz coexistence with legacy in the PHY

Neighboring legacy APs and clients may be present in the primary or secondary channels of the 40 MHz channel. A common way to reserve the medium for transmission is to send an RTS/CTS. The neighboring APs and clients will respect management frames. To improve efficiency for a 40 MHz device, it is desirable to make the reservation on both the primary and secondary channels simultaneously.

The non-HT duplicate format is designed to be receivable by legacy (and HT) devices in both the primary and secondary channels simultaneously. In essence a packet is transmitted in legacy OFDM format in the upper and lower halves of the 40 MHz channel simultaneously.

The L-STF, L-LTF, and L-SIG are transmitted identically as given by Eq. (5.1). This is followed immediately by the Data field which is transmitted in legacy format similar to the L-SIG. The Data field may be transmitted at any legacy data rate. The equation for the waveform of the Data field is given below:

$$
r_{\text{LEG-DUP}}^{(i_{TX})}(t) = \frac{1}{\sqrt{104}} \sum_{n=0}^{N_{SYM}-1} \left[\sum_{k=-26}^{26} \left(D_{k,n} + p_{n+1} P_k \right) \exp \left(j\, 2\pi\, (k - 32)\, \Delta_F \right. \right.
$$
$$
\times \left(t - n \cdot T_{SYM} - T_{GI} - T_{CS}^{i_{TX}} \right) \right) + j \sum_{k=-26}^{26} \left(D_{k,n} + p_{n+1} P_k \right)
$$
$$
\times \exp \left(j\, 2\pi\, (k + 32)\, \Delta_F \left(t - n \cdot T_{SYM} - T_{GI} - T_{CS}^{i_{TX}} \right) \right) \right]
\tag{5.11}
$$

where N_{SYM} is the number of OFDM symbols defined in Eq. (4.16); $\left(D_{k,n} + p_{n+1} P_k \right)$ are the data and pilots, defined in Eq. (4.15); $\Delta_F = 312.5$ kHz (40 MHz/128); $T_{SYM} = 4$ μs symbol time, composed of 3.2 μs OFDM symbol time and 0.8 μs guard interval; $T_{GI} = 0.8$ μs guard interval; and $T_{CS}^{i_{TX}}$ is the cyclic shift for transmit antenna chain i_{TX} with values given by Table 4.2.

5.1.7   Performance improvement with 40 MHz

Just based on maximum data rate from the MCS tables, 40 MHz provides more than double the data rate as 20 MHz. In the absence of the required SNR of the individual data rates, it is difficult to determine the actual benefit in a real system. Figure 5.9 illustrates simulation results for over-the-air throughput as a function of range. Over-the-air throughput is defined as the data rate multiplied by one minus the packet error rate. MAC overhead is not accounted for in this analysis. The link budget used to convert from SNR to range follows Table 5.2.

Note that the maximum throughput plateaus at the short range. This artifact is due to a maximum SNR of 35 dB. SNRs above 35 dB become more difficult to achieve with

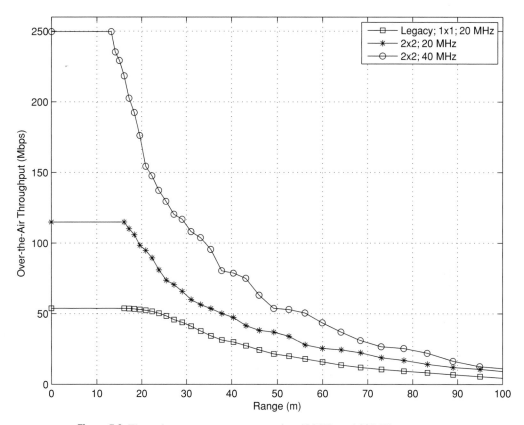

Figure 5.9 Throughput versus range comparing 40 MHz and 20 MHz.

low cost hardware. Therefore throughput values requiring greater than 35 dB SNR were not considered.

We see that in channel model D, the maximum throughput is achieved at less than 15 m at which point 40 MHz provides more than double the data rate as 20 MHz and more than quadruple the data rate of legacy 802.11a/g. Note that this is not true throughput provided by the wireless device. Actual throughput includes overhead due to PHY preambles and the MAC.

We also observe that 40 MHz has an unexpected benefit of range improvement at moderate data rates. For example, if we examine the ranges at 50 Mbps in Figure 5.9, we see that the range is approximately 25 m for legacy, 40 m for 20 MHz HT, and 55 m for 40 MHz HT. Similarly, at 100 Mbps, the approximate range for 20 MHz HT is 20 m and for 40 MHz HT is 35 m. At 100 Mbps, the 20 MHz HT operates with MCS 14. On the other hand, 40 MHz HT could operate with MCS 5 composed of a single spatial stream. This enables receive diversity gain for 40 MHz HT and increases range. Receive diversity was briefly introduced in Section 3.2, and is discussed in more detail in Section 6.1.

For legacy, the simulation uses one transmit antenna and one receive antenna, typical of 802.11a/g devices. At 50 Mbps, legacy operates with 64 QAM, $r = 3/4$, requiring a

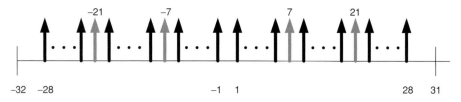

Figure 5.10 20 MHz subcarrier design.

high SNR with the 1×1 system, which then results in a low range. The HT devices operating at 50 Mbps benefit from the diversity gain equivalent to a 1×2 system.

5.2   20 MHz enhancements: Additional data subcarriers

As alluded to in Chapter 4, the number of data subcarriers in 20 MHz was increased to 52, which improves the data rate of 802.11n over legacy 802.11a/g for the same MCS even with single stream. As discussed in Section 4.1, legacy 802.11a/g has 48 data subcarriers. This results in an 8% increase in data rate.

Two data subcarriers are added to each end of the spectrum. The total number of subcarriers is 56. The data subcarriers occupy subcarriers -28 to -22, -20 to -8, -7 to -1, 1 to 6, 8 to 20, and 22 to 28. The pilot subcarriers are located in the same subcarriers as legacy, -21, -7, 7, 21. Again, as with legacy, there is still a null at DC. The new subcarrier design for 20 MHz HT operation is given in Figure 5.10.

When it was proposed to add two subcarriers at either edge of the band, measurements were performed which demonstrated that the signal still fits within the 802.11a spectral mask with very little extra PA backoff or filtering. The measurements are given in Coffey *et al.* (2004, Slides 26–27). Furthermore, analysis was given in Hedberg (2005), which demonstrated that the adjacent channel interference and Tx and Rx filter requirements are comparable between 56 subcarriers and 52 subcarriers. However, to improve the adjacent channel interference in alternate channels, the floor of the 802.11a mask was tightened to -45 dBr. The new 20 MHz HT spectral mask is given in Figure 5.11.

5.3   MCS enhancements: Spatial streams and code rate

For completeness of the section on high throughput, we must mention the obvious that with MIMO we achieve large data rate increases over 802.11a/g with two, three, and four spatial streams. MIMO provides a date rate increase over 802.11a/g proportional to the number of spatial streams.

In addition, a higher code rate (R) of 5/6 was added to further increase the data rate by 11%. The SNR requirement with 64-QAM, rate 5/6 is higher than the highest data rate in 802.11a/g, which only uses 64-QAM, $R = 3/4$. However, with the added robustness

Table 5.3 MCS definition (Perahia *et al.*, 2006; IEEE, 2007)

MCS indices for 1/2/3/4 spatial streams	Modulation	FEC code rate	Minimum sensitivity (dBm) (20 MHz channel spacing)	Minimum sensitivity (dBm) (40 MHz channel spacing)
0/8/16/24	BPSK	½	−82	−79
1/9/17/25	QPSK	½	−79	−76
2/10/18/26	QPSK	¾	−77	−74
3/11/19/27	16 QAM	½	−74	−71
4/12/20/28	16 QAM	¾	−70	−67
5/13/21/29	64 QAM	⅔	−66	−63
6/14/22/30	64 QAM	¾	−65	−62
7/15/23/31	64 QAM	⅚	−64	−61

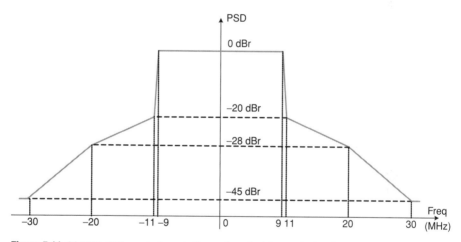

Figure 5.11 20 MHz HT spectral mask. Reproduced with permission from IEEE (2007) © IEEE.

features in 802.11n, the MCSs with 64-QAM, R = 5/6 are functional at a reasonable range.

The MCSs for 20 MHz are given in Tables 4.7–4.10, and the MCSs for 40 MHz are given in Tables 5.14–5.18. A summary of the MCSs as they map to modulation and code rate and their corresponding minimum receiver sensitivity is given in Table 5.3. Minimum receiver sensitivity is measured in a conducted test environment at a PER of 10% with packets of length 4096 bytes.

To determine the implementation margin, we simulate the 802.11n MIMO system in an AWGN environment without impairments. This type of system model is designated as Comparison Criteria 59 in Stephens (2004). To generate AWGN results for an $N \times N$ MIMO system, a Fourier matrix is used for the channel. The expression for the Fourier matrix is given in Eq. (5.12) (Berke, 2004):

$$\mathbf{F}_N = \begin{pmatrix} 1 & 1 & 1 & \cdots & 1 \\ 1 & W^1 & W^2 & \cdots & W^{(N-1)} \\ 1 & W^2 & W^4 & \cdots & W^{2(N-1)} \\ \vdots & \vdots & \vdots & & \vdots \\ 1 & W^{(N-1)} & W^{(N-1)2} & \cdots & W^{(N-1)^2} \end{pmatrix} \tag{5.12}$$

where $W = \exp(-j2\pi/N)$, and can be expressed explicitly as

$$\mathbf{F}_1 = 1$$

$$\mathbf{F}_2 = \begin{pmatrix} 1 & 1 \\ 1 & -1 \end{pmatrix}$$

$$\mathbf{F}_3 = \begin{pmatrix} 1 & 1 & 1 \\ 1 & e^{-j2\pi/3} & e^{-j4\pi/3} \\ 1 & e^{-j4\pi/3} & e^{-j8\pi/3} \end{pmatrix} \tag{5.13}$$

$$\mathbf{F}_4 = \begin{pmatrix} 1 & 1 & 1 & 1 \\ 1 & -j & -1 & j \\ 1 & -1 & 1 & -1 \\ 1 & j & -1 & -j \end{pmatrix}$$

In this manner all receive antennas are coupled with the transmit antennas, yet the matrix is orthogonal allowing for complete separation of the spatial streams without noise enhancement. The capacity of such a system can be derived from Eq. (3.10) as

$$C(\text{bps/Hz}) = N_{SS} \cdot \log_2 [1 + \text{SNR}] \tag{5.14}$$

and the capacity per spatial stream (SS) is given as

$$C(\text{bps/Hz/SS}) = \log_2 [1 + \text{SNR}] \tag{5.15}$$

Since the capacity per spatial stream is solely a function of SNR, the PER curves are only dependent on modulation, code rate, and SNR. The PER performance with no impairments with packets of length 4096 bytes for each modulation and code rate, as a function of SNR, is given in Figure 5.12.

Extracting the required SNR at 10% PER from Figure 5.12 and assuming a receiver noise figure of 6 dB, we can calculate the margin relative to the minimum sensitivity as given below:

$$\begin{aligned} \text{Margin (dB)} = {}& \text{RecSen(dBm)} - 30 \\ & - \left[\text{SNR(dB)} + \left(10 \cdot \log(1.38 \times 10^{-23} \cdot 290) \right. \right. \\ & \left. \left. + 10 \cdot \log(\text{BW(Hz)}) + \text{NF(dB)} \right) \right] \end{aligned} \tag{5.16}$$

where RecSen is the minimum receive sensitivity from Table 5.3. The margin ranges from 11 to 12 dB, as given in Table 5.4. The margin is the same for 20 MHz and 40 MHz since the 3 dB difference in bandwidth is equivalent to the difference in minimum sensitivity. Such a large margin may seem excessive; however, algorithmic degradations and hardware impairments degrade performance and require sufficient margin for cost effective implementations.

Table 5.4 Margin relative to minimum sensitivity

Modulation	FEC code rate	Required SNR (dB) @ 10% PER	Margin relative to minimum sensitivity (dB)
BPSK	1/2	1	12.0
QPSK	1/2	4	12.0
QPSK	3/4	6.5	11.5
16-QAM	1/2	9.75	11.2
16-QAM	3/4	13	12.0
64-QAM	2/3	17.25	11.7
64-QAM	3/4	18.75	11.2
64-QAM	5/6	19.75	11.2

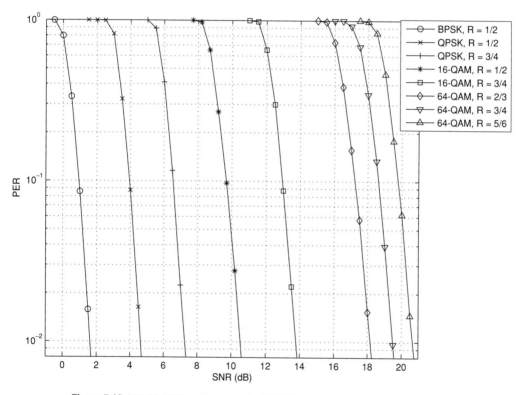

Figure 5.12 MIMO PER performance in AWGN.

The above analysis only provides reference with respect to measurements in a conductive test environment. To get a true sense of the required SNR in a frequency selective fading MIMO environment, we simulate the MCSs in the channel models described in Section 3.5 and with physical layer impairments described in Section 3.5.4. The equalizer is based on MMSE. Synchronization, channel estimation, and phase tracking are included in the simulation to give an accurate model for algorithm degradations.

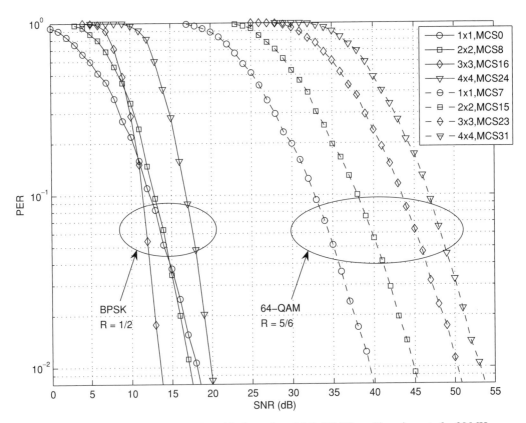

Figure 5.13 Simulated results with channel model B, NLOS, and impairments for 20 MHz.

Figures 5.13–5.15 illustrate the simulated results for 20 MHz in channel models B, D, and E, respectively, in an NLOS environment. The curves with circles are one spatial steam. The curves with squares are two spatial streams. The curves with diamonds are three spatial streams. Finally, the curves with triangles are four spatial streams. All curves have transmit and receive antennas equal to the number of spatial streams. The solid lines correspond to MCSs with BPSK and R = ½. The dotted lines correspond to MCSs with 64-QAM and R = 5/6.

We should note that obtaining an SNR above 35 dB is difficult to achieve with reasonable cost hardware since it exceeds the noise figure, impairments, and implementation losses of typical transmitters and receivers, particularly in RF circuits and ADC/DAC. Required SNRs beyond 35 dB are provided in these figures for comparison purposes only and are not intended to indicate realistic hardware capabilities. In general, links with MCSs of 64-QAM, R = 5/6 are difficult to establish without additional means to reduce the required SNR.

We see that for BPSK, R = 1/2 the curves are tightly clustered, whereas for 64-QAM, R = 5/6 the required SNR substantially increases with the number of spatial streams. With higher order modulation, the system is more sensitive to impairments as the number of spatial streams increases.

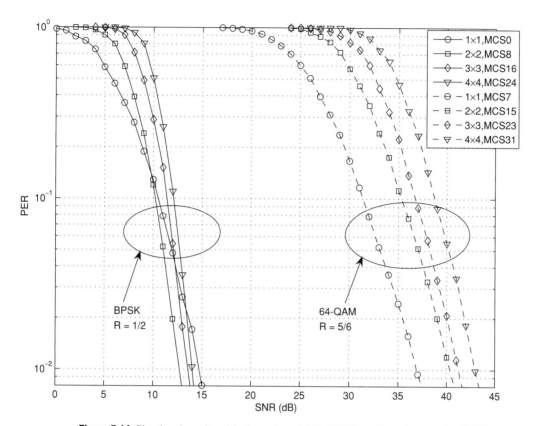

Figure 5.14 Simulated results with channel model D, NLOS, and impairments for 20 MHz.

The required SNR with channel model B is higher than with channel model D. Channel model D has a longer delay spread than channel model B, and therefore more frequency selective fading and less flat fading. The benefit of coded OFDM is its ability to mitigate frequency selective fading, resulting in better performance in channel model D. As seen in Figure 5.15, the performance degrades a little in channel model E as compared to channel model D. The longer delay spread of channel model E begins to effect synchronization and tracking.

5.4 Greenfield (GF) preamble

The MF preamble trades throughput efficiency for interoperability between HT and legacy devices by including the L-STF, L-LTF, and L-SIG in the preamble. During the development of the 802.11n standard amendment, many were of the opinion that network conditions would exist that would benefit from a more efficient preamble. To improve efficiency, a preamble without legacy fields which is not compatible with legacy OFDM devices was adopted.

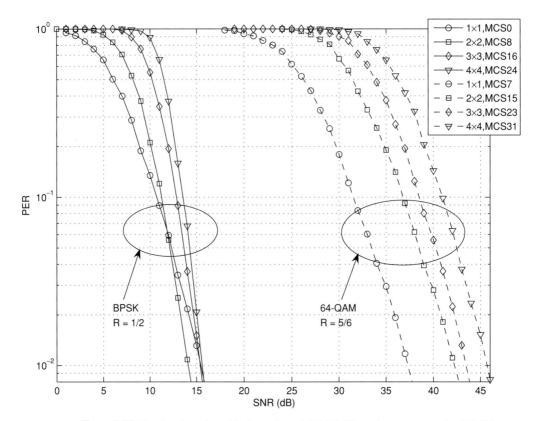

Figure 5.15 Simulated results with channel model E, NLOS, and impairments for 20 MHz.

Environments which do not require legacy compatibility are termed "Greenfield" deployments (from which the name of the preamble was derived). At initial development of 802.11n, adoption and deployment of 802.11a in 5 GHz was slow in much of the world and at that time much of the band was unused. Furthermore, new installations of WLANs in office buildings where no legacy devices were present would not require legacy compatibility. As another example, single family dwellings with no neighbors utilizing legacy devices would also not require legacy compatibility (van Nee and Jones, 2006).

Proponents of GF preamble format also proposed the use of the GF preamble even in the presence of legacy devices. If access to the medium is controlled through MAC protection mechanisms (legacy RTS/CTS, for example), then within the reserved time legacy compatibility is not required. During the reserved time, GF may be used to reduce overhead. MAC protection mechanisms and using the NAV to reserve time are discussed in Section 10.6.

5.4.1   Format of the GF preamble

The fields of the GF preamble are illustrated in Figure 5.16. The MF preamble is also included in Figure 5.16 for direct comparison.

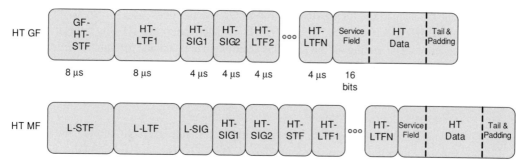

Figure 5.16 Greenfield format preamble.

A new HT-STF for GF (GF-HT-STF) replaces the L-STF in the MF preamble. The waveform for GF-HT-STF is derived in the same manner as the MF HT-STF given in Eq. (4.17), except that it is 8 μs in length. The increase in length matches the L-STF to provide enough time for start-of-packet detection, AGC setting, frequency estimation, and time synchronization. The HT cyclic shifts given in Table 4.3 are applied to the GF-HT-STF and all subsequent fields in a GF packet. Furthermore, if spatial expansion or beamforming is applied to the spatial streams, the spreading matrix, Q, is applied to the GF-HT-STF and all subsequent fields in a GF packet as defined in Eq. (4.31) for 20 MHz and Eq. (5.2) for 40 MHz.

Since we are not concerned about legacy compatibility with the GF preamble, long HT cyclic shifts are applied to the spatial streams of the GF-HT-STF. With the long cyclic shifts, accurate acquisition and AGC setting required by MIMO receivers is attained. There is no need for a subsequent Short Training field as is required in the MF preamble. This saves 4 μs compared to the MF preamble.

Without the requirement of legacy compatibility, there is also no need for the L-SIG. All information necessary for 11n is conveyed in the HT-SIG$_1$ and HT-SIG$_2$. This saves an additional 4 μs compared to the MF preamble. With the absence of the L-SIG, the L-LTF is not required. Therefore, HT-LTF1 replaces the L-LTF as compared to MF.

In MF, the two long training symbols of the L-LTF were used for accurate frequency and timing as described in Section 4.1.2. For comparable frequency and time estimation capability, the GF preamble also requires repetition of the long training symbols. Therefore for GF the same Long Training field is used as in MF (given in Eq. (4.19) for 20 MHz and Eq. (5.2) for 40 MHz), except that HT-LTF1 is twice as long as the other HT-LTFs. HT-LTF1 consists of two periods of the long training symbols, preceded by a 1.6 μs cyclic prefix (IEEE, 2007). This results in a length of 8 μs.

Following HT-LTF1 in GF are the two HT-SIG fields. The content is identical to that of MF, given in Section 4.2.2.1. The waveform is similar to that of MF (defined by Eq. (4.16) for 20 MHz and Eq. (5.1) for 40 MHz); however, there are some differences which are outlined as follows.

To match the HT-SIGs in GF to HT-LTF1, the HT-SIGs must include the orthogonal mapping matrix. The first column of the orthogonal mapping matrix is applied to the HT-SIGs. Furthermore, the scaling of HT-LTF is based on 56 subcarriers for 20 MHz and 114 subcarriers for 40 MHz. However, the HT-SIG fields only utilize 52 subcarriers in 20 MHz and 104 subcarriers in 40 MHz to match that of the HT-SIG fields in the MF preamble. The scaling for the HT-SIGs is modified to match the per-subcarrier scaling of the HT-SIG in the GF preamble to HT-LTF1. In addition, in the GF preamble the scaling is based on the number of spatial streams rather than the number of transmit antennas. Therefore in the GF preamble the scaling is $1/\sqrt{N_{SS} \cdot 56}$ for 20 MHz and $1/\sqrt{N_{SS} \cdot 114}$ for 40 MHz for the HT-SIG.

In MF, the first value of the pilot polarity is applied to the L-SIG. In GF, the first two values of the pilot polarity are applied to the HT-SIGs. The same pilot polarity sequence is used as in MF.

The 20 MHz waveform for the GF HT-SIG (including spatial expansion) is given in the equation below:

$$
r_{HT-SIG}^{(i_{TX})}(t) = \frac{1}{\sqrt{N_{SS} \cdot 56}} \sum_{n=0}^{1} \sum_{k=-26}^{26} \sum_{i_{SS}=1}^{N_{SS}} \left[[Q_k]_{i_{TX}, i_{SS}} \cdot [P_{HTLTF}]_{i_{SS},1} \cdot (jD_{k,n} + p_n P_k) \right.
$$
$$
\left. \times \exp \left(j 2\pi \, k \Delta_F \left(t - n \cdot T_{SYM} - T_{GI} - T_{CS}^{i_{SS}} \right) \right) \right] \tag{5.17}
$$

where N_{SS} is the number of spatial streams; Q is the spatial mapping matrix, discussed further in Section 4.2.3.8; P_{HTLTF} is the orthogonal mapping matrix, as defined by Eq. (4.20); $D_{k,n}$ is the mapping of complex data values to subcarrier locations, which is the same as L-SIG defined by Eq. (4.15); p_n is the pilot polarity defined to be the same as legacy OFDM and given by Eq. (4.5); P_k are the pilot subcarriers defined to be the same as legacy OFDM and given by Eq. (4.4); $\Delta_F = 312.5$ kHz; $T_{SYM} = 4$ μs symbol time, composed of 3.2 μs OFDM symbol time and 0.8 μs guard interval; $T_{GI} = 0.8$ μs guard interval; and $T_{CS}^{i_{SS}}$ represents the cyclic shift for spatial stream i_{SS} with values given by Table 4.3.

The 40 MHz waveform for the GF HT-SIG (including spatial expansion) is given in the equation below, with the same parameter definitions as in Eq. (5.17):

$$
r_{HT-SIG}^{(i_{TX})}(t) = \frac{1}{\sqrt{N_{SS} \cdot 114}} \left\{ \sum_{n=0}^{1} \sum_{k=-26}^{26} \sum_{i_{SS}=1}^{N_{SS}} \left[[Q_{k-32}]_{i_{TX}, i_{SS}} \cdot [P_{HTLTF}]_{i_{SS},1} \cdot (jD_{k,n} + p_n P_k) \right. \right.
$$
$$
\left. \times \exp \left(j 2\pi \, (k-32) \Delta_F \left(t - n \cdot T_{SYM} - T_{GI} - T_{CS}^{i_{SS}} \right) \right) \right]
$$
$$
+ \sum_{n=0}^{1} \sum_{k=-26}^{26} \sum_{i_{SS}=1}^{N_{SS}} \left[[Q_{k+32}]_{i_{TX}, i_{SS}} \cdot [P_{HTLTF}]_{i_{SS},1} \cdot (jD_{k,n} + p_n P_k) \right.
$$
$$
\left. \left. \times \exp \left(j 2\pi \, (k+32) \Delta_F \left(t - n \cdot T_{SYM} - T_{GI} - T_{CS}^{i_{SS}} \right) \right) \right] \right\} \tag{5.18}
$$

Since HT-LTF1 in GF replaced the L-LTF in MF, only the remaining HT-LTFs for two or more spatial streams are needed following the HT-SIG. These are identical to the HT-LTFs in the MF, the expressions of which are given in Eq. (4.19) for 20 MHz and Eq. (5.2) for 40 MHz, respectively. The additional HT-LTFs are 4 μs in length. The total number of HT-LTFs as a function of the number of spatial streams is the same as MF, and given in Eq. (4.21).

Table 5.5 Preamble length

Number of spatial streams	MF (μs)	GF (μs)
1	36	24
2	40	28
3	44	32
4	48	36

For two or more spatial streams, the Data field follows the additional HT-LTFs. For a single spatial stream, there is only the HT-LTF1, therefore the Data field follows the HT-SIGs. The Data field in GF is the same as in MF, as described in Section 4.2.3.

5.4.2 PHY efficiency

The GF preamble reduces the overhead of the PHY preamble by 12 μs as compared to the MF preamble. The preamble length, as a function of number of spatial streams, is given in Table 5.5.

During the 802.11n standard development process, the proponents of GF claimed substantial improvements in efficiency over MF, especially with higher data rates and shorter packet lengths. The efficiency improvement in the PHY is illustrated in Figure 5.17. Several data rates ranging from 6.5 Mbps with 20 MHz and one spatial steam (SS) to 540 Mbps with 40 MHz and four spatial streams are given. For applications like voice over IP (VoIP), which uses short data packets (of the order of 100 bytes), the benefit of GF preamble would be considerable. Additionally, the time on air is shorter, which means the system can support more callers. This is a benefit since the handheld devices that are the typical VoIP clients are very sensitive to battery life. Shorter time on air reduces the time that the transmitter and receiver are active, which decreases power consumption and increases battery life.

5.4.3 Issues with GF

The detractors of the GF preamble argued during the 802.11n standard development process that the GF preamble has many issues and is not necessary (McFarland *et al.*, 2006). These issues include small network efficiency improvement, lack of interoperability with legacy devices, implementation issues, and the added burden of auto-detection between multiple 11n preamble types. In the end, a compromise was reached resulting in the inclusion of the GF preamble in 802.11n, but as an optional preamble format.

5.4.3.1 Network efficiency

During the proposal phase of the 802.11n standard development, a MAC simulation was performed to demonstrate the sensitivity of network throughput to the PHY preamble length (Stephens *et al.*, 2005). The MAC simulations were run over system scenarios for residential, large enterprise, and hot spot as described in Chapter 1 and defined in

Table 5.6 Network throughput versus PHY overhead duration (Stephens *et al.*, 2005)

System scenario	2 TX × 2 RX × 20 MHz			2 TX × 2 RX × 40 MHz		
	25 μs preamble	50 μs preamble	Efficiency improvement	25 μs preamble	50 μs preamble	Efficiency improvement
Residential	109 Mbps	106 Mbps	3%	169 Mbps	160 Mbps	5%
Large enterprise	103 Mbps	98 Mbps	5%	228 Mbps	226 Mbps	1%
Hot spot	86 Mbps	80 Mbps	7%	194 Mbps	184 Mbps	5%

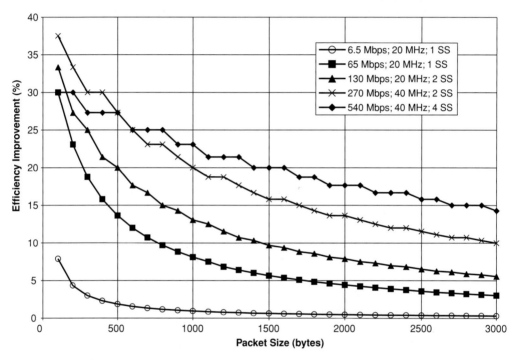

Figure 5.17 PHY efficiency improvement of GF over MF.

Stephens *et al.* (2004a). Separate simulations were performed for 20 MHz and 40 MHz, although both assumed devices with two transmit and two receive antennas. The simulations we repeated for different PHY preamble lengths. Table 5.6 provides the results for 25 and 50 μs preambles, which encompass the range of the length for the MF and GF preambles. We see that the network efficiency only improves by a very small amount.

Further analysis given in McFarland *et al.* (2006) demonstrated small throughput improvement of GF compared to MF preamble with UDP and TCP traffic. The assumptions of the analysis were one AP and one client, ideal traffic patterns in the sense that there was packet bursting with unlimited TXOP lengths, and there was no PHY or MAC packet loss. Results were given for a range of packet lengths, data rates, and both aggregated and un-aggregated. For a detailed discussion on aggregation, see Section 8.2.

Table 5.7 UDP throughput improvement of GF compared to MF without aggregation (McFarland *et al.*, 2006)

Data rate	Single packet exchange		Bursting	
	64 byte packet	1500 byte packet	64 byte packet	1500 byte packet
6.5 Mbps; 20 MHz; 1 SS	3%	0%	4%	0%
65 Mbps; 20 MHz; 1 SS	6%	3%	13%	5%
130 Mbps; 20 MHz; 2 SS	7%	5%	14%	7%
300 Mbps; 40 MHz; 2 SS; Short GI	7%	4%	15%	10%

Table 5.8 Throughput improvement of GF compared to MF with aggregation (McFarland *et al.*, 2006)

Data rate	UDP		TCP	
	20 subframes per packet; 64 bytes per subframe	20 subframes per packet; 1500 bytes per subframe	20 subframes per packet; 64 bytes per subframe	20 subframes per packet; 1500 bytes per subframe
6.5 Mbps; 20 MHz; 1 SS	0%	2%	5%	2%
65 Mbps; 20 MHz; 1 SS	2%	1%	4%	1%
130 Mbps; 20 MHz; 2 SS	3%	0%	2%	1%
300 Mbps; 40 MHz; 2 SS; Short GI	4%	0%	0%	0%

Table 5.7 illustrates the small improvement to UDP traffic without aggregation. Four data rates are given ranging from 6.5 Mbps with 20 MHz and a single spatial stream to 300 Mbps with 40 MHz, two spatial streams, and short guard interval. Two types of simulations are given. The first is a single packet exchange. The second is bursting of data/ACK sequences separated by SIFS with an infinite TXOP. Even without aggregation, the maximum benefit of GF is only 15% throughput improvement.

One of the key MAC features in 802.11n is aggregation to improve efficiency. Table 5.8 illustrates the further reduced benefit of GF compared to MF for UDP and TCP with A-MPDU aggregation. Results are given for two packet lengths. Each packet consists of 20 subframes. Each subframe contains either 64 bytes or 1500 bytes. With aggregation, the maximum benefit of GF compared to MF reduces to 5%.

5.4.3.2 Interoperability issues with legacy

Obviously a legacy device cannot process a GF packet. Therefore, in the presence of legacy devices in the BSS, it is required to protect GF packets. However, this protection does not extend to neighboring BSSs. A GF capable AP with only HT devices associated to it could allow unprotected GF transmissions. Legacy devices in neighboring BSSs would not be protected from GF transmissions.

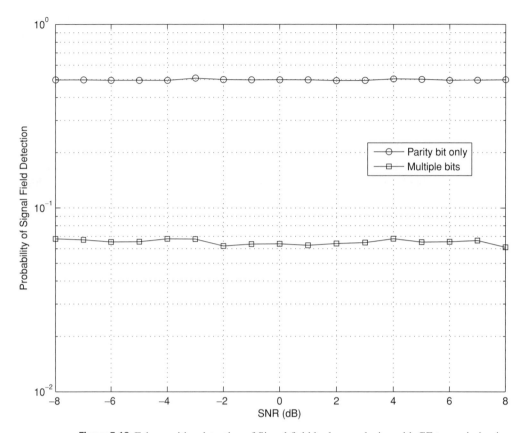

Figure 5.18 False positive detection of Signal field by legacy device with GF transmission in channel model D, NLOS.

Several situations can occur when a legacy device receives a GF packet. Both the GF and legacy preambles have an 8 μs STF and an 8 μs LTF followed by the SIG. Since the legacy and GF preambles are similar when the legacy device receives a GF preamble, there is a high likelihood of false detects. This is especially true since there is only a single parity bit in the legacy Signal field, which leads to false positives. Fortunately, many legacy devices use extra fields in the Signal field to reduce the false detect rate. Some possible techniques are described in Section 4.1.3. Figure 5.18 illustrates simulation results that demonstrate the probability that a legacy device falsely detects a valid Signal field when receiving a GF transmission. When only using the single parity bit, the false detect probability is 50%. However, when checking additional bits in the Signal field for valid conditions the probability is reduced to 7%.

False detects lead to legacy devices staying off the air for a random amount of time. If the amount of time is too short, the legacy device interferes with the HT GF device. If the amount of time is too long, the legacy device is unfairly disadvantaged.

Even when the legacy device properly determines that the received signal is not a legacy packet, it only relies on the energy level of the received signal to determine how long to stay off the air. Based on the standard, the legacy device is only required to remain off the air for a received level of -62 dBm or greater. Therefore, HT GF devices will most likely be interfered with by legacy devices in neighboring BSSs.

5.4.3.3   Implementation issues

There are some minor implementation issues with the GF preamble. Since with one spatial stream the Data field follows the HT-SIG field and with two or more spatial streams the HT-LTFs follows the HT-SIG, it is necessary to rapidly decode the HT-SIG to determine the number of HT-LTFs which dictates what field comes next (McFarland *et al.*, 2006). In addition, the GF preamble uses long cyclic shifts on the STF, which requires extra care in start-of-packet and time synchronization algorithms (McFarland *et al.*, 2006).

5.4.4   Preamble auto-detection

Having two HT preamble formats requires auto-detection between MF and GF formats. Since MF is mandatory and GF is optional, obviously a GF-capable device must auto-detect between MF and GF. However an MF-only device is also required to partially process a GF packet.

The 802.11n standard amendment states, "An HT STA that does not support the reception of an HT Greenfield format packet shall be able to detect that an HT Greenfield format packet is an HT transmission (as opposed to a non-HT transmission). In this case the receiver shall decode the HT-SIG and determine if the HT-SIG cyclic redundancy check (CRC) passes" (IEEE, 2007). Therefore, even a device that does not support GF must process a GF packet up to the HT-SIG and check the CRC. All HT devices need to auto-detect for legacy, HT MF, and HT GF.

The timing of the fields of the legacy, MF, and GF preambles is given in Figure 5.19. Upon receiving a packet, a device must parallel process all the different types of preambles until the correct format is determined. During the time from 0 to 16 µs, a device must parallel process the legacy STF/LTF for legacy and MF and the HT STF/LTF for GF. From 16 to 20 µs, the device must process the legacy SIG for legacy and MF preambles. From 16 to 24 µs, the device must process the HT-SIGs for a GF preamble.

As was described in Section 4.2.2.1, a 90 degree rotation between the L-SIG and HT-SIG$_1$ may be used to auto-detect between legacy and HT MF. For legacy and GF there is no rotation between 16 and 20 µs and 20 and 24 µs. Between 16 and 24 µs, the HT device must process the HT-SIG$_1$ and HT-SIG$_2$ in the case of a GF preamble. Similarly, between 20 and 28 µs, the HT device must process the HT-SIG$_1$ and HT-SIG$_2$ in the case of MF. The combinations of rotations, parity check on L-SIG, and CRC checks on HT-SIG at different times allows differentiation between legacy, HT MF, and HT GF.

If it has been determined that the packet is neither MF nor GF (failed CRC checks and improper rotations), but the parity check of the L-SIG passes, the packet is processed as

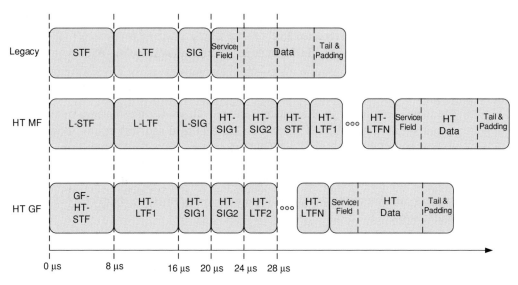

Figure 5.19 Timing of the preamble fields in legacy, MF, and GF.

legacy OFDM. On the other hand, if it has been determined that the packet is GF, but the device is non-GF capable, the device must still properly defer transmission.

The proponents of GF wanted to make it mandatory, thereby ensuring interoperability between all HT devices. The opponents to GF did not feel that the marginal benefit of GF warranted the added complexity of an additional preamble type. As a compromise, a non-GF capable device is given two options when receiving a GF packet. In the first option, the device may extract the Signal field parameters and calculate the transmit time (TXTIME) of the packet. The PHY of the device indicates to the MAC that the channel is busy for this duration. This is the clear channel assessment (CCA). The second option is that the PHY of the non-GF capable device maintains CCA busy until the received level drops below −72 dBm for 20 MHz and −69 dBm for 40 MHz. These thresholds are derived from the receiver minimum input level sensitivity (−82 dBm for 20 MHz and −79 dBm for 40 MHz) plus 10 dB. This allows the non-GF capable device to not have to extract the contents of the HT-SIG when it occurs during a GF packet. A non-GF capable device is only required to check the CRC of a GF packet, which does not require extracting the contents.

A GF capable device must receive all packet formats, and maintain CCA busy based on TXTIME calculated from the contents of the Signal field. The correct duration must be calculated even for rates and modes that are not supported by the device. The TXTIME equation for the GF preamble is given below:

$$\text{TXTIME} = T_{\text{GF-HT-STF}} + T_{\text{HT-LTF1}} + T_{\text{HT-SIG}}$$
$$+ (N_{\text{LTF}} - 1) \cdot T_{\text{HT-LTFs}} + T_{\text{SYM}} \cdot N_{\text{SYM}} \qquad (5.19)$$

where $T_{\text{HT-GF-STF}}, T_{\text{HT-LTF1}}, T_{\text{HT-SIG}} = 8\,\mu\text{s}$ duration; $T_{\text{HT-LTFs}} = 4\,\mu\text{s}$ duration; $T_{\text{SYM}} = 4\,\mu\text{s}$ duration for 800 ns guard interval operation; N_{SYM} is the number of OFDM symbols in the Data field defined by Eq. (4.22) for one BCC encoder and Eq. (5.6) for two BCC encoders; and N_{LTF} is the number of long training symbols, determined from the MCS – the number of long training symbols is based on the number of spatial streams indicated by the MCS (excluding a sounding packet).

The computation of TXTIME changes for short guard interval, sounding packet, LDPC, and STBC, which are described in Sections 5.5, 12.6, 6.4, and 6.3, respectively.

There are some exceptions, which also apply to a non-GF capable device receiving a MF packet. An exception to the TXTIME computation is if the reserved bit is decoded as 0. If the HT-SIG CRC is valid, then a reserved bit value of 0 indicates an invalid condition or a false positive of the CRC computation. Furthermore, if the MCS field is in the range 77–127 or the STBC field has a value of 3, TXTIME can not be computed. In these conditions, the device must not access the medium for as long as the received signal level remains above −62 dBm for 20 MHz and −59 dBm for 40 MHz, rather than based on TXTIME computation. These thresholds are derived from the receiver minimum input level sensitivity (−82 dBm for 20 MHz and −79 dBm for 40 MHz) plus 20 dB.

For further information on the preamble detection, refer to Section 4.2.4.3.

5.5   Short guard interval

The standard OFDM data symbol for 802.11n has a 4 μs duration, comprising a 0.8 μs guard interval (GI) and 3.2 μs of data. This is the same as it is for legacy OFDM symbols. To further increase the data rate in 802.11n, an optional short GI was adopted. With the short GI, the guard interval of the data symbols is reduced to 0.4 μs and the overall symbol length is 3.6 μs. The guard interval of the preamble was not modified. This corresponds to an increase in data rate of 11%. With the short GI, the maximum data rate for 802.11n with 20 MHz is 289 Mbps and with 40 MHz is 600 Mbps.

The waveform equations for the Data field must be changed such that the data symbol duration, T_{SYM}, is reduced to 3.6 μs and the guard interval time, T_{GI}, is reduced to 0.4 μs. This occurs in Eq. (4.30) for MF and 20 MHz operation. Then with spatial expansion in Eq. (4.31), the same changes are made, but only for the Data field. Similarly for 40 MHz, the same changes (limited to the Data field) are made to Eq. (5.2). And finally, Eq. (5.10) for the HT duplicate is also modified with reduced values when short GI is utilized. Correspondingly, the TXTIME computation for the MF preamble in Eq. (4.32) and the GF preamble in Eq. (5.19) is modified by also reducing T_{SYM} to 3.6 μs.

There is one packet format in which short GI may not be used. With the GF preamble and one spatial stream, the data symbols immediately follow the HT-SIG. However, the receiver does not know that this situation has occurred until it parses the HT-SIG. In a typical design, the receiver continues processing the subsequent symbol in parallel to extracting the information from the HT-SIG. The first step is to remove the guard interval before performing the Fourier transform. In order to receive a GF with one spatial stream and short GI, the receiver would have to buffer the signal until it determined the number of spatial streams and the length of the GI from the HT-SIG. In order to avoid this

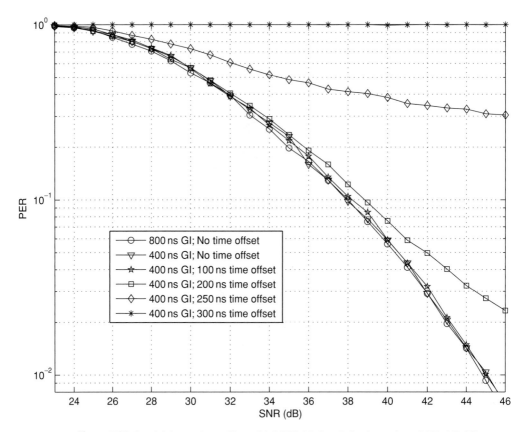

Figure 5.20 Sensitivity to time offset with MCS 15, 2 × 2, in channel model B, NLOS.

complication, the use of short GI with GF with one spatial stream is not permitted by the 802.11n standard.

As discussed in Section 2.1, the guard interval is used to mitigate inter-symbol interference due to the delay spread of the channel environment. By shortening the GI, the signal is more susceptible to longer delay spread channels. Delay spread is also caused by filtering in the transmitter and receiver. Transmit filters are used to reduce out-of-band emissions. The larger the reduction of interference to adjacent bands, the longer the delay spread of the filter. Similarly on receive, the more there is rejection of adjacent bands, the greater the delay spread that is incurred from the receive filter.

By shortening the guard interval, the receiver requires much more precise time synchronization, especially in channels with longer delay spread. Therefore, one way to quantify the receiver sensitivity to shortening of the guard interval is to model the receiver performance with increasing fixed time offsets in the time synchronization. We compare the performance with 800 ns GI to 400 ns GI. Figure 5.20 illustrates the performance for MCS 15 with two transmit antennas and two receiver antennas in channel model B, with NLOS conditions. Figures 5.21 and 5.22 illustrate results with the same configuration,

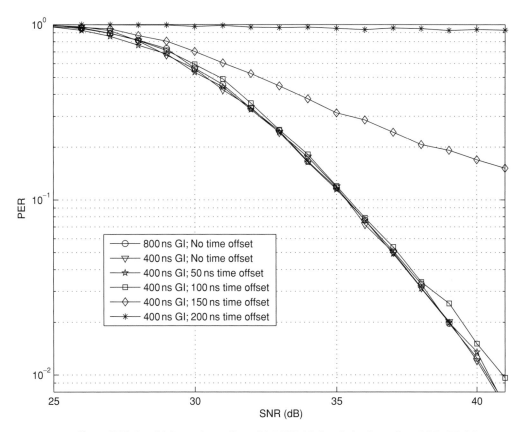

Figure 5.21 Sensitivity to time offset with MCS 15, 2×2, in channel model D, NLOS.

except that the simulation is performed in channel models D and E, respectively, with NLOS conditions.

Though not illustrated in the figures, packets transmitted with 800 ns GI experience no degradation with time offsets up to 300 ns with channel models B, D, and E. On the other hand, with 400 ns GI the performance degradation is directly coupled to the delay spread of the channel. In channel model B, which has an RMS delay spread of 15 ns, there is no degradation up to a time offset of 100 ns. As the offset is increased to 200 ns, slight degradation is experienced. The receiver fails when the time offset is increased to 250 ns or more. In channel model D which has an RMS delay spread of 50 ns, there is no degradation with a time offset of 50 ns. There is a slight degradation with a time offset of 100 ns. There is a large degradation with a time offset of 150 ns, and the receiver fails with an offset greater than 150 ns. In channel model E which has an RMS delay spread of 100 ns, there is also no degradation with a time offset of 50 ns. However with a time offset of 100 ns there is more degradation than in channel model D. There is a large degradation with a time offset of 150 ns, and the receiver fails with offset greater than 200 ns.

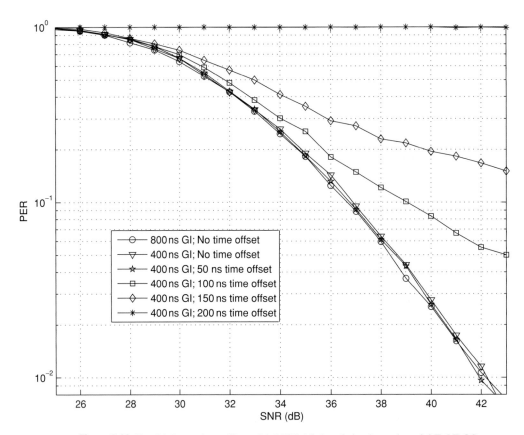

Figure 5.22 Sensitivity to time offset with MCS 15, 2×2, in channel model E, NLOS.

To summarize, noticeable degradation occurs at time offsets which decrease from 200 ns for channel model B to 150 ns for channel model D to 100 ns for channel model E. This decrement roughly corresponds to the increase in RMS delay spread between the channels. When critically sampled, a time offset of 100 ns for channel model E is only two samples in time, which requires much more accuracy in the time synchronization and tracking algorithm in the receiver.

Short GI adds another dimension to rate adaptation beyond modulation, code rate, and number of spatial streams. With each added dimension, the complexity of rate adaptation increases significantly. In addition, short GI only provides an incremental improvement in data rate in shorter delay spread channels. Similar incremental improvements may be attained by changing the MCS. Therefore, a typical approach to the use of short GI is to only try to switch to short GI when the highest data rate of the device with long GI has been achieved. The highest MCS is indicative of a good channel environment and no further increases in data rate can be achieved by modulation, code rate, and number of spatial streams. As a final rate adaptation step, the device can switch to short GI for the maximum data rate of the device. If acknowledgements are received to the transmitted packets, the device knows that the channel supports short GI and the maximum data rate. With added complexity, one may determine the delay spread of the channel from the channel estimate to try to use short GI in more situations.

References

Berke, B. (2004). *TGn Comparison Criteria – Proposal for CC59*, IEEE 802.11–04/0176r0.

Coffey, S., Jones, V. K., Trachewsky, J., *et al.* (2004). *WWiSE IEEE 802.11n Proposal*, IEEE 802.11–04/0935r4.

Hedberg, D. (2005). *Adjacent Channel Interference and Filtering for 56-carrier Signals*, IEEE 802.11–04/1579r1.

IEEE (2007). *IEEE P802.11n(tm)/D3.00, Draft Amendment to STANDARD for Information Technology–Telecommunications and Information Exchange Between Systems – Local and Metropolitan Networks – Specific Requirements. Part 11: Wireless LAN Medium Access Control (MAC) and Physical Layer (PHY). Amendment 4: Enhancements for Higher Throughput.*

McFarland, B., Nakao, S., Takeda, D., *et al.* (2006). *Green Field Analysis*, IEEE 802.11–06/452r1.

Perahia, E., Aldana, C., Kim, J., *et al.* (2006). *Joint Proposal Team PHY Simulation Results*, IEEE 802.11–06/0067r2.

Stephens, A. (2004), *IEEE 802.11 TGn Comparison Criteria*. IEEE 802.11–03/814r31.

Stephens, A., Bjerke, B., Jechoux, B., *et al.* (2004). *Usage Models*, IEEE 802.11–03/802r23.

Stephens, A., Akhmetov, D., and Shtin, S. (2005). *TGn Sync Simulation Results – Goodput versus PHY Overhead*, IEEE 802.11–05/479r0.

van Nee, R. and Jones, V. K. (2006). *Mandatory Greenfield Preamble*, IEEE 802.11–6/971r1.

Appendix 5.1: Channel allocation

Channel center frequencies are defined at every integral multiple of 5 MHz beginning from the channel starting frequency for the specific band, as follows:

$$\text{Channel center frequency} = \text{channel starting frequency} + 5 \cdot n_{ch} \text{ (MHz)} \qquad (5.20)$$

where n_{ch} is the channel number ($= 0, 1, \ldots, 200$ for 5 GHz, and $= 1, 2, \ldots, 13$ for 2.4 GHz). For the 5 GHz band, the channel starting frequency is equal to 5000. For the 2.4 GHz band, the channel starting frequency is equal to 2407.

Tables 5.9 and 5.10 give the 20 MHz channel allocations for USA and Europe, respectively. Tables 5.11 and 5.12 give the 40 MHz channel allocations for USA and Europe, respectively. The 40 MHz channel allocations also include both conditions where the secondary 20 MHz channel is below or above the primary 20 MHz channel. This is indicated by the secondary field.

Table 5.9 20 MHz channel allocation for USA

Frequency band (GHz)	Regulatory class	Channel number	Center frequency (MHz)	Transmit power limit (mW)
ISM 2.4–2.483	12	1, 2, ..., 11	2412, 2417, ..., 2462	1000
U-NII lower 5.15–5.25	1	36, 40, 44, 48	5180, 5200, 5220, 5240	40
U-NII middle 5.25–5.35	2	52, 56, 60, 64	5260, 5280, 5300, 5320	200
Additional U-NII 5.47–5.725	4	100, 104, ..., 140	5500, 5520, ..., 5700	200
Upper U-NII 5.725–5.85	3	149, 153, 157, 161	5745, 5765, 5785, 5805	800
	5	149, 153, 157, 161, 165	5745, 5765, 5785, 5805, 5825	1000

Table 5.10 20 MHz channel allocation for Europe

Frequency band (GHz)	Regulatory class	Channel number	Center frequency (MHz)	Transmit power limit (mW)
ISM 2.4–2.483	4	1, 2, ..., 13	2412, 2417, ..., 2472	100
5.15–5.25	1	36, 40, 44, 48	5180, 5200, 5220, 5240	200
5.25–5.35	2	52, 56, 60, 64	5260, 5280, 5300, 5320	200
5.47–5.725	3	100, 104, ..., 140	5500, 5520, ..., 5700	1000

Table 5.11 40 MHz channel allocation for USA

Frequency band (GHz)	Regulatory class	Primary channel number	Secondary field	Secondary channel number
ISM 2.4–483	32	1	1	5
		2	1	6
		3	1	7
		4	1	8
		5	1	9
		6	1	10
		7	1	11
	33	5	−1	1
		6	−1	2
		7	−1	3
		8	−1	4
		9	−1	5
		10	−1	6
		11	−1	7
U-NII lower 5.15–5.25	22	36	1	40
		44	1	48
	27	40	−1	36
		48	−1	44
U-NII middle 5.25–5.35	23	52	1	56
		60	1	64
	28	56	−1	52
		64	−1	60
Additional U-NII 5.47–5.725	24	100	1	104
		108	1	112
		116	1	120
		124	1	128
		132	1	136
	29	104	−1	100
		112	−1	108
		120	−1	116
		128	−1	124
		136	−1	132
Upper U-NII 5.725–5.85	25 (for 800 MW) or	149	1	153
	26 (for 1000 MW)	157	1	161
	30 (for 800 MW) or	153	−1	149
	31 (for 1000 MW)	161	−1	157

Table 5.12 40 MHz channel allocation for Europe

Frequency band (GHz)	Regulatory class	Primary channel number	Secondary field	Secondary channel number
ISM 2.4–2.483	11	1	1	5
		2	1	6
		3	1	7
		4	1	8
		5	1	9
		6	1	10
		7	1	11
		8	1	12
		9	1	13
	12	5	−1	1
		6	−1	2
		7	−1	3
		8	−1	4
		9	−1	5
		10	−1	6
		11	−1	7
		12	−1	8
		13	−1	9
5.15–5.25	5	36	1	40
		44	1	48
	8	40	−1	36
		48	−1	44
5.25–5.35	6	52	1	56
		60	1	64
	9	56	−1	52
		64	−1	60
5.47–5.725	7	100	1	104
		108	1	112
		116	1	120
		124	1	128
		132	1	136
	10	104	−1	100
		112	−1	108
		120	−1	116
		128	−1	124
		136	−1	132

Table 5.13 Symbols used for MCS parameters (IEEE, 2007)

Symbol	Explanation
R	Code rate
N_{BPSCS}	Number of coded bits per single carrier for each spatial stream (or modulation order)
N_{CBPS}	Number of coded bits per OFDM symbol
N_{DBPS}	Number of data bits per OFDM symbol
N_{ES}	Number of BCC encoders

Table 5.14 40 MHz MCS parameters for single spatial stream (IEEE, 2007)

MCS index	Modulation	R	N_{ES}	N_{BPSCS}	N_{CBPS}	N_{DBPS}	Data rate (Mbps)
0	BPSK	$^1/_2$	1	1	108	54	13.5
1	QPSK	$^1/_2$	1	2	216	108	27.0
2	QPSK	$^3/_4$	1	2	216	162	40.5
3	16-QAM	$^1/_2$	1	4	432	216	54.0
4	16-QAM	$^3/_4$	1	4	432	324	81.0
5	64-QAM	$^2/_3$	1	6	648	432	108.0
6	64-QAM	$^3/_4$	1	6	648	486	121.5
7	64-QAM	$^5/_6$	1	6	648	540	135.0

Appendix 5.2: 40 MHz basic MCS tables

The following tables consist of the MCSs for 40 MHz "basic" rates. As defined herein, the basic rates include equal modulation and the mandatory 800 ns guard interval. Unequal modulation is described in Section 12.4. Short guard interval is described in Section 5.5.

For 40 MHz, the number of data subcarriers is 108 and the number of pilot subcarriers is 6. The only exception is for HT duplicate format (MCS 32), where the number of data subcarriers is 96 and the number of pilot subcarriers is 8.

Table 5.13 describes the symbols used in the subsequent tables.

Table 5.14 defines the 40 MHz MCS parameters and data rates for single spatial stream transmission. The data rate is calculated by dividing N_{DBPS} by the symbol time of 4 μs (3.2 μs OFDM symbol plus 0.8 μs guard interval).

Table 5.15 defines the 40 MHz MCS parameters and data rates for two spatial stream transmission.

Table 5.16 defines the 40 MHz MCS parameters and data rates for three spatial stream transmission.

Table 5.17 defines the 40 MHz MCS parameters and data rates for four spatial stream transmission.

Table 5.18 defines the 40 MHz MCS parameters and data rates for HT duplicate format.

Table 5.15 40 MHz MCS parameters for two spatial streams (IEEE, 2007)

MCS index	Modulation	R	N_{ES}	N_{BPSCS}	N_{CBPS}	N_{DBPS}	Data rate (Mbps)
8	BPSK	$^1/_2$	1	1	216	108	27.0
9	QPSK	$^1/_2$	1	2	432	216	54.0
10	QPSK	$^3/_4$	1	2	432	324	81.0
11	16-QAM	$^1/_2$	1	4	864	432	108.0
12	16-QAM	$^3/_4$	1	4	864	648	162.0
13	64-QAM	$^2/_3$	1	6	1296	864	216.0
14	64-QAM	$^3/_4$	1	6	1296	972	243.0
15	64-QAM	$^5/_6$	1	6	1296	1080	270.0

Table 5.16 40 MHz MCS parameters for three spatial streams (IEEE, 2007)

MCS index	Modulation	R	N_{ES}	N_{BPSCS}	N_{CBPS}	N_{DBPS}	Data rate (Mbps)
16	BPSK	$^1/_2$	1	1	324	162	40.5
17	QPSK	$^1/_2$	1	2	648	324	81.0
18	QPSK	$^3/_4$	1	2	648	486	121.5
19	16-QAM	$^1/_2$	1	4	1296	648	162.0
20	16-QAM	$^3/_4$	1	4	1296	972	243.0
21	64-QAM	$^2/_3$	2	6	1944	1296	324.0
22	64-QAM	$^3/_4$	2	6	1944	1458	364.5
23	64-QAM	$^5/_6$	2	6	1944	1620	405.0

Table 5.17 40 MHz MCS parameters for four spatial streams (IEEE, 2007)

MCS index	Modulation	R	N_{ES}	N_{BPSCS}	N_{CBPS}	N_{DBPS}	Data rate (Mbps)
24	BPSK	$^1/_2$	1	1	432	216	54.0
25	QPSK	$^1/_2$	1	2	864	432	108.0
26	QPSK	$^3/_4$	1	2	864	648	162.0
27	16-QAM	$^1/_2$	1	4	1728	864	216.0
28	16-QAM	$^3/_4$	2	4	1728	1296	324.0
29	64-QAM	$^2/_3$	2	6	2592	1728	432.0
30	64-QAM	$^3/_4$	2	6	2592	1944	486.0
31	64-QAM	$^5/_6$	2	6	2592	2160	540.0

Table 5.18 40 MHz MCS parameters for HT duplicate format (IEEE, 2007)

MCS index	Modulation	R	N_{ES}	N_{BPSCS}	N_{CBPS}	N_{DBPS}	Data rate (Mbps)
32	BPSK	$^1/_2$	1	1	48	24	6.0

Appendix 5.3: Physical layer waveform parameters

Table 5.19 provides a summary and comparison of waveform parameters for 802.11a, 80.11n 20 MHz and 40 MHz.

Table 5.19 Physical layer waveform parameters

	802.11a	802.11n 20 MHz	802.11n 40 MHz	802.11n 40 MHz MCS 32	802.11n non-HT duplicate
Data subcarriers	48	52	108	96	96
Unique modulated data values	48	52	108	48	48
Pilot subcarriers	4	4	6	8	8
Subcarrier frequency spacing	312.5 kHz (20 MHz/64)	312.5 kHz (20 MHz/64)	312.5 kHz (40 MHz/128)	312.5 kHz (40 MHz/128)	312.5 kHz (40 MHz/128)
IDFT/DFT period	3.2 μs	3.2 μs	3.2 μs	3.2 μs	3.2 μs
Guard interval of data symbols	0.8 μs	Normal: 0.8 μs Short: 0.4 μs	Normal: 0.8 μs Short: 0.4 μs	Normal: 0.8 μs Short: 0.4 μs	0.8 μs

6 Robust performance

With the addition of MIMO to IEEE 802.11, many new WLAN devices will have multiple antennas. Though an important benefit of multiple antennas is increased data rate with multiple spatial streams, multiple antennas may be also used to significantly improve the robustness of the system. Multiple antennas enable optional features such as receive diversity, spatial expansion, transmit beamforming, and space-time block coding (STBC). The topic of transmit beamforming is addressed in Chapter 12.

Advanced coding has also been added to 802.11n to further improve link robustness with the inclusion of the optional low density parity check (LDPC) codes and STBC. STBC combines multiple antennas with coding.

To simplify notation in the following sections, the system descriptions for receive diversity, STBC, and spatial expansion are given in the frequency domain for a single subcarrier. This is done since each technique is in fact applied to each subcarrier in the frequency band. It is assumed that, at the transmitter, the frequency domain data is transformed into a time domain waveform as described in Section 4.2. Furthermore, the receive procedure to generate frequency domain samples is described in Section 4.2.4.

To quantify the benefits of these features, this chapter contains simulation results modeling each technique. For each function, the simulation results include physical layer impairments, as described in Section 3.5. The equalizer is based on MMSE. Synchronization, channel estimation, and phase tracking are included in the simulation. For receive diversity, spatial expansion, and STBC, the simulations are performed using channel model B with NLOS conditions, as described in Section 3.5. For LDPC, the simulation is performed using channel model D with NLOS conditions.

6.1 Receive diversity

The discussion of maximal-ratio combining (MRC) in Section 3.2 gave an introduction to receive diversity with a SISO system. In general, additional receive antennas can be added to the receiver in order to increase the diversity order. This same principle also applies to a MIMO receiver. If the receiver has more antennas than spatial streams in the transmitted signal, the diversity order increases by the difference between the number of receive antennas and the number of transmitted spatial streams. A MIMO system with two transmit antennas (and two spatial streams) and three receive antennas is illustrated in Figure 6.1.

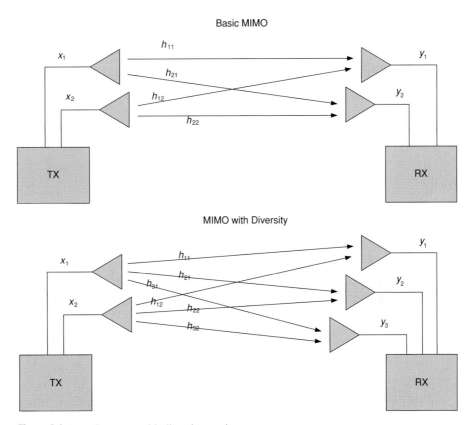

Basic MIMO

MIMO with Diversity

Figure 6.1 MIMO system with diversity receiver.

Figure 6.1 shows both a basic MIMO system in addition to a MIMO system with diversity. In the basic MIMO case the number of transmit antennas, spatial streams, and receive antennas are all equal to two. In the receive diversity case, however, there are two transmit antennas and spatial streams, but three receive antennas. The figure also illustrates all possible channel paths.

6.1.1 Maximal ratio combining basics

An $M \times N$ MIMO/SDM system was described in Section 3.3, and is repeated in the equation below:

$$Y_N = \sqrt{\rho/M} \cdot H_{N \times M} X_M + Z_N \tag{6.1}$$

where X_M is the transmitted data; $H_{N \times M}$ is the channel fading matrix; Z_N is additive white Gaussian noise (AWGN) defined as Normal(0,1); ρ is the average signal-to-noise ratio (SNR); and Y is the received signal. There are N receive antennas and M transmit

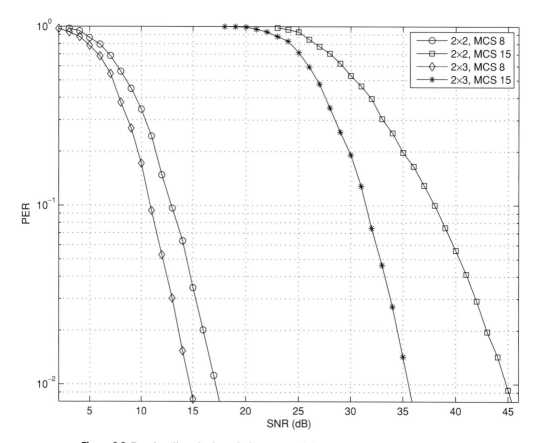

Figure 6.2 Receive diversity benefit for two spatial streams with channel model B, NLOS, and 20 MHz.

antennas. Similar to Section 3.2, an MRC receiver is given as follows:

$$R = H^*Y$$
$$= \sqrt{\rho/M} \cdot H^*HX + H^*Z \tag{6.2}$$

where H^* is the Hermitian (complex conjugate transpose) of H. After MRC is applied, the received signal may be equalized as follows:

$$\hat{X} = \sqrt{M/\rho} \cdot (H^*H)^{-1} R$$
$$= X + \sqrt{M/\rho} \cdot (H^*H)^{-1} H^T Z \tag{6.3}$$

6.1.2   MIMO performance improvement with receive diversity

The extra receive antennas provide substantial improvement in SNR especially at higher modulation and code rates. Figure 6.2 illustrates the benefit with two spatial streams and

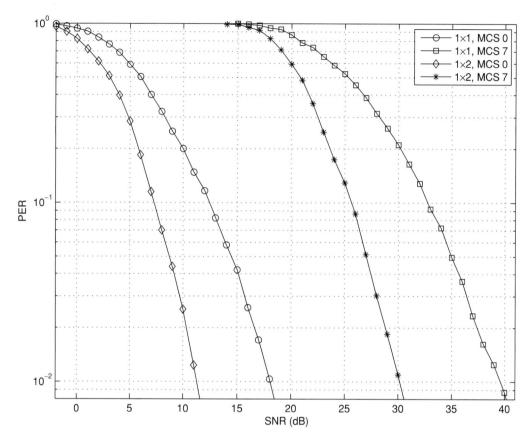

Figure 6.3 Receive diversity benefit for one spatial stream with channel model B, NLOS, and 20 MHz.

three receive antennas, as compared to two receive antennas. The simulation environment is described at the beginning of the chapter.

As illustrated in Figure 6.2, with MCS 15, at a PER of 1% the required SNR decreases by 9 dB with the extra receive antenna. With MCS 8, the required SNR decreases by 2.5 dB. With multiple spatial streams the diversity benefit is much more significant with higher modulation orders as these are more sensitive to fading.

An important aspect with MCS 15 is that the required SNR drops to 36 dB, which is within the capability of moderate cost transmitters and receivers. With channel model B and only two receive antennas, however, a required SNR of 45 dB greatly exceeds the noise figure, impairments, and implementation losses of current typical transmitters and receivers.

MIMO receivers with two receive antennas provide diversity gain over single antenna 802.11a/g receivers. In this situation, a single stream transmission received by a two antenna receiver has performance comparable to a 1×2 diversity system. The benefit over single antenna receivers is illustrated by Figure 6.3. We see that with single spatial

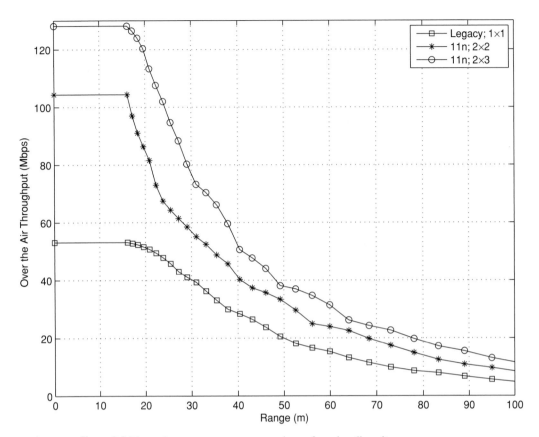

Figure 6.4 Throughput versus range comparison of receive diversity.

stream, the receive diversity gain is roughly 7 dB with low modulation order and increases to 9 dB with high modulation order.

Added robustness from receive diversity enables longer ranges at a given throughput. This is demonstrated in Figure 6.4, which illustrates simulation results for over-the-air throughput as a function of range. Over-the-air throughput is defined as the data rate multiplied by one minus the packet error rate. The results incorporate those illustrated in Figure 6.2 in addition to simulations for MCS 9–14. Range is derived from the required SNR read from the PER waterfall curves and the link budget parameters outlined in Table 5.2.

Note that the maximum throughput plateaus at a shorter range. This artifact is due to a maximum SNR of 35 dB. SNRs above 35 dB become increasingly difficult to achieve with moderate cost hardware. Therefore, throughput values requiring greater than 35 dB SNR were not considered.

As can be seen in Figure 6.4, 50 Mbps data rates are limited to a maximum of 20 m with a legacy single antenna device. With a 2 × 2 802.11n system, 50 Mbps can be sustained up to 35 m. With a 2 × 3 802.11n system, the range is further increased to

40 m. At a data rate of 50 Mbps, MIMO systems may rate adapt to single stream modes, equivalent to a 1×2 diversity system with the 2×2 802.11n system and a 1×3 diversity system with the 2×3 802.11n system. Furthermore, the range at which 100 Mbps can be sustained increases from approximately 17 m to 25 m with a 2×3 802.11n system over a 2×2 system.

With diversity, the required SNR of the peak data rate MCS is substantially reduced. As illustrated in Figure 6.4, at 15 m the peak data rate for 2×3 increases to almost 130 Mbps. This data rate is not achievable with the 2×2 system.

6.1.3   Selection diversity

Selection diversity provides a means for a modest amount of diversity combining gain with reduced hardware complexity. With selection diversity, there are fewer RF chains than antennas. A subset of antennas with the best channel are selected and connected to the RF chains, reducing the necessary number of RF chains. The gain with selection diversity is less than maximal ratio combining as demonstrated by Jakes (1974). For example with a single transmit antenna and N receive antennas in an uncorrelated flat fading Rayleigh channel, selection diversity improves the average output SNR by a factor of $\sum_{k=1}^{N} 1/k$. With the same conditions, maximal ratio combining improves the average output SNR by a factor of N.

In order to select the best antennas, the 802.11n standard provides a mechanism to train up to eight antennas with four RF chains. Training for antenna selection is specified on both transmit and receive. Antenna training is performed with sounding packets. Refer to Chapter 12 for information on channel sounding and sounding packets.

6.2   Spatial expansion

Spatial expansion (SE) is described in detail in Section 4.2.3.8 as a simple means to transmit fewer spatial streams with more antennas. With the appropriate selection of cyclic shifts, as given in Table 4.5, spatial expansion provides a small transmit diversity benefit in channels with more pronounced flat fading (e.g. channel model B). Figure 6.5 illustrates the performance of SE with a single spatial stream. Figure 6.6 illustrates the performance of SE with two spatial streams. Refer to the beginning of the chapter for simulation conditions. SE provides 2 dB diversity gain at a PER of 1%. However, with 64-QAM, $R = 5/6$ (MCS 7 and MCS 15) there is no gain at a PER of 10%. Furthermore, with larger delay spread channels, which cause more frequency selective fading but less flat fading, the gain is reduced.

6.3   Space-time block coding

Space-time block coding (STBC) using the Alamouti algorithm (Alamouti, 1998) is a simple optional transmit diversity scheme in 802.11n which provides the same diversity

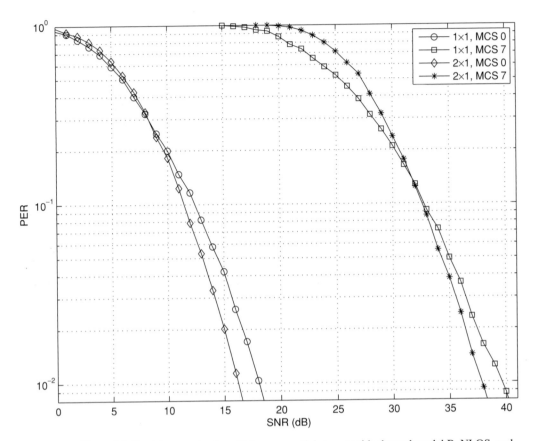

Figure 6.5 Spatial expansion benefit for one spatial stream with channel model B, NLOS, and 20 MHz.

gain as maximal-ratio receiver combining when comparing a 2×1 system to a 1×2 system. The advantage of such a scheme is that robust link performance is achieved with low cost, small form factor devices (which make multiple antennas costly and less effective) and do not require high data rates (e.g. handheld devices). A possible scenario is one where two antennas are used on transmit and receive in the AP, and only one antenna is used for transmit and receive in the handheld device. The downlink from the AP to the handheld device is 2×1 STBC, and the uplink from the device to the AP is 1×2 MRC. Furthermore, STBC is completely open loop and does not require any feedback or additional system complexity, unlike transmit beamforming.

However, when comparing STBC to MRC, if the same total radiated power is considered with both schemes, STBC is disadvantaged. For example, a single transmit antenna device typically has an output power of ~17 dBm. A two antenna device with the same total output power has an output power of 14 dBm from each antenna. As such, an STBC system has a 3 dB power penalty with respect to MRC. In addition, STBC requires additional Long Training fields for channel estimation, which reduces efficiency.

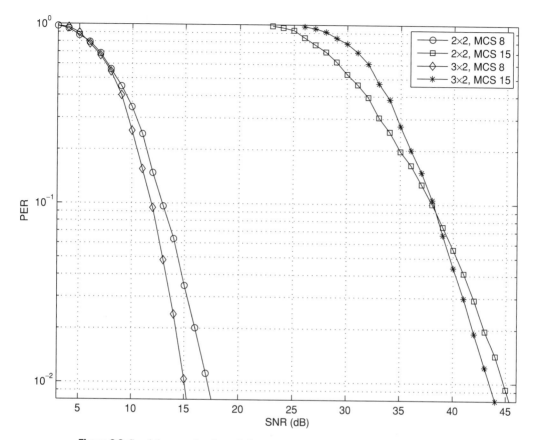

Figure 6.6 Spatial expansion benefit for two spatial streams with channel model B, NLOS, and 20 MHz.

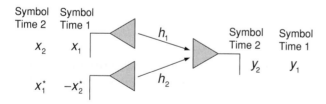

Figure 6.7 Alamouti scheme.

6.3.1   Alamouti scheme background

MRC is described in Section 3.2 and Section 6.1.1. The received signal at each antenna is multiplied by the conjugate of the channel weight, as given in Eq. (3.6). The received signal results in transmitted data scaled by the sum of the magnitude squared of the channel taps.

By comparison, with the Alamouti scheme two data samples are coded and transmitted over two antennas and two symbol periods, as illustrated in Figure 6.7.

At symbol time one, x_1 is transmitted from antenna one and $-x_2^*$ is transmitted from antenna two. The signal y_1 is received at symbol time one. At symbol time two, x_2 is transmitted from antenna one and x_1^* is transmitted from antenna two. The signal y_2 is received at symbol time two. Following the notation given in Section 3.1, the received signals are given as follows (note that the output transmitted power is normalized to be equivalent to that in Eq. (3.5), as given by dividing by $\sqrt{2}$):

$$y_1 = \sqrt{P/2} \cdot h_1 \cdot x_1 + \sqrt{P/2} \cdot h_2 \cdot -x_2^* + z_1$$
$$y_2 = \sqrt{P/2} \cdot h_1 \cdot x_2 + \sqrt{P/2} \cdot h_2 \cdot x_1^* + z_2$$

(6.4)

After two symbols are received, the received signals are combined to extract the transmitted data as follows:

$$r_1 = h_1^* \cdot y_1 + h_2 \cdot y_2^*$$
$$r_2 = -h_2^* \cdot y_1 + h_1 \cdot y_2^*$$

(6.5)

The resulting combined signals match the solution for MRC in Eq. (3.6), except for dividing by $\sqrt{2}$ (the power disadvantage), as follows:

$$r_1 = \sqrt{P/2} \cdot (|h_1|^2 + |h_2|^2) \cdot x_1 + h_1 z_1 + h_2 z_2^*$$
$$r_2 = \sqrt{P/2} \cdot (|h_1|^2 + |h_2|^2) \cdot x_2 - h_2 z_1^* + h_1^* z_2$$

(6.6)

Figure 6.8 illustrates the performance of STBC compared to other system configurations for MCS 7 (single stream, 64-QAM, $R = 5/6$), with simulation conditions as described at the beginning of the chapter.

For comparison, Figure 6.8 includes curves for 1×1 (denoted by circles), 2×1 spatial expansion (denoted by squares), 1×2 MRC (denoted by diamonds), and 2×1 STBC (denoted by stars). At a PER of 1%, STBC provides a substantial amount of gain, approximately 5.5 dB over 1×1 and 4 dB over SE. However, STBC is demonstrated to perform 4 dB worse than MRC. As explained, STBC experiences a 3 dB power penalty relative to MRC. Furthermore, STBC is more susceptible to impairments. Unlike the data, the pilot subcarriers are not space-time block coded and do not experience diversity gain, degrading frequency and phase tracking. Received signals are combined between two consecutive symbols in time. Without extra measures to compensate frequency offset and phase drift between the two symbols additional degradation can occur. Furthermore, STBC frames must be a multiple of two OFDM symbols, thereby adding extra padding overhead.

Figure 6.9 illustrates the performance of STBC with MCS 0 (single stream, BPSK, $R = 1/2$). In the case of MCS 0, STBC provides marginal benefit over spatial expansion. As described in Section 6.1.2, MRC gain decreases from 9 dB with MCS 7 to 7 dB with MCS 0. Coupling this with the 3 dB power loss and more than 1 dB impairment loss reduces the gain of STBC to 2.25 dB, which is only 0.5 dB greater than SE. STBC has a clear benefit in extending the range of the higher data rates. However, at the lowest data rate which sets the maximum coverage area, STBC does not provide a measurable improvement in coverage over SE.

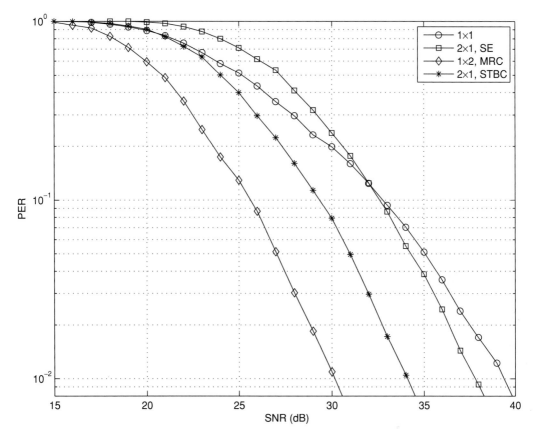

Figure 6.8 STBC performance with MCS 7, 20 MHz, and channel model B, NLOS.

6.3.2   Additional STBC antenna configurations

In 802.11n, antenna configurations beyond 2×1 are implemented. For two spatial streams (MCS 8–15), STBC can be used with either three or four transmit antennas and a minimum of two receive antennas. With three transmit antennas, one spatial stream is transmitted with the 2×1 STBC configuration, and the second spatial stream is transmitted without STBC. This is illustrated by Figure 6.10 and given in Eq. (6.7):

$$y_1 = \sqrt{P/3} \cdot h_{11} \cdot x_1 + \sqrt{P/3} \cdot h_{12} \cdot -x_2^* + \sqrt{P/3} \cdot h_{13} \cdot x_3 + z_1$$

$$y_2 = \sqrt{P/3} \cdot h_{11} \cdot x_2 + \sqrt{P/3} \cdot h_{12} \cdot x_1^* + \sqrt{P/3} \cdot h_{13} \cdot x_4 + z_2$$

$$y_3 = \sqrt{P/3} \cdot h_{21} \cdot x_1 + \sqrt{P/3} \cdot h_{22} \cdot -x_2^* + \sqrt{P/3} \cdot h_{23} \cdot x_3 + z_3 \qquad (6.7)$$

$$y_4 = \sqrt{P/3} \cdot h_{21} \cdot x_2 + \sqrt{P/3} \cdot h_{22} \cdot x_1^* + \sqrt{P/3} \cdot h_{23} \cdot x_4 + z_4$$

With four antennas and two spatial streams, each spatial stream is separately coded with the 2×1 STBC mapping. This is illustrated by Figure 6.11 and given

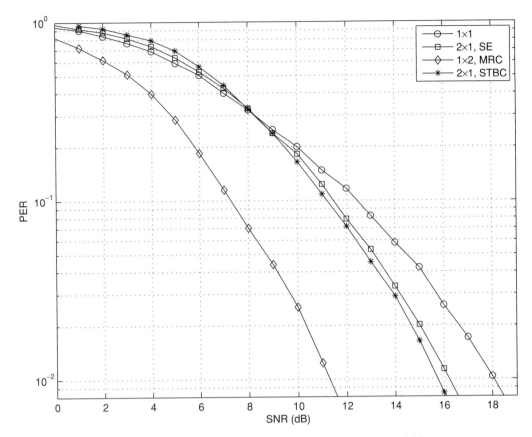

Figure 6.9 STBC performance with MCS 0, 20 MHz, and channel model B.

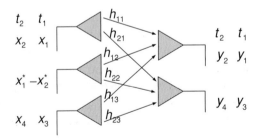

Figure 6.10 3×2 STBC.

in Eq. (6.8):

$$y_1 = \sqrt{P/4} \cdot h_{11} \cdot x_1 + \sqrt{P/4} \cdot h_{12} \cdot -x_2^* + \sqrt{P/4} \cdot h_{13} \cdot x_3 + \sqrt{P/4} \cdot h_{14} \cdot -x_4^* + z_1$$

$$y_2 = \sqrt{P/4} \cdot h_{11} \cdot x_2 + \sqrt{P/4} \cdot h_{12} \cdot x_1^* + \sqrt{P/4} \cdot h_{13} \cdot x_4 + \sqrt{P/4} \cdot h_{14} \cdot x_3^* + z_2$$

$$y_3 = \sqrt{P/4} \cdot h_{21} \cdot x_1 + \sqrt{P/4} \cdot h_{22} \cdot -x_2^* + \sqrt{P/4} \cdot h_{23} \cdot x_3 + \sqrt{P/4} \cdot h_{24} \cdot -x_4^* + z_3$$

$$y_4 = \sqrt{P/4} \cdot h_{21} \cdot x_2 + \sqrt{P/4} \cdot h_{22} \cdot x_1^* + \sqrt{P/4} \cdot h_{23} \cdot x_4 + \sqrt{P/4} \cdot h_{24} \cdot x_3^* + z_4$$

$$(6.8)$$

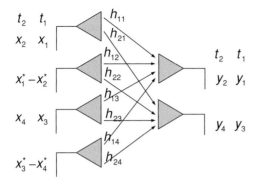

Figure 6.11 4×2 STBC.

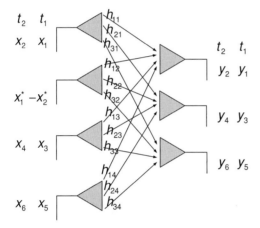

Figure 6.12 4×3 STBC.

With four transmit antennas and three spatial streams, one spatial stream is transmitted with the standard 2×1 configuration, and the second and third spatial streams are transmitted without STBC. This is illustrated by Figure 6.12 and given in Eq. (6.9):

$$y_1 = \sqrt{P/4} \cdot h_{11} \cdot x_1 + \sqrt{P/4} \cdot h_{12} \cdot -x_2^* + \sqrt{P/4} \cdot h_{13} \cdot x_3 + \sqrt{P/4} \cdot h_{14} \cdot x_5 + z_1$$

$$y_2 = \sqrt{P/4} \cdot h_{11} \cdot x_2 + \sqrt{P/4} \cdot h_{12} \cdot x_1^* + \sqrt{P/4} \cdot h_{13} \cdot x_4 + \sqrt{P/4} \cdot h_{14} \cdot x_6 + z_2$$

$$y_3 = \sqrt{P/4} \cdot h_{21} \cdot x_1 + \sqrt{P/4} \cdot h_{22} \cdot -x_2^* + \sqrt{P/4} \cdot h_{23} \cdot x_3 + \sqrt{P/4} \cdot h_{24} \cdot x_5 + z_3$$

$$y_4 = \sqrt{P/4} \cdot h_{21} \cdot x_2 + \sqrt{P/4} \cdot h_{22} \cdot x_1^* + \sqrt{P/4} \cdot h_{23} \cdot x_4 + \sqrt{P/4} \cdot h_{24} \cdot x_6 + z_4$$

$$y_5 = \sqrt{P/4} \cdot h_{31} \cdot x_1 + \sqrt{P/4} \cdot h_{32} \cdot -x_2^* + \sqrt{P/4} \cdot h_{33} \cdot x_3 + \sqrt{P/4} \cdot h_{34} \cdot x_5 + z_5$$

$$y_6 = \sqrt{P/4} \cdot h_{31} \cdot x_2 + \sqrt{P/4} \cdot h_{32} \cdot x_1^* + \sqrt{P/4} \cdot h_{33} \cdot x_4 + \sqrt{P/4} \cdot h_{34} \cdot x_6 + z_6$$

$$(6.9)$$

Other antenna configurations can be achieved by combining STBC with SE. For example, a system with three transmit antennas, one spatial stream, and one receive antenna utilizes a 2×1 STBC configuration followed by the (3 Tx, 2 SS) spatial expansion matrix (given in Table 4.5) to map the two output STBC streams to three antennas.

6.3.3   STBC receiver and equalization

The STBC system descriptions given in Eqs. (6.4) and (6.7)–(6.9) may be reformulated in vector/matrix notation with the structure $Y = HX + Z$. In this format, we can utilize the basic ZF or MMSE receiver described in Section 3.6. For 2×1,

$$\begin{bmatrix} y_1 \\ y_2^* \end{bmatrix} = \sqrt{\rho/2} \cdot \begin{bmatrix} h_1 & -h_2 \\ h_2^* & h_1^* \end{bmatrix} \cdot \begin{bmatrix} x_1 \\ x_2^* \end{bmatrix} + \begin{bmatrix} z_1 \\ z_2^* \end{bmatrix} \tag{6.10}$$

The estimate of the data is attained by equalization and taking the conjugate of the estimate of x_2. The other STBC system configurations may be rewritten as follows. For 3×2,

$$\begin{bmatrix} y_1 \\ y_2^* \\ y_3 \\ y_4^* \end{bmatrix} = \sqrt{\rho/3} \cdot \begin{bmatrix} h_{11} & -h_{12} & h_{13} & 0 \\ h_{12}^* & h_{11}^* & 0 & h_{13}^* \\ h_{21} & -h_{22} & h_{23} & 0 \\ h_{22}^* & h_{21}^* & 0 & h_{23}^* \end{bmatrix} \cdot \begin{bmatrix} x_1 \\ x_2^* \\ x_3 \\ x_4^* \end{bmatrix} + \begin{bmatrix} z_1 \\ z_2^* \\ z_3 \\ z_4^* \end{bmatrix} \tag{6.11}$$

For 4×2,

$$\begin{bmatrix} y_1 \\ y_2^* \\ y_3 \\ y_4^* \end{bmatrix} = \sqrt{\rho/4} \cdot \begin{bmatrix} h_{11} & -h_{12} & h_{13} & -h_{14} \\ h_{12}^* & h_{11}^* & h_{14}^* & h_{13}^* \\ h_{21} & -h_{22} & h_{23} & -h_{24} \\ h_{22}^* & h_{21}^* & h_{24}^* & h_{23}^* \end{bmatrix} \cdot \begin{bmatrix} x_1 \\ x_2^* \\ x_3 \\ x_4^* \end{bmatrix} + \begin{bmatrix} z_1 \\ z_2^* \\ z_3 \\ z_4^* \end{bmatrix} \tag{6.12}$$

For 4×3,

$$\begin{bmatrix} y_1 \\ y_2^* \\ y_3 \\ y_4^* \\ y_5 \\ y_6^* \end{bmatrix} = \sqrt{\rho/4} \cdot \begin{bmatrix} h_{11} & -h_{12} & h_{13} & 0 & h_{14} & 0 \\ h_{12}^* & h_{11}^* & 0 & h_{13}^* & 0 & h_{14}^* \\ h_{21} & -h_{22} & h_{23} & 0 & h_{24} & 0 \\ h_{22}^* & h_{21}^* & 0 & h_{23}^* & 0 & h_{24}^* \\ h_{31} & -h_{32} & h_{33} & 0 & h_{34} & 0 \\ h_{32}^* & h_{31}^* & 0 & h_{33}^* & 0 & h_{34}^* \end{bmatrix} \cdot \begin{bmatrix} x_1 \\ x_2^* \\ x_3 \\ x_4^* \\ x_5 \\ x_6^* \end{bmatrix} + \begin{bmatrix} z_1 \\ z_2^* \\ z_3 \\ z_4^* \\ z_5 \\ z_6^* \end{bmatrix} \tag{6.13}$$

STBC may also be combined with receive diversity for extra diversity gain. For example, a 2×1 STBC system with an extra receive antenna is illustrated in Figure 6.13.

The STBC system with an additional receive antenna is expressed as follows:

$$\begin{bmatrix} y_1 \\ y_2^* \\ y_3 \\ y_4^* \end{bmatrix} = \sqrt{\rho/2} \cdot \begin{bmatrix} h_{11} & -h_{12} \\ h_{12}^* & h_{11}^* \\ h_{21} & -h_{22} \\ h_{22}^* & h_{21}^* \end{bmatrix} \cdot \begin{bmatrix} x_1 \\ x_2^* \end{bmatrix} + \begin{bmatrix} z_1 \\ z_2^* \\ z_3 \\ z_4^* \end{bmatrix} \tag{6.14}$$

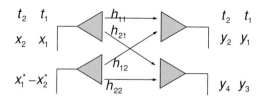

Figure 6.13 2×1 STBC with extra receive antenna.

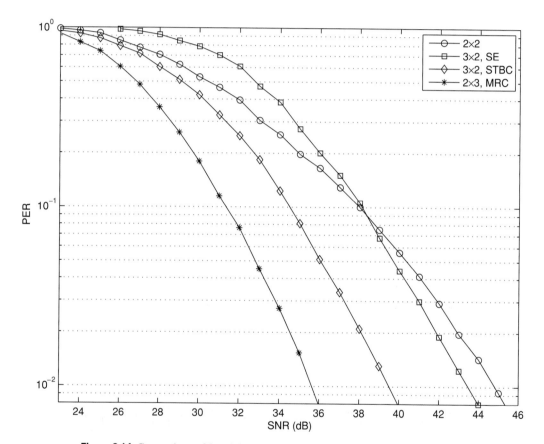

Figure 6.14 Comparison of 3×2 STBC with SE and MRC for MCS 15 and channel model B, NLOS.

Simulation results for the additional STBC configurations are given in Figure 6.14 for 3×2 STBC, MCS 15, Figure 6.15 for 4×2 STBC, MCS 15, and Figure 6.16 for 4×3 STBC, MCS 23. The simulation conditions are described at the beginning of this chapter.

The simulation results in Figure 6.14 for 3×2 STBC illustrate that this configuration may support MCS 15 at a PER of 10%, but will not support 1% with a typical receiver due to the high required SNR in channel model B. At a PER of 10%, STBC provides

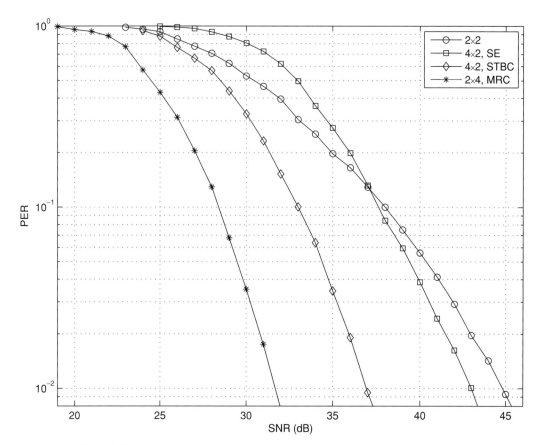

Figure 6.15 Comparison of 4 × 2 STBC with SE and MRC for MCS 15 and channel model B, NLOS.

3.5 dB gain over a 2 × 2 system and a 3 × 2 SE system. On the other hand, STBC results in a 3 dB worse performance than a 2 × 3 MRC system.

The simulation results in Figure 6.15 for 4 × 2 STBC illustrate that it still may not support MCS 15 at a PER of 1% even with the two extra transmit diversity antennas. However, at a PER of 10%, 4 × 2 STBC performs 5 dB better than a 2 × 2 system and a 4 × 2 SE system. However, a 2 × 4 MRC system provides a 4.5 dB better performance than 4 × 2 STBC.

The final configuration, illustrated in Figure 6.16, demonstrates that 4 × 3 STBC may not support MCS 23 in channel model B. With two streams uncoded, the system does not provide enough diversity for MCS 23 in channel model B. In fact, even 3 × 4 MRC barely supports MCS 23 in channel model B.

6.3.4   Transmission and packet encoding process with STBC

In the full 802.11n packet encoding process, STBC encoding is applied after the constellation mapper, but before cyclic shifts and spatial mapping are applied. With the

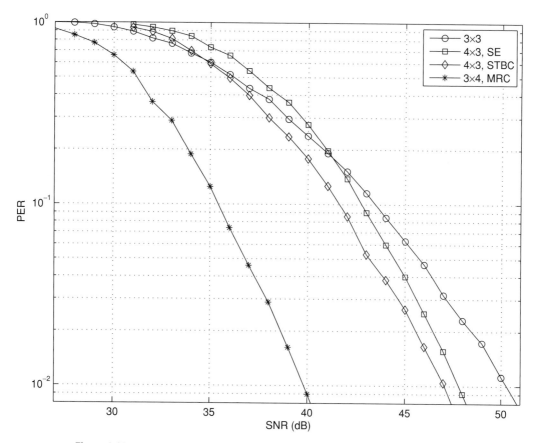

Figure 6.16 Comparison of 4×3 STBC with SE and MRC for MCS 23 and channel model B, NLOS.

addition of STBC, we need to differentiate between spatial streams and the output from STBC encoding. The output of the STBC encoding is termed space-time streams. Figure 6.17 illustrates a transmitter block diagram with two spatial streams, 3×2 STBC with an output of three space-time streams, and spatial mapping to four transmit chains.

To transmit a high throughput (HT) packet with STBC, the STBC field of the HT-SIG must be set to indicate the difference between the number of space-time streams and the number of spatial streams, as given in Table 6.1. The number of spatial streams is indicated by the selected MCS. For example, MCS 15 has two spatial streams.

When comparing notation in Section 4.2 for the mixed format high throughput preamble, waveforms for fields that are transmitted in HT format are modified to account for the number of space-time streams rather than spatial streams. The first field is the HT-STF whose waveform is given by Eq. (4.17). In that equation, N_{SS} and i_{SS} are replaced by N_{STS} and i_{STS}, respectively. Furthermore, the cyclic shift values, given in Table 4.3, are indexed based on N_{STS} and i_{STS}.

Table 6.1 HT-SIG STBC field value

$N_{STS} \times N_{SS}$	STBC field value
2×1	1
3×2	1
4×2	2
4×3	1
No STBC	0

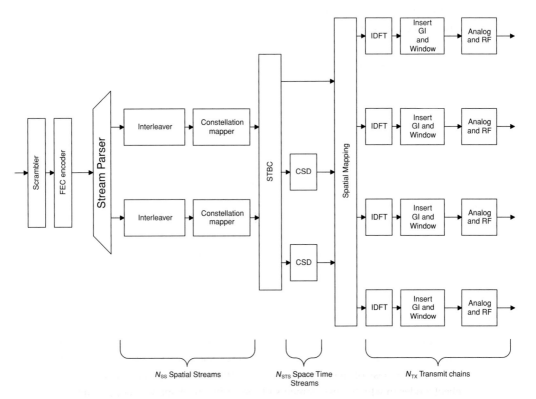

Figure 6.17 Transmitter block diagram with STBC. Reproduced with permission from IEEE (2007) © IEEE.

STBC requires a long training symbol for every space-time stream. Therefore the waveform equations for HT-LTF given by Eq. (4.19) are similarly modified by replacing N_{SS} and i_{SS} by N_{STS} and i_{STS}, respectively. Upon reception, the total number of long training symbols is calculated by N_{SS} (determined from the MCS field) plus the STBC field value.

When transmitting the Data field with STBC, the number of symbols must be even to allow for encoding over two symbols. Therefore the number of OFDM symbols in the

Data field is modified as given by Eq. (6.15) (IEEE, 2007b):

$$N_{SYM} = m_{STBC} \cdot \left\lceil \frac{8 \cdot length + 16 + 6 \cdot N_{ES}}{m_{STBC} \cdot N_{DBPS}} \right\rceil \qquad (6.15)$$

where $m_{STBC} = 2$ for STBC, 1 otherwise; length is the value of the Length field in octets in the HT-SIG field illustrated in Figure 4.16; N_{DBPS} is the number of data bits per OFDM symbol determined by the selected MCS; and N_{ES} is the number of encoders, described in Section 5.1.4. The symbol $\lceil x \rceil$ denotes the smallest integer greater than or equal to x, corresponding to the *ceiling* function. The number of pad bits is therefore $N_{SYM} \cdot N_{DBPS} - 8 \cdot length - 16 - 6 \cdot N_{ES}$.

As a note, upon reception, TXTIME is calculated based on the contents of the HT-SIG. Since the number of OFDM symbols is modified based on STBC, which is part of the TXTIME calculation, this correspondingly impacts the duration of TXTIME.

To create the data waveform, STBC is applied to the data symbols after the constellation mapping, as illustrated in Figure 6.17. Pilots are then inserted at the appropriate subcarrier locations for each symbol for each space-time stream. As mentioned the pilots are sent without STBC. Therefore, in Section 4.2.3.6, N_{SS} and i_{SS} are replaced by N_{STS} and i_{STS}, respectively. This is followed by similarly modifying the HT-DATA waveform in Eq. (4.30). If the number of transmit antennas exceeds the number of space-time streams, the waveform for spatial expansion in Eq. (4.31) is also modified by replacing N_{SS} and i_{SS} by N_{STS} and i_{STS}, respectively.

To transmit STBC with the Greenfield format preamble, the same modifications are made as with MF, except for the HT-SIG waveform. With GF preamble, the HT-SIG is also transmitted in HT format. Therefore, in Eqs. (5.17) and (5.18) the HT-SIG is also modified by replacing N_{SS} and i_{SS} by N_{STS} and i_{STS}, respectively.

6.4 Low density parity check codes

As an optional mode, 802.11n introduced advanced coding using low density parity check (LDPC) codes. LDPC codes were discovered by Gallager (1962). LDPC codes are a special class of linear block codes. Furthermore, they are a special case of parity check codes in which most elements of the parity check matrix are zeros and only a few matrix elements contain ones. The "low density" portion of the LDPC name was termed from this property.

LDPC codes were rediscovered in the 1990s when Turbo coding made iterative decoding systems popular. Since then much research into LDPC codes has demonstrated that these codes achieve near Shannon capacity performance while still maintaining relatively low decoding complexity. A further simplification in the 802.11n system design is that the frequency interleaver is not necessary, due to the intrinsic randomness of LDPC codes.

This section focuses on the specifics of the 802.11n LDPC encoding scheme. The initial steps to generate the code word involve selecting the code word length and determining the number of code words. This is followed by computing the amount of shortening bits and then producing the parity bits. If necessary, puncturing or repetition is then performed.

Readers interested in more details relating to LDPC coding theory, code design, and iterative decoding techniques are encouraged to pursue further investigation by examining two introductory articles. An overview of LDPC codes is provided in Richardson and Urbanke (2003). An overview of the architecture of iterative decoders is given in Yeo and Anantharam (2003). Reference lists in both articles give the reader further resources on LDPC codes and decoding algorithms. Also of interest to the reader is Tanner (1981), in which a graphical representation of the LDPC decoding process, Tanner graphs, was first presented. Much of the advantage of LDPC comes from reduced complexity, sub-optimal iterative decoding techniques which are still very close in performance to Shannon capacity. A number of these techniques are given in Fossorier *et al.* (1999).

6.4.1   LDPC encoding process

A critical part of the LDPC encoding process in 802.11n is fitting the payload bits simultaneously into both an integer number of OFDM symbols and an integer number of LDPC code words. Part of this process is to determine which one of three different LDPC code word lengths to use: 648 bits, 1296 bits, or 1944 bits. Each code word contains information bits and parity bits. The number of information bits is determined based on the rate of the code from the selected MCS. For example, with a code word size of 1944 bits and an MCS with a coding rate of 5/6, the number of information bits is 1620 and the number of parity bits is 324 bits.

For packets shorter than 322 bytes, we need to determine which code word size to use. For packets longer than 322 bytes, a code word size of 1944 bits is always used. Shortening, puncturing, and repetition techniques are used to adjust the number of coded bits. The encoding process is illustrated in Figure 6.18 and outlined in the following steps.

In describing the encoding process we provide examples based on MCS 15 and 20 MHz mode. We illustrate the values of parameters with two sizes of data packets: one of length 3000 bytes and the other of length 3053 bytes.

6.4.1.1   Step 1: Calculating the minimum number of OFDM symbols

The formula for the number of OFDM symbols, N_{SYM}, is slightly modified from Eq. (6.16), since we do not include the six tail bits:

$$N_{SYM} = m_{STBC} \cdot \left\lceil \frac{8 \cdot \text{length} + 16}{m_{STBC} \cdot N_{DBPS}} \right\rceil \tag{6.16}$$

where length is the value of the Length field in octets in the HT-SIG field, illustrated in Figure 4.16; N_{DBPS} is the number of data bits per OFDM symbol determined by the selected MCS; and $m_{STBC} = 2$ for STBC, 1 otherwise. The symbol $\lceil x \rceil$ denotes the smallest integer greater than or equal to x, corresponding to the *ceiling* function.

In this step, we have computed the minimum number of OFDM symbols. If excessive puncturing is indicated in later steps, the number of OFDM symbols will be increased.

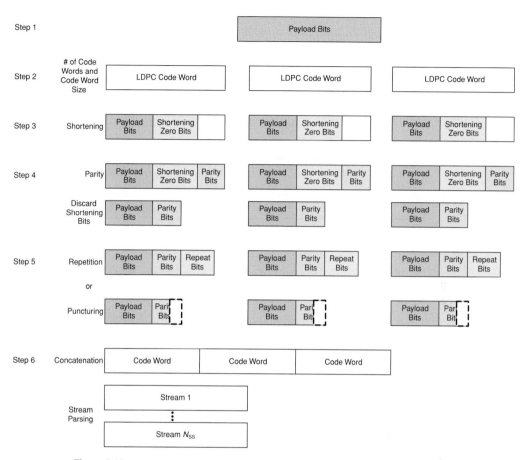

Figure 6.18 LDPC encoding process.

We also define the number of payload bits as the number of data bits and the number of bits in the SERVICE field, which is part of Eq. (6.16):

$$N_{\text{pld}} = 8 \cdot \text{length} + 16 \tag{6.17}$$

With a data packet of length of 3000 bytes (24 000 bits), the number of payload bits, N_{pld}, is equal to $8 \cdot 3000 + 16 = 24\,016$ bits. With MCS 15 and 20 MHz as an example, the number of data bits per OFDM symbol is equal to 520. Therefore the minimum number of OFDM symbols is 47. With a data packet of length of 3053 bytes (24 424 bits), N_{pld} is equal to 24 440 bits. However, even with the longer length of the payload, N_{SYM} is still equal to 47.

6.4.1.2 Step 2: Determining the code word size and number of code words

To determine the LDPC code word size and the number of code words, we must first calculate the total number of coded bits that will fill all the OFDM symbols. Given the number of OFDM symbols, we calculate the total number of coded bits that will fill all

Table 6.2 LDPC code word length and number of code words (IEEE, 2007)

Range (bits) of N_{TCB}	Number of LDPC code words (N_{CW})	LDPC code word length in bits (L_{LDPC})
$N_{TCB} \leq 648$	1	1296, if $N_{TCB} \geq N_{pld} + 912 \cdot (1 - R)$ 648, otherwise
$648 < N_{TCB} \leq 1296$	1	1944, if $N_{TCB} \geq N_{pld} + 1464 \cdot (1 - R)$ 1296, otherwise
$1296 < N_{TCB} \leq 1944$	1	1944
$1944 < N_{TCB} \leq 2592$	2	1944, if $N_{TCB} \geq N_{pld} + 2916 \cdot (1 - R)$ 1296, otherwise
$2592 < N_{TCB}$	$\left\lceil \dfrac{N_{pld}}{1944 \cdot R} \right\rceil$	1944

the symbols, as follows:

$$N_{TCB} = N_{CBPS} \cdot N_{SYM} \tag{6.18}$$

where N_{CBPS} is the number of coded bits per OFDM symbol determined by the selected MCS.

In our examples, N_{CBPS} is equal to 624 with MCS 15 and 20 MHz mode. With N_{SYM} equal to 47, N_{TCB} is equal to 29 328 bits for both data packet lengths.

The LDPC coding performance improves with increasing code word size. Therefore the objective in choosing the code word length is to maximize the code word size. However, if with small payload sizes a code word length is selected that exceeds N_{TCB}, puncturing of bits is needed. Since puncturing typically degrades performance, a balance must be struck between maximizing the code word length and minimizing the number of punctured bits. The computation of the number of LDPC code words (N_{CW}) and selection of the LDPC code word length (L_{LDPC}) is given in Table 6.2, based on N_{TCB}, N_{pld}, and the code rate, R, determined by the selected MCS.

In our examples, the code word length is 1944. The number of code words is 15 with a 3000 byte data packet. The number of code words is 16 with a 3053 byte data packet. As further illustration of the code word length selection, with MCS 15 and 20 MHz mode, when the data packet size is between 1 and 57 bytes, inclusive, the code word length is 1296. When the data packet size is between 58 and 63, inclusive, the code word length is 648.

When we discuss puncturing in Step 5, we will further explore the implications of the selection of these code word lengths.

Figure 6.19 demonstrates the LDPC code word length as a function of data packet length for MCS 15 and 20 MHz mode. We see the effect of the logical conditions in Table 6.2. The code word length begins with 1296 bits, then as the data packet length increases the code word length drops to 648. As the data packet length further increases,

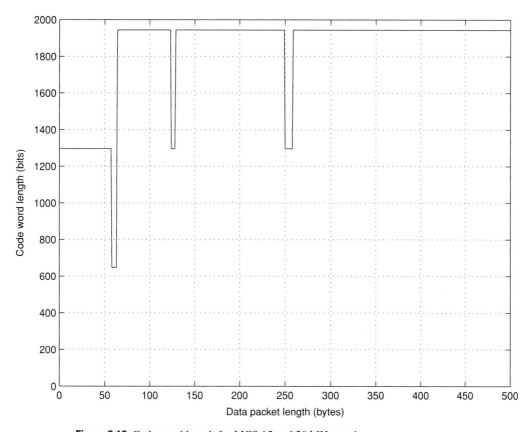

Figure 6.19 Code word length for MCS 15 and 20 MHz mode.

the code word length jumps back up to 1944 bits. Then two conditions cause the code word length to drop back down to 1296 for a short range of data packet lengths. For data packet lengths of 259 bytes and longer, an LDPC code word length of 1944 bits is always used.

6.4.1.3　Step 3: Determining the number of shortening zero bits

For most combinations of payload size and selected MCS, there are not enough information bits to fill the information part of all the code words. Shortening is the process of adding zeros in the information portion of the code word prior to computing the parity bits. After computing the parity bits the zeros are discarded. Next, the parity bits are concatenated to the information bits.

The total number of coded bits that are carried by all the LDPC code words is simply the number of code words multiplied by the code word length. If we multiply that product by the code rate, we know the total number of information bits carried by all the LDPC code words. If the total number of information bits carried in the LDPC code words is greater than the number of payload bits, then N_{shrt} shortening zero bits are required:

$$N_{shrt} = \max\left(0, (N_{CW} \cdot N_{LDPC} \cdot R) - N_{pld}\right) \tag{6.19}$$

If shortening is necessary, the shortening zero bits are distributed over all N_{CW} code words. It is not always possible to evenly distribute the payload bits and shortening zero bits in each LDPC code word. Each code word contains $N_{LDPC} \cdot R$ information bits comprised of payload bits and shortening zero bits. The first $\mathrm{rem}(N_{shrt}, N_{CW})$ code words contains one more shortening zero bit and one fewer payload bit than the remaining code words.

We define the information block of an LDCP code as $(i_0, i_1, \ldots, i_{N_{LDPC}-1})$. Furthermore, we define the minimum number of shortening bits per code word as follows:

$$N_{spcw} = \lfloor N_{shrt}/N_{CW} \rfloor \tag{6.20}$$

where the symbol $\lfloor x \rfloor$ denotes the largest integer less than or equal to x, corresponding to the *floor* function. Therefore, in the first $\mathrm{rem}(N_{shrt}, N_{CW})$ code words, the subblock of information bits $(i_0, i_1, \ldots, i_{N_{LDPC} \cdot R - N_{spcw}-2})$ contain payload bits, and the subblock of information bits $(i_{N_{LDPC} \cdot R - N_{spcw}-1}, i_{N_{LDPC} \cdot R - N_{spcw}}, \ldots, i_{N_{LDPC} \cdot R-1})$ are set to zero. In the remaining code words, the subblock of information bits $(i_0, i_1, \ldots, i_{N_{LDPC} \cdot, R - N_{spcw}-1})$ contain payload bits, and the subblock of information bits $(i_{N_{LDPC} \cdot R - N_{spcw}}, i_{N_{LDPC} \cdot R - N_{spcw}+1}, \ldots, i_{N_{LDPC} \cdot R-1})$ are set to zero.

In the example with a 3000 byte data packet, we require $15 \cdot 1944 \cdot 5/6 - 24\,016 = 284$ shortening zero bits. The minimum number of shortening bits per code word, N_{spcw}, is equal to $\lfloor 284/15 \rfloor = 18$. Therefore, the first $\mathrm{rem}(284, 15) = 14$ code words have 19 zeros appended to $1944 \cdot 5/6 - 18 - 1 = 1601$ payload bits and the 15th code word has 18 zeros appended to the 1602 payload bits.

In the example with a 3053 byte data packet, we require $16 \cdot 1944 \cdot 5/6 - 24\,440 = 1480$ shortening zero bits. The minimum number of shortening bits per code word, N_{spcw}, is equal to $\lfloor 1480/16 \rfloor = 92$. Therefore, the first $\mathrm{rem}(1480, 16) = 8$ code words have 93 zeros appended to $1944 \cdot 5/6 - 92 - 1 = 1527$ payload bits and code words 9 through 16 have 92 zeros appended to the 1528 payload bits.

Figure 6.20 illustrates the number of shortening bits per code word for data payload lengths up to 3500 bytes, with MCS 15 and 20 MHz mode. The general trend is such that the number of shortening bits per code word decreases as the data payload length increases. This indicates that, as the data payload length increases, the payload bits fit better into the code words.

6.4.1.4   Step 4: Generating the parity bits

The LDPC encoder used in 802.11n is systematic, whereby it encodes an information block of $N_{LDPC} \cdot R$ bits into a code word of size N_{LDPC} by appending $N_{LDPC} \cdot (1 - R)$ parity bits to the original information bits. The code word c is described as follows, where i represents the information bits (which may include shortening) and p represents the parity bits:

$$c = (i_0, i_1, \ldots, i_{N_{LDPC} \cdot R-1}, p_0, p_1, \ldots, p_{N_{LDPC} \cdot (1-R)-1}) \tag{6.21}$$

There are N_{CW} code words for the payload bits, as determined in Step 2.

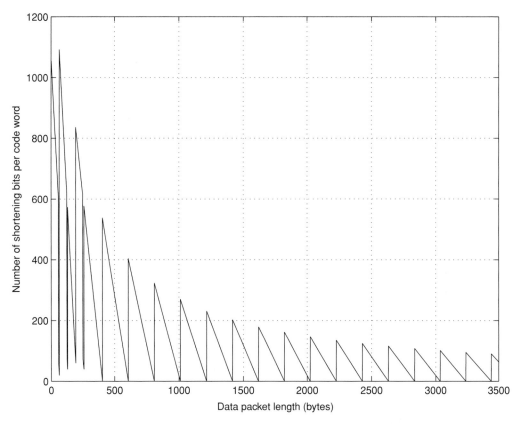

Figure 6.20 Number of shortening bits per code word for MCS 15 and 20 MHz mode.

The 802.11n standard defines 12 different parity check matrices (H), one for each of the combination of three code word sizes (648, 1296, and 1944 bits) and four coding rates (1/2, 2/3, 3/4, and 5/6). The parity check matrices are of size $N_{\mathrm{LDPC}} \cdot (1 - R) \times N_{\mathrm{LDPC}}$ bits. The parity check matrices are provided in Appendix 6.1. A particular parity check matrix is selected based on the code word size and the code rate from the MCS. By definition, any code word c is orthogonal to every row of the parity check matrix H (Proakis, 1989). Note that the matrix computations are modulo-2 operations:

$$H \cdot c^{\mathrm{T}} = \mathbf{0} \tag{6.22}$$

One approach to generating the parity bits is given as follows. First we segment the parity check matrix and the code word:

$$[H_i \mid H_p] \cdot \begin{bmatrix} i_0 \\ \vdots \\ i_{N_{\mathrm{LDPC}} \cdot R - 1} \\ -\;-\;-\;-\;-\;- \\ p_0 \\ \vdots \\ p_{N_{\mathrm{LDPC}} \cdot (1-R) - 1} \end{bmatrix} = \mathbf{0} \tag{6.23}$$

where the dimension of H_i is $N_{\text{LDPC}} \cdot (1 - R) \times N_{\text{LDPC}} \cdot R$ and the dimension of H_p is $N_{\text{LDPC}} \cdot (1 - R) \times N_{\text{LDPC}} \cdot (1 - R)$. Equation (6.23) is expanded into separate information and parity matrices:

$$H_i \cdot \begin{bmatrix} i_0 \\ \vdots \\ i_{N_{\text{LDPC}} \cdot R-1} \end{bmatrix} + H_p \cdot \begin{bmatrix} p_0 \\ \vdots \\ p_{N_{\text{LDPC}} \cdot (1-R)-1} \end{bmatrix} = \mathbf{0} \tag{6.24}$$

With modulo-2 based operations, Eq. (6.24) is equivalent to the following:

$$H_p \cdot \begin{bmatrix} p_0 \\ \vdots \\ p_{N_{\text{LDPC}} \cdot (1-R)-1} \end{bmatrix} = H_i \cdot \begin{bmatrix} i_0 \\ \vdots \\ i_{N_{\text{LDPC}} \cdot R-1} \end{bmatrix} \tag{6.25}$$

Finally, the following equation is in a form which can be used to generate the parity bits from the parity check matrix and the information bits for each of N_{CW} code words:

$$\begin{bmatrix} p_0 \\ \vdots \\ p_{N_{\text{LDPC}} \cdot (1-R)-1} \end{bmatrix} = H_p^{-1} \cdot H_i \cdot \begin{bmatrix} i_0 \\ \vdots \\ i_{N_{\text{LDPC}} \cdot R-1} \end{bmatrix} \tag{6.26}$$

If the information bits contained shortening zero bits, the shortening zero bits are discarded from each code word. In each code word, the parity bits are concatenated to the end of the remaining information bits.

6.4.1.5   Step 5: Packing into OFDM symbols

After creating each code word by concatenating the information bits with the parity bits, the total number of coded bits is equal to the number of payload bits plus the total number of parity bits, and the total number of parity bits is computed by $N_{\text{CW}} \cdot N_{\text{LDPC}} \cdot (1 - R)$. The total number of coded bits is $N_{\text{pld}} + N_{\text{CW}} \cdot N_{\text{LDPC}} \cdot (1 - R)$. In most circumstances this does not equal the total number of bits needed to fill the OFDM symbols, N_{TCB}.

Therefore, if $N_{\text{pld}} + N_{\text{CW}} \cdot N_{\text{LDPC}} \cdot (1 - R)$ is greater than N_{TCB}, then it is necessary to puncture the coded bits such that the coded bits fit into the OFDM symbols. However, puncturing of parity bits degrades the performance of the code. Therefore, if the number of parity bits to be punctured is too large, an extra OFDM symbol is added and N_{TCB} is recomputed to the larger size. This reduces the number of punctured bits or eliminates puncturing entirely.

On the other hand, if $N_{\text{pld}} + N_{\text{CW}} \cdot N_{\text{LDPC}} \cdot (1 - R)$ is less than N_{TCB} (where N_{TCB} may have been recomputed due to the puncturing test), then it is necessary to repeat coded bits to fill in all the necessary bits in the OFDM symbols.

In the example with the 3000 byte data packet, the total number of coded bits is equal to $24\,016 + 15 \cdot 1944 \cdot (1 - 5/6) = 28\,876$. The value for N_{TCB} is $29\,328$. Therefore, this configuration requires repetition. In the example with the 3053 byte data packet, the total number of coded bits is equal to $24\,440 + 16 \cdot 1944 \cdot (1 - 5/6) = 29\,624$. The value for N_{TCB} is the same as with the 3000 byte data packet. Since the total number of coded bits exceeds N_{TCB}, puncturing is required.

Puncturing

The number of bits to be punctured is given as follows:

$$N_{punc} = [N_{pld} + N_{CW} \cdot N_{LDPC} \cdot (1 - R)] - N_{TCB} \qquad (6.27)$$

If the number of parity bits to be punctured is too large, coding performance suffers. Therefore, we check for two conditions. The first condition is if N_{punc} exceeds 30% of the total number of parity bits, and is expressed as follows:

$$N_{punc} > 0.3 \cdot N_{CW} \cdot N_{LDPC} \cdot (1 - R) \qquad (6.28)$$

The second condition is to determine whether

$$N_{punc} > 0.1 \cdot N_{CW} \cdot N_{LDPC} \cdot (1 - R) \quad \text{AND} \quad N_{shrt} > 1.2 \cdot N_{punc} \cdot R/(1 - R) \qquad (6.29)$$

is true. If either of the conditions in Eqs. (6.28) or (6.29) is true, then N_{TCB} is increased by another OFDM symbol. The increase to N_{TCB} is given as an update to the previous value as follows:

$$N_{TCB} = N_{TCB} + N_{CBPS} \cdot m_{STBC}$$

The number of OFDM symbols is recomputed as follows:

$$N_{SYM} = \frac{N_{TCB}}{N_{CBPS}} \qquad (6.30)$$

Finally, N_{punc} is recomputed using Eq. (6.27) with the new value of N_{TCB}. Even if the new value of N_{punc} still results in either of the conditions in Eqs. (6.28) or (6.29) being true, no further increase to N_{TCB} is performed.

Once the value of N_{punc} is computed, the number of punctured bits per code word is determined. The puncturing is distributed over all N_{CW} code words. It is not always possible to evenly distribute the puncturing in each LDPC code word. When necessary, the first rem(N_{punc}, N_{CW}) code words contain one more punctured parity bit than the remaining code words. The minimum number of punctured parity bits per code word is as follows:

$$N_{ppcw} = \lfloor N_{punc}/N_{CW} \rfloor \qquad (6.31)$$

In the first rem(N_{punc}, N_{CW}) code words, the parity check bits $p_{N_{LDPC} \cdot (1-R)-N_{ppcw}-1}, \ldots, p_{N_{LDPC} \cdot (1-R)-1}$ are discarded. In the remaining code words, the parity check bits $p_{N_{LDPC} \cdot (1-R)-N_{ppcw}}, \ldots, p_{N_{LDPC} \cdot (1-R)-1}$ are discarded.

In the example with the 3053 byte data packet, which requires puncturing, the total number of punctured parity bits is $[24\,440 + 16 \cdot 1944 \cdot (1 - 5/6)] - 29\,328 = 296$. The first rem(296, 16) = 8 code words are punctured by $\lfloor 296/16 \rfloor + 1 = 19$ parity bits. The remaining 8 code words are punctured by 18 parity bits.

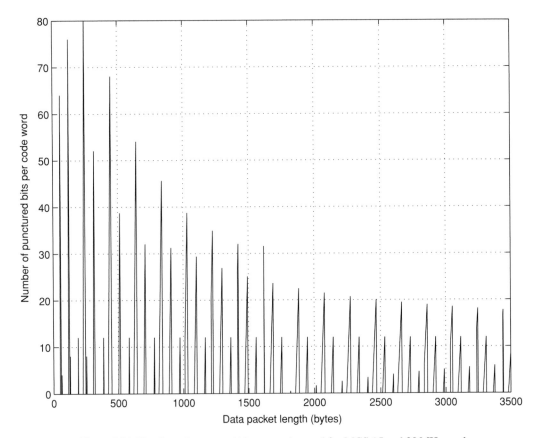

Figure 6.21 Number of punctured bits per code word for MCS 15 and 20 MHz mode.

Figure 6.21 illustrates the number of punctured parity bits per code word for payload lengths up to 3500 bytes. The general trend is that the number of punctured bits per code word decreases as the payload length increases. This indicates that as the payload length increases, the payload and parity bits fit better into the code words.

Repetition
After checking for whether puncturing is required and a possible increase in N_{TCB} and N_{SYM}, we determine whether repetition of coded bits is necessary. The number of coded bits to be repeated is given as follows:

$$N_{rep} = N_{TCB} - \left[N_{pld} + N_{CW} \cdot N_{LDPC} \cdot (1 - R) \right] \tag{6.32}$$

The repeated bits are distributed over all N_{CW} code words. It is not always possible to evenly distribute the repeated bits in each LDPC code word. When necessary, the first $\text{rem}(N_{rep}, N_{CW})$ code words contains one more repeated bit than the remaining code words. The minimum number of repeated bits per code word is as follows:

$$N_{rpcw} = \lfloor N_{rep} / N_{CW} \rfloor \tag{6.33}$$

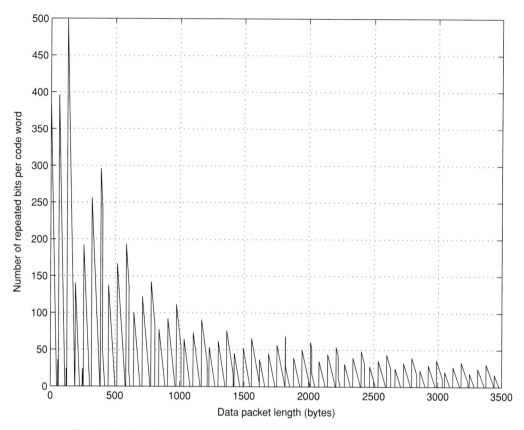

Figure 6.22 Number of repeated bits per code word for MCS 15 and 20 MHz mode.

For each code word, the repeated bits are generated by copying bits for the code word itself, starting from the information bit i_0 and continuing sequentially through to the parity bits as necessary. Furthermore, if additional repeated bits are necessary, bits are copied beginning again with the information bit i_0 and continuing sequentially through to the parity bits as necessary.

In the example with the 3000 byte data packet, which requires repetition, the number of repetition bits is $29\,328 - [24\,016 + 15 \cdot 1944 \cdot (1 - 5/6)] = 452$. The first $\mathrm{rem}(452, 15) = 2$ code words contains $\lfloor 452/15 \rfloor + 1 = 31$ repetition bits. The remaining 13 code words contains 30 repetition bits.

Figure 6.22 illustrates the number of repeated bits per code word for payload lengths up to 3500 bytes. The general trend is such that the number of repeated bits per code word decreases as the payload length increases. This again indicates that as the payload length increases, the payload and parity bits fit better into the code words.

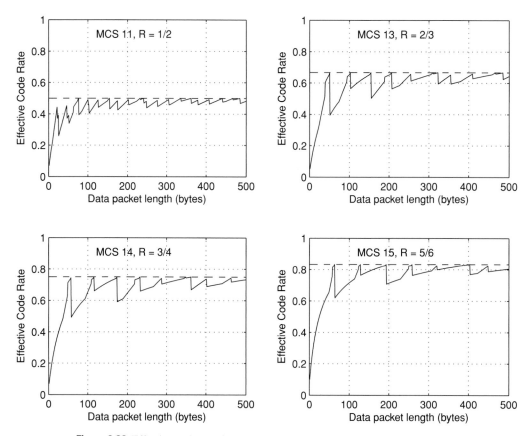

Figure 6.23 Effective code rate for LDPC for 20 MHz mode.

6.4.1.6   Step 6: Stream parsing

The resulting code words are concatenated together. Parsing of the LDPC encoded bit string into spatial streams is identical to BCC, as described in Section 4.2.3.3. A single LDPC encoder is always used, so no encoder parsing is necessary. Furthermore, interleaving is not necessary with LDPC, so the interleaver steps are not performed. Thus with LDPC, the next step after stream parsing is modulation mapping.

6.4.2   Effective code rate

With BCC, the selected MCS determines the code rate. With LDPC, the rate of the selected MCS determines the number of information bits and parity bits in the code word. However, with the operations of shortening, puncturing, and repetition, the actual numbers of payload bits and parity bits used to create the code words are different than the native rate of the code word. To compare the coding gain between BCC and LDPC we derive a new metric for LDPC, termed effective code rate.

Shortening fills in the missing information bits in the LDPC code word, but they are discarded before transmission. Therefore, the payload bits are a better measure

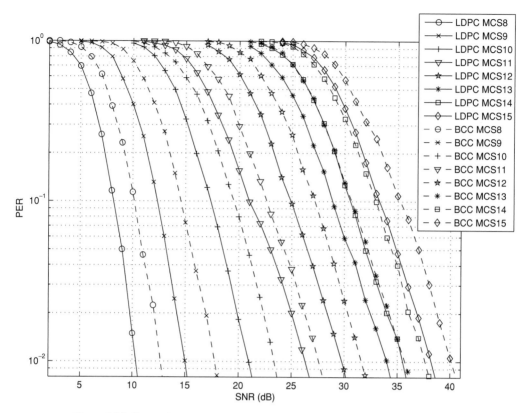

Figure 6.24 Comparison between LDPC and BCC for 2 × 2, 20 MHz, channel model D, NLOS, and payload length of 970 bytes.

of the actual transmitted information than the number of information code word bits. In many cases the parity bits are punctured, which increases the effective code rate, and must be accounted for in the metric. Repetition in LDPC is very similar to padding in BCC, so we do not account for repetition in the new metric when comparing BCC and LDCP performance. The new metric for effective code rate is given as follows:

$$\tilde{R} = \frac{N_{\text{pld}}}{N_{\text{pld}} + N_{\text{CW}} \cdot N_{\text{LDPC}} \cdot (1 - R) - N_{\text{punc}}} \tag{6.34}$$

Figure 6.23 illustrates the effective code rate for LDPC for MCS 11, 13, 14, and 15 for 20 MHz mode for a range of data packet lengths. The code rates for these MCSs are 1/2, 2/3, 3/4, and 5/6, respectively. In all four cases, the effective code rate converges to the MCS code rate by a data packet length of 500 bytes. Shorter packets differ the most between the effective code rate and the MCS code rate for all four cases. As we see in Figure 6.20, shorter packets have a much larger number of shortening bits. Therefore, the ratio between payload bits and information bits in a code word is much smaller than the native rate of the code.

6.4.3   LDPC coding gain

Since BCC is the mandatory coding scheme of 802.11n, a device is required to have a BCC encoder and decoder. Using the optional LDPC code means implementing both the BCC and LDPC encoders and decoders. Therefore the basic questions regarding LDPC coding are: (a) how much additional gain does it provide to the system and (b) is it worth having a second encoder/decoder type in the device?

To give some insight into answering these questions, we provide simulation results for LDPC and BCC with two transmit antennas, two receive antennas and 20 MHz mode. The simulations are performed in the conditions detailed at the beginning of the chapter.

For a fair comparison between BCC and LDPC, a payload length of 970 bytes was selected such that the effective code rate of LDPC is very close to the rate of the MCS. The PER versus SNR waterfall curves are illustrated in Figure 6.24. The solid lines are the LDPC curves, and the dashed lines are the BCC curves. The gain of LDPC over BCC ranges from 1.5 dB to 3 dB, at a PER of 1%, depending on the MCS.

References

Alamouti, S. (1998). A simple transmit diversity technique for wireless communications. *IEEE JSAC*, **16**(8), 1451–8.

Fossorier, M. P. C., Mihaljević, M., and Imai, H. (1999). Reduced complexity iterative decoding of low-density parity check codes based on belief propagation. *IEEE Transactions on Communications*, **47**(5), 673–80.

Gallager, R. (1962). Low-density parity-check codes. *IRE Transactions on Information Theory*, January, 21–8.

IEEE (2007). *IEEE P802.11n™/D3.00, Draft Amendment to STANDARD for Information Technology – Telecommunications and Information Exchange Between Systems – Local and Metropolitan Networks – Specific Requirements. Part 11: Wireless LAN Medium Access Control (MAC) and Physical Layer (PHY). Amendment 4: Enhancements for Higher Throughput.*

Jakes, W. C. (1974). *Microwave Mobile Communications*. New York: Wiley.

Proakis, J. G. (1989). *Digital Communications*. New York: McGraw-Hill.

Richardson, T. and Urbanke, R. (2003), The renaissance of Gallager's low-density parity-check codes. *IEEE Communications Magazine*, August, 126–31.

Tanner, R. M. (1981). A recursive approach to low complexity codes. *IEEE Transactions on Information Theory*, **IT-27**(5), 533–47.

Yeo, E. and Anantharam, V. (2003). Iterative decoder architectures. *IEEE Communications Magazine*, August, 132–40.

Appendix 6.1: Parity check matrices

Since the parity check matrices are extremely large and difficult to display in print, certain notational short cuts are taken in IEEE (2007) to represent the matrices. This notation does not have anything to do with the design of the parity check matrices.

$$P_0 = \begin{bmatrix} 1 & 0 & 0 & 0 \\ 0 & 1 & 0 & 0 \\ 0 & 0 & 1 & 0 \\ 0 & 0 & 0 & 1 \end{bmatrix}, P_1 = \begin{bmatrix} 0 & 1 & 0 & 0 \\ 0 & 0 & 1 & 0 \\ 0 & 0 & 0 & 1 \\ 1 & 0 & 0 & 0 \end{bmatrix}, P_2 = \begin{bmatrix} 0 & 0 & 1 & 0 \\ 0 & 0 & 0 & 1 \\ 1 & 0 & 0 & 0 \\ 0 & 1 & 0 & 0 \end{bmatrix}, P_3 = \begin{bmatrix} 0 & 0 & 0 & 1 \\ 1 & 0 & 0 & 0 \\ 0 & 1 & 0 & 0 \\ 0 & 0 & 1 & 0 \end{bmatrix}$$

Figure 6.25 Cyclic permutation matrices with dimension 4×4.

The first notation simplification is to define a cyclic permutation matrix P_i of dimension $Z \times Z$. The matrix P_0 is an identity matrix of dimension $Z \times Z$. The matrix P_1 is an identity matrix of dimension $Z \times Z$ with its columns cyclically shifted to the right by 1 element. The matrix P_i is an identity matrix of dimension $Z \times Z$ with its columns cyclically shifted to the right by i elements. Some examples of cyclic permutation matrices with dimension 4×4 are illustrated in Figure 6.25.

The parity check matrices are partitioned into sub-matrices consisting of the cyclic permutation matrices and all-zero matrices. The all-zero matrices of dimension $Z \times Z$ are represented by **0**. The dimensions of the parity check matrix is determined by the combination of (code word size, code rate, Z) which are used as a subscript of H in the following equations.

For a code word size of 648 bits, Z is 27, which means that the cyclic permutation matrices and all-zero matrices have the dimensions 27×27. For a code rate of 1/2, the parity check matrix has the dimensions 324×648 bits and is given in Eq. (6.35) (IEEE, 2007):

$H_{648,1/2,27}$

$$= \begin{bmatrix}
P_0 & 0 & 0 & 0 & P_0 & P_0 & 0 & 0 & P_0 & 0 & 0 & P_0 & P_1 & P_0 & 0 & 0 & 0 & 0 & 0 & 0 & 0 & 0 & 0 & 0 \\
P_{22} & P_0 & 0 & 0 & P_{17} & 0 & P_0 & P_0 & P_{12} & 0 & 0 & 0 & 0 & P_0 & P_0 & 0 & 0 & 0 & 0 & 0 & 0 & 0 & 0 & 0 \\
P_6 & 0 & P_0 & 0 & P_{10} & 0 & 0 & 0 & P_{24} & 0 & P_0 & 0 & 0 & 0 & P_0 & P_0 & 0 & 0 & 0 & 0 & 0 & 0 & 0 & 0 \\
P_2 & 0 & 0 & P_0 & P_{20} & 0 & 0 & 0 & P_{25} & P_0 & 0 & 0 & 0 & 0 & 0 & P_0 & P_0 & 0 & 0 & 0 & 0 & 0 & 0 & 0 \\
P_{23} & 0 & 0 & 0 & P_3 & 0 & 0 & 0 & P_0 & 0 & P_9 & P_{11} & 0 & 0 & 0 & 0 & P_0 & P_0 & 0 & 0 & 0 & 0 & 0 & 0 \\
P_{24} & 0 & P_{23} & P_1 & P_{17} & P_3 & 0 & P_{10} & 0 & 0 & 0 & 0 & 0 & 0 & 0 & 0 & P_0 & P_0 & 0 & 0 & 0 & 0 & 0 & 0 \\
P_{25} & 0 & 0 & 0 & P_8 & 0 & 0 & P_7 & P_{18} & 0 & 0 & P_0 & 0 & 0 & 0 & 0 & 0 & P_0 & P_0 & 0 & 0 & 0 & 0 & 0 \\
P_{13} & P_{24} & 0 & 0 & P_0 & 0 & P_8 & 0 & P_6 & 0 & 0 & 0 & 0 & 0 & 0 & 0 & 0 & 0 & P_0 & P_0 & 0 & 0 & 0 & 0 \\
P_7 & P_{20} & 0 & P_{16} & P_{22} & P_{10} & 0 & 0 & P_{23} & 0 & 0 & 0 & 0 & 0 & 0 & 0 & 0 & 0 & 0 & P_0 & P_0 & 0 & 0 & 0 \\
P_{11} & 0 & 0 & 0 & P_{19} & 0 & 0 & 0 & P_{13} & 0 & P_3 & P_{17} & 0 & 0 & 0 & 0 & 0 & 0 & 0 & 0 & P_0 & P_0 & 0 & 0 \\
P_{25} & 0 & P_8 & 0 & P_{23} & P_{18} & 0 & P_{14} & P_9 & 0 & 0 & 0 & 0 & 0 & 0 & 0 & 0 & 0 & 0 & 0 & 0 & P_0 & P_0 & 0 \\
P_3 & 0 & 0 & 0 & P_{16} & 0 & 0 & P_2 & P_{25} & P_5 & 0 & 0 & P_1 & 0 & 0 & 0 & 0 & 0 & 0 & 0 & 0 & 0 & P_0 & P_0
\end{bmatrix}$$

(6.35)

For a code rate of 2/3, the parity check matrix has the dimensions 216×648 bits and is given in Eq. (6.36) (IEEE, 2007):

$H_{648,2/3,27}$

$$
=
\begin{bmatrix}
P_{25} & P_{26} & P_{14} & 0 & P_{20} & 0 & P_2 & 0 & P_4 & 0 & 0 & P_8 & 0 & P_{16} & 0 & P_{18} & P_1 & P_0 & 0 & 0 & 0 & 0 & 0 & 0 \\
P_{10} & P_9 & P_{15} & P_{11} & 0 & P_0 & 0 & P_1 & 0 & 0 & P_{18} & 0 & P_8 & 0 & P_{10} & 0 & 0 & P_0 & P_0 & 0 & 0 & 0 & 0 & 0 \\
P_{16} & P_2 & P_{20} & P_{26} & P_{21} & 0 & P_6 & 0 & P_1 & P_{26} & 0 & P_7 & 0 & 0 & 0 & 0 & 0 & 0 & P_0 & P_0 & 0 & 0 & 0 & 0 \\
P_{10} & P_{13} & P_5 & P_0 & 0 & P_3 & 0 & P_7 & 0 & 0 & P_{26} & 0 & P_{13} & 0 & P_{16} & 0 & 0 & 0 & P_0 & P_0 & 0 & 0 & 0 & 0 \\
P_{23} & P_{14} & P_{24} & 0 & P_{12} & 0 & P_{19} & 0 & P_{17} & 0 & 0 & 0 & P_{20} & 0 & P_{21} & 0 & P_0 & 0 & 0 & 0 & P_0 & P_0 & 0 & 0 \\
P_6 & P_{22} & P_9 & P_{20} & 0 & P_{25} & 0 & P_{17} & 0 & P_8 & 0 & P_{14} & 0 & P_{18} & 0 & 0 & 0 & 0 & 0 & 0 & P_0 & P_0 & 0 & 0 \\
P_{14} & P_{23} & P_{21} & P_{11} & P_{20} & 0 & P_{24} & 0 & P_{18} & 0 & P_{19} & 0 & 0 & 0 & 0 & P_{22} & 0 & 0 & 0 & 0 & 0 & 0 & P_0 & P_0 \\
P_{17} & P_{11} & P_{11} & P_{20} & 0 & P_{21} & 0 & P_{26} & 0 & P_3 & 0 & 0 & P_{18} & 0 & P_{26} & 0 & P_1 & 0 & 0 & 0 & 0 & 0 & 0 & P_0
\end{bmatrix}
\tag{6.36}
$$

For a code rate of 3/4, the parity check matrix has the dimensions 162×648 and is given in Eq. (6.37) (IEEE, 2007):

$H_{648,3/4,27}$

$$
=
\begin{bmatrix}
P_{16} & P_{17} & P_{22} & P_{24} & P_9 & P_3 & P_{14} & 0 & P_4 & P_2 & P_7 & 0 & P_{26} & 0 & P_2 & 0 & P_{21} & 0 & P_1 & P_0 & 0 & 0 & 0 & 0 \\
P_{25} & P_{12} & P_{12} & P_3 & P_3 & P_{26} & P_6 & P_{21} & 0 & P_{15} & P_{22} & 0 & P_{15} & 0 & P_4 & 0 & 0 & P_{16} & 0 & P_0 & P_0 & 0 & 0 & 0 \\
P_{25} & P_{18} & P_{26} & P_{16} & P_{22} & P_{23} & P_9 & 0 & P_0 & 0 & P_4 & 0 & P_4 & 0 & P_8 & P_{23} & P_{11} & 0 & 0 & 0 & P_0 & P_0 & 0 & 0 \\
P_9 & P_7 & P_0 & P_1 & P_{17} & 0 & 0 & P_7 & P_3 & 0 & P_3 & P_{23} & 0 & P_{16} & 0 & 0 & P_{21} & 0 & P_0 & 0 & 0 & P_0 & P_0 & 0 \\
P_{24} & P_5 & P_{26} & P_7 & P_1 & 0 & 0 & P_{15} & P_{24} & P_{15} & 0 & P_8 & 0 & P_{13} & 0 & P_{13} & 0 & P_{11} & 0 & 0 & 0 & 0 & P_0 & P_0 \\
P_2 & P_2 & P_{19} & P_{14} & P_{24} & P_1 & P_{15} & P_{19} & 0 & P_{21} & 0 & P_2 & 0 & P_{24} & 0 & P_3 & 0 & P_2 & P_1 & 0 & 0 & 0 & 0 & P_0
\end{bmatrix}
\tag{6.37}
$$

For a code rate of 5/6, the parity check matrix has the dimensions 108×648 and is given in Eq. (6.38) (IEEE, 2007):

$H_{648,5/6,27}$

$$
=
\begin{bmatrix}
P_{17} & P_{13} & P_8 & P_{21} & P_9 & P_3 & P_{18} & P_{12} & P_{10} & P_0 & P_4 & P_{15} & P_{19} & P_2 & P_5 & P_{10} & P_{26} & P_{19} & P_{13} & P_{13} & P_1 & P_0 & 0 & 0 \\
P_3 & P_{12} & P_{11} & P_{14} & P_{11} & P_{25} & P_5 & P_{18} & P_0 & P_9 & P_2 & P_{26} & P_{26} & P_{10} & P_{24} & P_7 & P_{14} & P_{20} & P_4 & P_2 & 0 & P_0 & P_0 & 0 \\
P_{22} & P_{16} & P_4 & P_3 & P_{10} & P_{21} & P_{12} & P_5 & P_{21} & P_{14} & P_{19} & P_5 & 0 & P_8 & P_5 & P_{18} & P_{11} & P_5 & P_5 & P_{15} & P_0 & 0 & P_0 & P_0 \\
P_7 & P_7 & P_{14} & P_{14} & P_4 & P_{16} & P_{16} & P_{24} & P_{24} & P_{10} & P_1 & P_7 & P_{15} & P_6 & P_{10} & P_{26} & P_8 & P_{18} & P_{21} & P_{14} & P_1 & 0 & 0 & P_0
\end{bmatrix}
\tag{6.38}
$$

For a code word size of 1296 bits, Z is 54, which means that the cyclic permutation matrices and all-zero matrices have the dimensions 54×54. For a code rate of 1/2, the parity check matrix has the dimensions 648×1296 bits and is given in Eq. (6.39)

(IEEE, 2007):

$H_{1296,1/2,54}$

$$
= \begin{bmatrix}
P_{40} & 0 & 0 & 0 & P_{22} & 0 & P_{49} & P_{23} & P_{43} & 0 & 0 & 0 & P_1 & P_0 & 0 & 0 & 0 & 0 & 0 & 0 & 0 & 0 & 0 & 0 \\
P_{50} & P_1 & 0 & 0 & P_{48} & P_{35} & 0 & 0 & P_{13} & 0 & P_{30} & 0 & 0 & P_0 & P_0 & 0 & 0 & 0 & 0 & 0 & 0 & 0 & 0 & 0 \\
P_{39} & P_{50} & 0 & 0 & P_4 & 0 & P_2 & 0 & 0 & 0 & 0 & P_{49} & 0 & 0 & P_0 & P_0 & 0 & 0 & 0 & 0 & 0 & 0 & 0 & 0 \\
P_{33} & 0 & 0 & P_{38} & P_{37} & 0 & 0 & P_4 & P_1 & 0 & 0 & 0 & 0 & 0 & 0 & P_0 & P_0 & 0 & 0 & 0 & 0 & 0 & 0 & 0 \\
P_{45} & 0 & 0 & 0 & P_0 & P_{22} & 0 & 0 & P_{20} & P_{42} & 0 & 0 & 0 & 0 & 0 & 0 & P_0 & P_0 & 0 & 0 & 0 & 0 & 0 & 0 \\
P_{51} & 0 & 0 & P_{48} & P_{35} & 0 & 0 & 0 & P_{44} & 0 & P_{18} & 0 & 0 & 0 & 0 & 0 & 0 & P_0 & P_0 & 0 & 0 & 0 & 0 & 0 \\
P_{47} & P_{11} & 0 & 0 & 0 & P_{17} & 0 & 0 & P_{51} & 0 & 0 & 0 & P_0 & 0 & 0 & 0 & 0 & 0 & P_0 & P_0 & 0 & 0 & 0 & 0 \\
P_5 & 0 & P_{25} & 0 & P_6 & 0 & P_{45} & 0 & P_{13} & P_{40} & 0 & 0 & 0 & 0 & 0 & 0 & 0 & 0 & 0 & P_0 & P_0 & 0 & 0 & 0 \\
P_{33} & 0 & 0 & P_{34} & P_{24} & 0 & 0 & 0 & P_{23} & 0 & 0 & P_{46} & 0 & 0 & 0 & 0 & 0 & 0 & 0 & 0 & P_0 & P_0 & 0 & 0 \\
P_1 & 0 & P_{27} & 0 & P_1 & 0 & 0 & 0 & P_{38} & 0 & P_{44} & 0 & 0 & 0 & 0 & 0 & 0 & 0 & 0 & 0 & 0 & P_0 & P_0 & 0 \\
0 & P_{18} & 0 & 0 & P_{23} & 0 & P_8 & P_0 & P_{35} & 0 & 0 & 0 & 0 & 0 & 0 & 0 & 0 & 0 & 0 & 0 & 0 & 0 & P_0 & P_0 \\
P_{49} & 0 & P_{17} & 0 & P_{30} & 0 & 0 & 0 & P_{34} & 0 & 0 & P_{19} & P_1 & 0 & 0 & 0 & 0 & 0 & 0 & 0 & 0 & 0 & 0 & P_0
\end{bmatrix}
$$

(6.39)

For a code rate of 2/3, the parity check matrix has the dimensions 432×1296 bits and is given in Eq. (6.40) (IEEE, 2007):

$H_{1296,2/3,54}$

$$
= \begin{bmatrix}
P_{39} & P_{31} & P_{22} & P_{43} & 0 & P_{40} & P_4 & 0 & P_{11} & 0 & 0 & P_{50} & 0 & 0 & 0 & P_6 & P_1 & P_0 & 0 & 0 & 0 & 0 & 0 & 0 \\
P_{25} & P_{52} & P_{41} & P_2 & P_6 & 0 & P_{14} & 0 & P_{34} & 0 & 0 & 0 & P_{24} & 0 & P_{37} & 0 & 0 & P_0 & P_0 & 0 & 0 & 0 & 0 & 0 \\
P_{43} & P_{31} & P_{29} & P_0 & P_{21} & 0 & P_{28} & 0 & 0 & P_2 & 0 & 0 & P_7 & 0 & P_{17} & 0 & 0 & 0 & P_0 & P_0 & 0 & 0 & 0 & 0 \\
P_{20} & P_{33} & P_{48} & 0 & P_4 & P_{13} & 0 & P_{26} & 0 & 0 & P_{22} & 0 & 0 & P_{46} & P_{42} & 0 & 0 & 0 & 0 & P_0 & P_0 & 0 & 0 & 0 \\
P_{45} & P_7 & P_{18} & P_{51} & P_{12} & P_{25} & 0 & 0 & 0 & P_{50} & 0 & 0 & P_5 & 0 & 0 & 0 & P_0 & 0 & 0 & 0 & P_0 & P_0 & 0 & 0 \\
P_{35} & P_{40} & P_{32} & P_{16} & P_5 & 0 & 0 & P_{18} & 0 & 0 & P_{43} & P_{51} & 0 & P_{32} & 0 & 0 & 0 & 0 & 0 & 0 & 0 & P_0 & P_0 & 0 \\
P_9 & P_{24} & P_{13} & P_{22} & P_{28} & 0 & 0 & P_{37} & 0 & 0 & P_{25} & 0 & 0 & P_{52} & 0 & P_{13} & 0 & 0 & 0 & 0 & 0 & 0 & P_0 & P_0 \\
P_{32} & P_{22} & P_4 & P_{21} & P_{16} & 0 & 0 & 0 & P_{27} & P_{28} & 0 & 0 & P_{38} & 0 & 0 & 0 & P_8 & P_1 & 0 & 0 & 0 & 0 & 0 & P_0
\end{bmatrix}
$$

(6.40)

For a code rate of 3/4, the parity check matrix has the dimensions 324×1296 and is given in Eq. (6.41) (IEEE, 2007):

$H_{1296,3/4,54}$

$$
= \begin{bmatrix}
P_{39} & P_{40} & P_{51} & P_{41} & P_3 & P_{29} & P_8 & P_{36} & 0 & P_{14} & 0 & P_6 & 0 & P_{33} & 0 & P_{11} & 0 & P_4 & P_1 & P_0 & 0 & 0 & 0 \\
P_{48} & P_{21} & P_{47} & P_9 & P_{48} & P_{35} & P_{51} & 0 & P_{38} & 0 & P_{28} & 0 & P_{34} & 0 & P_{50} & 0 & P_{50} & 0 & 0 & P_0 & P_0 & 0 & 0 \\
P_{30} & P_{39} & P_{28} & P_{42} & P_{50} & P_{39} & P_5 & P_{17} & 0 & P_6 & 0 & P_{18} & 0 & P_{20} & 0 & P_{15} & 0 & P_{40} & 0 & 0 & P_0 & P_0 & 0 \\
P_{29} & P_0 & P_1 & P_{43} & P_{36} & P_{30} & P_{47} & 0 & P_{49} & 0 & P_{47} & 0 & P_3 & 0 & P_{35} & 0 & P_{34} & 0 & P_0 & 0 & 0 & P_0 & P_0 \\
P_1 & P_{32} & P_{11} & P_{23} & P_{10} & P_{44} & P_{12} & P_7 & 0 & P_{48} & 0 & P_4 & 0 & P_9 & 0 & P_{17} & 0 & P_{16} & 0 & 0 & 0 & P_0 & P_0 \\
P_{13} & P_7 & P_{15} & P_{47} & P_{23} & P_{16} & P_{47} & 0 & P_{43} & 0 & P_{29} & 0 & P_{52} & 0 & P_2 & 0 & P_{53} & 0 & P_1 & 0 & 0 & 0 & P_0
\end{bmatrix}
$$

(6.41)

For a code rate of 5/6, the parity check matrix has the dimensions 216×1296 and is given in Eq. (6.42) (IEEE, 2007):

$$H_{1296,5/6,54}$$

$$= \begin{bmatrix} P_{48}\,P_{29}\,P_{37}\,P_{52}\,P_2 & P_{16}\,P_6 & P_{14}\,P_{53}\,P_{31}\,P_{34}\,P_5 & P_{18}\,P_{42}\,P_{53}\,P_{31}\,P_{45}\mathbf{0} & P_{46}\,P_{52}\,P_1\,P_0\mathbf{0}\ \ \mathbf{0} \\ P_{17}\,P_4 & P_{30}\,P_7 & P_{43}\,P_{11}\,P_{24}\,P_6 & P_{14}\,P_{21}\,P_6 & P_{39}\,P_{17}\,P_{40}\,P_{47}\,P_7 & P_{15}\,P_{41}\,P_{19}\mathbf{0}\ \ \mathbf{0}\ \ P_0\,P_0\mathbf{0} \\ P_7\ \ P_2 & P_{51}\,P_{31}\,P_{46}\,P_{23} & P_{16}\,P_{11}\,P_{53}\,P_{40}\,P_{10}\,P_7 & P_{46}\,P_{53}\,P_{33}\,P_{35}\mathbf{0} & P_{25}\,P_{35}\,P_{38}\,P_0\mathbf{0}\ \ P_0\,P_0 \\ P_{19}\,P_{48}\,P_{41}\,P_1 & P_{10}\,P_7 & P_{36}\,P_{47}\,P_5 & P_{29}\,P_{52}\,P_{52}\,P_{31}\,P_{10}\,P_{26}\,P_6 & P_3\ \ P_2\ \ \mathbf{0}\ \ \ P_{51}\,P_1\mathbf{0}\ \ \mathbf{0}\ \ P_0 \end{bmatrix}$$

$$(6.42)$$

For a code word size of 1944 bits, Z is 81, which means that the cyclic permutation matrices and all-zero matrices have the dimensions 81×81. For a code rate of 1/2, the parity check matrix has the dimensions 972×1944 bits and is given in Eq. (6.43) (IEEE, 2007):

$$H_{1944,1/2,81}$$

$$= \begin{bmatrix} P_{57}\mathbf{0}\ \ \mathbf{0}\ \ \mathbf{0}\ \ P_{50}\mathbf{0}\ \ P_{11}\mathbf{0}\ \ P_{50}\mathbf{0}\ \ P_{79}\mathbf{0}\ \ P_1\,P_0\mathbf{0}\ \mathbf{0}\ \mathbf{0}\ \mathbf{0}\ \mathbf{0}\ \mathbf{0}\ \mathbf{0}\ \mathbf{0}\ \mathbf{0}\ \mathbf{0} \\ P_3\ \ \mathbf{0}\ \ P_{28}\mathbf{0}\ \ \ P_0\ \ \mathbf{0}\ \ \ \mathbf{0}\ \ \ \mathbf{0}\ \ P_{55}\,P_7\ \ \mathbf{0}\ \ \ \mathbf{0}\ \ \ P_0\,P_0\mathbf{0}\ \mathbf{0}\ \mathbf{0}\ \mathbf{0}\ \mathbf{0}\ \mathbf{0}\ \mathbf{0}\ \mathbf{0} \\ P_{30}\mathbf{0}\ \ \mathbf{0}\ \ \mathbf{0}\ \ P_{24}\,P_{37}\mathbf{0}\ \ \ \mathbf{0}\ \ P_{56}\,P_{14}\mathbf{0}\ \ \mathbf{0}\ \ \ \mathbf{0}\ \ \mathbf{0}\ \ P_0\,P_0\mathbf{0}\ \mathbf{0}\ \mathbf{0}\ \mathbf{0}\ \mathbf{0}\ \mathbf{0}\ \mathbf{0} \\ P_{62}\,P_{53}\mathbf{0}\ \ \ \mathbf{0}\ \ P_{53}\mathbf{0}\ \ \ \mathbf{0}\ \ \ P_3\ \ P_{35}\mathbf{0}\ \ \ \mathbf{0}\ \ \ \mathbf{0}\ \ \mathbf{0}\ \ \mathbf{0}\ \ P_0\,P_0\mathbf{0}\ \mathbf{0}\ \mathbf{0}\ \mathbf{0}\ \mathbf{0}\ \mathbf{0} \\ P_{40}\mathbf{0}\ \ \ \mathbf{0}\ \ P_{20}\,P_{66}\mathbf{0}\ \ \ \mathbf{0}\ \ P_{22}\,P_{28}\mathbf{0}\ \ \ \mathbf{0}\ \ \ \mathbf{0}\ \ \mathbf{0}\ \ \mathbf{0}\ \ P_0\,P_0\mathbf{0}\ \mathbf{0}\ \mathbf{0}\ \mathbf{0}\ \mathbf{0} \\ P_0\ \ \mathbf{0}\ \ \ \mathbf{0}\ \ \ \mathbf{0}\ \ \ P_8\ \ \mathbf{0}\ \ \ P_{42}\mathbf{0}\ \ \ P_{50}\mathbf{0}\ \ \ \mathbf{0}\ \ P_8\ \ \mathbf{0}\ \ \mathbf{0}\ \ \mathbf{0}\ \ \mathbf{0}\ \ P_0\,P_0\mathbf{0}\ \mathbf{0}\ \mathbf{0}\ \mathbf{0} \\ P_{69}\,P_{79}\,P_{79}\mathbf{0}\ \ \ \mathbf{0}\ \ \ \mathbf{0}\ \ P_{56}\mathbf{0}\ \ \ P_{52}\mathbf{0}\ \ \ \mathbf{0}\ \ \ \mathbf{0}\ \ P_0\mathbf{0}\ \ \mathbf{0}\ \ \mathbf{0}\ \ \mathbf{0}\ \ P_0\,P_0\mathbf{0}\ \mathbf{0}\ \mathbf{0}\ \mathbf{0} \\ P_{65}\mathbf{0}\ \ \ \mathbf{0}\ \ \ \mathbf{0}\ \ P_{38}\,P_{57}\mathbf{0}\ \ \ \mathbf{0}\ \ P_{72}\mathbf{0}\ \ \ P_{27}\mathbf{0}\ \ \ \mathbf{0}\ \ \mathbf{0}\ \ \mathbf{0}\ \ \mathbf{0}\ \ P_0\,P_0\mathbf{0}\ \mathbf{0}\ \mathbf{0} \\ P_{64}\mathbf{0}\ \ \ \mathbf{0}\ \ \ \mathbf{0}\ \ P_{14}\,P_{52}\mathbf{0}\ \ \ \mathbf{0}\ \ P_{30}\mathbf{0}\ \ \ \mathbf{0}\ \ P_{32}\mathbf{0}\ \ \ \mathbf{0}\ \ \mathbf{0}\ \ \mathbf{0}\ \ \mathbf{0}\ \ P_0\,P_0\mathbf{0}\ \mathbf{0} \\ \mathbf{0}\ \ P_{45}\mathbf{0}\ \ \ P_{70}\,P_0\ \ \mathbf{0}\ \ \ \mathbf{0}\ \ \ \mathbf{0}\ \ P_{77}\,P_9\ \ \mathbf{0}\ \ \ \mathbf{0}\ \ \ \mathbf{0}\ \ \mathbf{0}\ \ \mathbf{0}\ \ \mathbf{0}\ \ \mathbf{0}\ \ P_0\,P_0\mathbf{0} \\ P_2\ \ P_{56}\mathbf{0}\ \ \ P_{57}\,P_{35}\mathbf{0}\ \ \ \mathbf{0}\ \ \ \mathbf{0}\ \ \ \mathbf{0}\ \ P_{12}\mathbf{0}\ \ \ \mathbf{0}\ \ \ \mathbf{0}\ \ \mathbf{0}\ \ \mathbf{0}\ \ \mathbf{0}\ \ \mathbf{0}\ \ P_0\,P_0 \\ P_{24}\mathbf{0}\ \ \ P_{61}\mathbf{0}\ \ \ P_{60}\mathbf{0}\ \ \ \mathbf{0}\ \ P_{27}\,P_{51}\mathbf{0}\ \ \ \mathbf{0}\ \ P_{16}\,P\ \mathbf{0}\ \ \mathbf{0}\ \ \mathbf{0}\ \ \mathbf{0}\ \ \mathbf{0}\ \ \mathbf{0}\ \ P_0 \end{bmatrix}$$

$$(6.43)$$

For a code rate of 2/3, the parity check matrix has the dimensions 648×1944 bits and is given in Eq. (6.44) (IEEE, 2007):

$$H_{1944,2/3,81}$$

$$= \begin{bmatrix} P_{61}\,P_{75}\,P_4 & P_{63}\,P_{56}\mathbf{0}\ \ \mathbf{0}\ \ \mathbf{0}\ \ \mathbf{0}\ \ \mathbf{0}\ \ \mathbf{0}\ \ P_8\ \mathbf{0}\ \ \ P_2\ \ P_{17}\,P_{25}\,P_1\,P_0\mathbf{0}\ \mathbf{0}\ \mathbf{0}\ \mathbf{0}\ \mathbf{0} \\ P_{56}\,P_{74}\,P_{77}\,P_{20}\mathbf{0}\ \ \mathbf{0}\ \ \mathbf{0}\ \ P_{64}\,P_{24}\,P_4 & P_{67}\mathbf{0}\ \ P_7\ \ \mathbf{0}\ \ \ \mathbf{0}\ \ \ \mathbf{0}\ \ \ P_0\,P_0\mathbf{0}\ \mathbf{0}\ \mathbf{0}\ \mathbf{0} \\ P_{28}\,P_{21}\,P_{68}\,P_{10}\,P_7 & P_{14}\,P_{65}\mathbf{0}\ \ \mathbf{0}\ \ \mathbf{0}\ \ P_{23}\mathbf{0}\ \ \ \mathbf{0}\ \ \mathbf{0}\ \ P_{75}\mathbf{0}\ \ \mathbf{0}\ \ P_0\,P_0\mathbf{0}\ \mathbf{0}\ \mathbf{0} \\ P_{48}\,P_{38}\,P_{43}\,P_{78}\,P_{76}\mathbf{0}\ \ \mathbf{0}\ \ \mathbf{0}\ \ \mathbf{0}\ \ P_5\ P_{36}\mathbf{0}\ \ P_{15}\,P_{72}\mathbf{0}\ \ \mathbf{0}\ \ \mathbf{0}\ \ P_0\,P_0\mathbf{0}\ \mathbf{0} \\ P_{40}\,P_2 & P_{53}\,P_{25}\mathbf{0}\ \ P_{52}\,P_{62}\mathbf{0}\ \ P_{20}\mathbf{0}\ \ \mathbf{0}\ \ P_{44}\mathbf{0}\ \ \mathbf{0}\ \ \ \mathbf{0}\ \ P_0\mathbf{0}\ \ \mathbf{0}\ \ P_0\,P_0\mathbf{0}\ \mathbf{0} \\ P_{69}\,P_{23}\,P_{64}\,P_{10}\,P_{22}\mathbf{0}\ \ P_{21}\mathbf{0}\ \ \mathbf{0}\ \ \mathbf{0}\ \ \mathbf{0}\ \ P_{68}\,P_{23}\,P_{29}\mathbf{0}\ \ \mathbf{0}\ \ \mathbf{0}\ \ \mathbf{0}\ \ P_0\,P_0\mathbf{0} \\ P_{12}\,P_0 & P_{68}\,P_{20}\,P_{55}\,P_{61}\mathbf{0}\ \ P_{40}\mathbf{0}\ \ \mathbf{0}\ \ P_{52}\mathbf{0}\ \ \mathbf{0}\ \ P_{44}\mathbf{0}\ \ \mathbf{0}\ \ \mathbf{0}\ \ \mathbf{0}\ \ P_0\,P_0 \\ P_{58}\,P_8 & P_{34}\,P_{64}\,P_{78}\mathbf{0}\ \ \mathbf{0}\ \ P_{11}\,P_{78}\,P_{24}\mathbf{0}\ \ \mathbf{0}\ \ \mathbf{0}\ \ \mathbf{0}\ \ P_{58}\,P_1\mathbf{0}\ \mathbf{0}\ \mathbf{0}\ \mathbf{0}\ \mathbf{0}\ \ P_0 \end{bmatrix}$$

$$(6.44)$$

For a code rate of 3/4, the parity check matrix has the dimensions 486×1944 and is given in Eq. (6.45) (IEEE, 2007):

$H_{1944,3/4,81}$

$$
= \begin{bmatrix}
P_{48} P_{29} P_{28} P_{39} P_9 \ P_{61} \mathbf{0} \ \ \mathbf{0} \ \ \mathbf{0} \ \ P_{63} P_{45} P_{80} \mathbf{0} \ \ \mathbf{0} \ \ \mathbf{0} \ \ P_{37} P_{32} P_{22} P_1 P_0 \mathbf{0} \ \mathbf{0} \ \mathbf{0} \ \mathbf{0} \\
P_4 \ P_{49} P_{42} P_{48} P_{11} P_{30} \mathbf{0} \ \ \mathbf{0} \ \ \mathbf{0} \ \ P_{49} P_{17} P_{41} P_{37} P_{15} \mathbf{0} \ \ P_{54} \mathbf{0} \ \ \mathbf{0} \ \ \mathbf{0} \ \ P_0 P_0 \mathbf{0} \ \mathbf{0} \ \mathbf{0} \\
P_{35} P_{76} P_{78} P_{51} P_{37} P_{35} P_{21} \mathbf{0} \ \ P_{17} P_{64} \mathbf{0} \ \ \mathbf{0} \ \ \mathbf{0} \ \ P_{59} P_7 \ \mathbf{0} \ \ \mathbf{0} \ \ P_{32} \mathbf{0} \ \ \mathbf{0} \ \ P_0 P_0 \mathbf{0} \ \mathbf{0} \\
P_9 \ P_{65} P_{44} P_9 \ P_{54} P_{56} P_{73} P_{34} P_{42} \mathbf{0} \ \ \mathbf{0} \ \ \mathbf{0} \ \ P_{35} \mathbf{0} \ \ \mathbf{0} \ \ \mathbf{0} \ \ P_{46} P_{39} P_0 \mathbf{0} \ \ \mathbf{0} \ \ P_0 P_0 \mathbf{0} \\
P_3 \ P_{62} P_7 \ P_{80} P_{68} P_{26} \mathbf{0} \ \ P_{80} P_{55} \mathbf{0} \ \ P_{36} \mathbf{0} \ \ P_{26} \mathbf{0} \ \ P_9 \ \mathbf{0} \ \ P_{76} \mathbf{0} \ \ \mathbf{0} \ \ \mathbf{0} \ \ \mathbf{0} \ \ P_0 P_0 \\
P_{26} P_{75} P_{33} P_{21} P_{69} P_{59} P_3 \ P_{38} \mathbf{0} \ \ \mathbf{0} \ \ \mathbf{0} \ \ P_{35} \mathbf{0} \ \ P_{62} P_{36} P_{26} \mathbf{0} \ \ \mathbf{0} \ \ P_1 \mathbf{0} \ \mathbf{0} \ \mathbf{0} \ \mathbf{0} \ \ P_0
\end{bmatrix}
$$

$$(6.45)$$

For a code rate of 5/6, the parity check matrix has the dimensions 324×1944 and is given in Eq. (6.46) (IEEE, 2007):

$H_{1944,5/6,81}$

$$
= \begin{bmatrix}
P_{13} P_{48} P_{80} P_{66} P_4 \ P_{74} P_7 \ P_{30} P_{76} P_{52} P_{37} P_{60} \mathbf{0} \ \ P_{49} P_{73} P_{31} P_{74} P_{73} P_{23} \mathbf{0} \ \ P_1 P_0 \mathbf{0} \ \mathbf{0} \\
P_{69} P_{63} P_{74} P_{56} P_{64} P_{77} P_{57} P_{65} P_6 \ P_{16} P_{51} \mathbf{0} \ \ P_{64} \mathbf{0} \ \ P_{68} P_9 \ P_{48} P_{62} P_{54} P_{27} \mathbf{0} \ \ P_0 P_0 \mathbf{0} \\
P_{51} P_{15} P_0 \ P_{80} P_{24} P_{25} P_{42} P_{54} P_{44} P_{71} P_{71} P_9 \ P_{67} P_{35} \mathbf{0} \ \ P_{58} \mathbf{0} \ \ P_{29} \mathbf{0} \ \ P_{53} P_0 \mathbf{0} \ \ P_0 P_0 \\
P_{16} P_{29} P_{36} P_{41} P_{44} P_{56} P_{59} P_{37} P_{50} P_{24} \mathbf{0} \ \ P_{65} P_4 \ P_{65} P_{52} \mathbf{0} \ \ P_4 \ \mathbf{0} \ \ P_{73} P_{52} P_1 \mathbf{0} \ \mathbf{0} \ \ P_0
\end{bmatrix}
$$

$$(6.46)$$

Part II

Medium access control layer

7   Medium access control

The medium access control (MAC) layer provides, among other things, addressing and channel access control that makes it possible for multiple stations on a network to communicate. IEEE 802.11 is often referred to as wireless Ethernet and, in terms of addressing and channel access, 802.11 is indeed similar to Ethernet, which was standardized as IEEE 802.3. As a member of the IEEE 802 LAN family, IEEE 802.11 makes use of the IEEE 802 48-bit global address space, making it compatible with Ethernet at the link layer. The 802.11 MAC also supports shared access to the wireless medium through a technique called carrier sense multiple access with collision avoidance (CSMA/CA), which is similar to the original (shared medium) Ethernet's carrier sense multiple access with collision detect (CSMA/CD). With both techniques, if the channel is sensed to be "idle," the station is permitted to transmit, but if the channel is sensed to be "busy" then the station defers its transmission. However, the very different media over which Ethernet and 802.11 operate mean that there are some differences.

The Ethernet channel access protocol is essentially to wait for the medium to go "idle," begin transmitting and, if a collision is detected while transmitting, to stop transmitting and begin a random backoff period. It is not feasible for a transmitter to detect a collision while transmitting in a wireless medium; thus the 802.11 channel access protocol attempts to avoid collisions. Once the medium goes "idle," the station waits a random period during which it continues to sense the medium, and if at the end of that period the medium is still "idle," it begins transmitting. The random period reduces the chances of a collision since another station waiting to access the medium would likely choose a different period, hence the collision avoidance aspect of CSMA/CA.

The simple distributed, contention-based access protocol supported by the CSMA/CA technique is the basis for the 802.11 MAC protocol and also where the similarity to Ethernet ends. The wireless medium, being very different from the wired medium, necessitates a number of additional features:

- The wireless medium is prone to errors and benefits significantly from having a low latency, link level error recovery mechanism.
- In a wireless medium not all stations can "hear" all other stations. Some stations may "hear" the station on one end of an exchange but not the station at the far end (the hidden node problem).
- The data rate that a channel can support is affected greatly by distance and other environmental effects. Also, channel conditions may change with time due to station

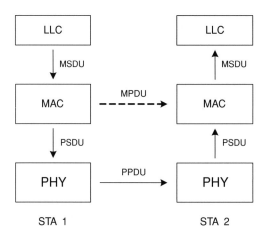

Figure 7.1 Protocol layering and messaging.

mobility or environmental changes. Stations need to continually adjust the data rate at which they exchange information to optimize throughput.

- Stations, often being mobile, need management mechanisms for associating with and disassociating from WLANs as they change location.

This chapter provides an overview of the 802.11 MAC prior to the enhancements introduced in 802.11n. After some background information on protocol layering, there is a brief overview of 802.11 management functions. This is followed by a more detailed overview of the channel access and data transfer aspects.

7.1   Protocol layering

Some basic concepts regarding protocol layering and messaging and illustrated in Figure 7.1 are needed in order to understand the MAC functionality. In this layered model, each entity, PHY and MAC, offers services to the entity in the layer immediately above it and user data is transferred between the layers as a service data unit (SDU). The MAC receives data from the logical link control (LLC) layer, and delivers data to the LLC layer through the MAC SDU (MSDU). The PHY receives data from the MAC and delivers data to the MAC in a PHY SDU[1] (PSDU).

A protocol is the means by which entities in the layered model exchange data and control information with their peer entities. This exchange takes place through protocol data units (PDUs). The MAC exchanges MAC PDUs (MPDUs) with its peer and the PHY exchanges PHY PDUs with its peer.

Another commonly used term in the 802.11 standard is station or STA, which refers to the MAC and PHY in the context of the device that incorporates these entities. In

---

[1] The IEEE 802.11 standard uses the term PLCP SDU instead of PHY SDU, where PLCP is the physical layer convergence procedure, a sublayer at the top of the PHY. The terms are equivalent.

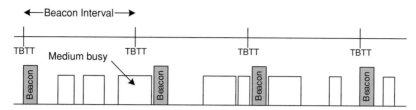

Figure 7.2 Beacon transmission on a busy network.

practice this is the network adaptor in a laptop or the communication subsystem in a mobile phone. An AP (access point) is a station with additional functions related to managing an infrastructure BSS and providing access to the distribution system.

7.2 Management functions

As described in Chapter 1, the BSS is the basic building block of an 802.11 WLAN. There are two types of BSS, the independent BSS (IBSS), which is an ad-hoc association of stations that communicate with one another directly, and the infrastructure BSS, which is anchored by an AP that may be connected to a distribution system (DS) and through which the majority of the data transfer takes place, both station to station and station to DS. In this book we are primarily concerned with the infrastructure BSS; however, much of the discussion applies equally to IBSSs.

A station becomes aware of the existence of a BSS through scanning, that is passively seeking Beacon transmissions or actively probing for the existence of an AP through a Probe Request/Response exchange.

A station's membership of a BSS is dynamic. The station may turn on and off, or the station may be mobile and move in or out of the area covered by the BSS. A station becomes a member of a BSS by becoming "associated" with the BSS. On leaving the BSS, a station becomes "disassociated." In an ESS comprised of multiple infrastructure BSSs, a station may migrate from one BSS to another BSS within the ESS through "reassociation."

7.2.1 Beacons

The AP in an infrastructure BSS periodically broadcasts Beacon frames. The Beacon period defines a fixed schedule of target beacon transmission time (TBTTs) and the Beacon frame itself is transmitted on or as close to the TBTT as possible subject to the medium being idle (Figure 7.2).

The Beacon frame carries regulatory information, capability information and information for managing the BSS.

7.2.2 Scanning

Scanning is the process by which a station discovers a BSS and the attributes associated with that BSS. Two forms of scanning are possible: passive scanning and active scanning.

Passive scanning is a receive only operation that is compatible with all regulatory domains. With passive scanning the station looks for Beacon transmissions and may switch channels to find these transmissions. Beacon frames include, among other things, information on the country code, maximum allowable transmit power, and the channels to be used for the regulatory domain. Once the station has discovered the AP through its Beacon transmission and has this regulatory information it may probe the AP directly for additional information using a Probe Request/Response exchange if that additional information is not present in the Beacon frame itself.

Active scanning may be used when it is permitted by the regulatory domain in which the station operates. With active scanning a station transmits Probe Request frames on each of the channels where it is seeking a BSS. Depending on the generality of the search, the Probe Request frame includes the following addressing information:

- **SSID (service set identifier).** The SSID in the Probe Request may be the SSID of the specific ESS for which the station is seeking BSSs or it may be the wildcard SSID.
- **BSSID (BSS identifier).** The BSSID in the Probe Request may be the BSSID of a specific BSS or it may be the wildcard BSSID.
- **DA (destination address).** The DA of the Probe Request frame is the broadcast address or the specific MAC address of the AP station.

An AP that receives a broadcast Probe Request sends a Probe Response to the station making the request if the following conditions are true:(a) the SSID is the wildcard SSID or matches the SSID of the ESS and (b) the BSSID is the wildcard BSSID or the AP's BSSID. Multiple APs may respond to a Probe Request using normal channel access procedures to avoid collisions.

7.2.3   Authentication

Authentication is the process by which two stations that wish to communicate establish their identity to a mutually acceptable level. The original 802.11 specification supported two authentication methods operating at the link level: open system authentication and shared key authentication. With the former any station may be admitted as a member of a BSS. With the latter, stations rely on the wired equivalent privacy (WEP) protocol to demonstrate knowledge of a shared encryption key.

WEP has been shown to be insecure and the newer security techniques in 802.11 support authentication through the IEEE 802.1X standard (IEEE, 2004). Authentication is performed prior to association; however, a detailed discussion on how this is performed is beyond the scope of this book.

7.2.4   Association

Before a station is allowed to send data via an AP it must become associated with the AP. Association provides a mapping between the station and AP that allows messages within the DS (distribution system) to reach the AP with which the station is associated and ultimately to the station itself. At any given instant a station may only be associated with a single AP.

Association is initiated by the station with the station sending an Association Request to the AP. If the station is admitted, the AP responds with an Association Response. With the Association Request and Response exchange, the station and AP exchange capability information (support for optional features) and the AP informs the station of specific operating parameters within the BSS.

7.2.5   Reassociation

Reassociation supports BSS-transition mobility, allowing a station to move from a current association with one AP to another within the same ESS. This keeps the DS informed of the current mapping between AP and station. Reassociation may also be performed to change attributes of the station association such as station capability information.

Reassociation is initiated by the station with the station sending a Reassociation Request to the AP. The AP responds with a Reassociation Response.

7.2.6   Disassociation

Disassociation terminates an existing association and may be performed by either the station or the AP. Stations should attempt to disassociate when they leave the network. However, because loss of communication may prevent this, a timeout mechanism allows the AP to disassociate the station without a message exchange should the station become unreachable.

To disassociate a station from the BSS, the AP or station sends a Disassociation frame. Disassociation is not a request, thus the other party merely acknowledges reception of the frame.

7.3   Distributed channel access

The specific CSMA/CA mechanism used in the 802.11 MAC is referred to as the distributed coordination function (DCF). A station that wishes to transmit first performs a clear channel assessment (CCA) by sensing the medium for a fixed duration, the DCF inter-frame space (DIFS). If the medium is idle then the station assumes that it may take ownership of the medium and begin a frame exchange sequence. If the medium is busy, the station waits for the medium to go idle, defers for DIFS, and waits for a further random backoff period. If the medium remains idle for the DIFS deferral and the backoff period, the station assumes that it may take ownership of the medium and begin a frame exchange sequence.

The random backoff period provides the collision avoidance aspect. When the network is loaded, multiple stations may be waiting for the medium to go idle having accumulated packets to send while the medium was busy. Since each station probabilistically selects a different backoff interval, collisions where more than one station begins transmission at the same time are unlikely.

Once a station has gained access to the medium, it maintains control of the medium by keeping a minimum gap, the short inter-frame space (SIFS), between frames in a sequence. Another station will not gain access to the medium during that sequence since it must defer for a fixed duration that is longer than SIFS. Rules limit the types of frame exchange sequences that are allowed and the duration of those sequences to prevent one station from monopolizing the medium.

Fundamental to CSMA/CA is the carrier sense. The DCF uses both physical and virtual carrier sense functions to determine the state of the medium. The physical carrier sense resides in the PHY and uses energy detect and preamble detect with frame length deferral to determine when the medium is busy. The virtual carrier sense resides in the MAC and uses reservation information carried in the Duration field of the MAC headers announcing impeding use of the medium. The virtual carrier sense mechanism is called the network allocation vector (NAV). The medium is determined to be idle only when both the physical and virtual carrier sense mechanisms indicate it to be so.

The DCF also makes use of the immediate feedback provided by the basic acknowledgement mechanism that has the responder send an ACK frame in response to the initiator's data or management frame. Not receiving the ACK response frame is a likely indication that the initiator's transmission was not correctly received, either due to collision or poor channel conditions at the time of the data transmission.

To further minimize the chance of collisions, and as a more robust collision detect mechanism, the initiating station may begin a sequence with a short control frame exchange using robustly modulated RTS and CTS frames. This sets the NAV in the stations surrounding both the initiator and responder, some of which may be hidden nodes unable to detect the more remote station's transmissions and thus only able to defer for frame transmissions from nearby nodes.

The DCF provides a distributed contention-based channel access function. Stations compete for channel access without the need for a central coordinator or arbiter. This mechanism is remarkable efficient and fairly apportions bandwidth among the active stations.

7.3.1   Basic channel access timing

Basic channel access timing from the original 802.11 specification is illustrated in Figure 7.3. The different inter-frame space (IFS) durations effectively provide access to the wireless medium at different priority levels.

7.3.1.1   SIFS

The short inter-frame space (SIFS) is used to separate a response frame from the frame that solicited the response, for example between a data frame and the ACK response. SIFS is designed to be as short as possible but still accommodate the latencies incurred in a reasonable implementation. These latencies include the decode latency in the PHY for demodulating the received frame, the MAC processing time for the received

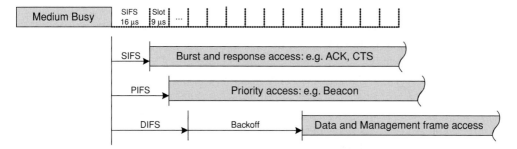

Figure 7.3 Basic channel access priorities with associated timing.

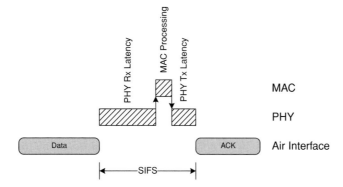

Figure 7.4 PHY and MAC latencies generating a response frame.

frame and building the response, and the transmitter startup time to send the response (Figure 7.4).

SIFS is also used to separate individual frames in a back-to-back data burst. Stations accessing the medium using SIFS timing do not check if the medium is busy, but simply switch to transmit mode (if not already in transmit mode) and begin transmission at the SIFS boundary.

The SIFS duration for a particular PHY is defined by the aSIFSTime parameter. For the 802.11a, 802.11g, and 802.11n PHYs the value is 16 μs.

7.3.1.2   Slot time

Timing for the other IFS durations is SIFS plus an integral number of slot times with transmission beginning on the slot boundary. In practice, propagation delays and, to a small extent, implementation inaccuracies mean that each station sees a slightly different boundary. The slot duration is designed to accommodate this variability and provide enough time for a transmitting station's preamble to be detected by neighboring stations before the next slot boundary. During each slot time, stations not yet transmitting remain in receive mode and check that the medium remains idle.

The slot time for a particular PHY is defined by the aSlotTime parameter. For the 802.11a, 802.11g, and 802.11n PHYs the value is 9 μs.

7.3.1.3   PIFS

The PCF inter-frame space (PIFS) defer provides the next highest access priority following SIFS and is used to gain priority access to the medium. PIFS is defined by the following equation:

$$PIFS = aSIFSTime + aSlotTime \tag{7.1}$$

The AP uses the PIFS defer to gain access to the medium to send a Beacon, start a contention free period, or to regain access to the medium if an expected response frame is not received during a contention free period. Despite its name, which reflects its original use for the point coordination function (see Section 9.1), PIFS is now also used for other priority operations, such as by a station that needs to send a Channel Switch Announcement frame (802.11h).

7.3.1.4   DIFS

The DCF inter-frame space (DIFS) is used by stations operating under the DCF to transmit data frames and management frames and is defined by the following equation:

$$DIFS = aSIFSTime + 2 \times aSlotTime \tag{7.2}$$

A station using the DCF is allowed to transmit if it determines that the medium is idle for the duration of the DIFS, or if it determines that the medium is idle for the duration of the DIFS plus the remaining backoff time following the reception of a correctly received frame.

7.3.1.5   Random backoff time

When the medium transitions from busy to idle, multiple stations may be ready to send data. To minimize collisions, stations wishing to initiate transfer select a random backoff count and defer for that number of slot times. The random backoff count is selected as a pseudo-random integer drawn from a uniform distribution over the interval [0, CW], where CW, an integer value, is the contention window.

The contention window (CW) parameter takes the initial value CWmin and effectively doubles on each unsuccessful MPDU transmit, for example each time an ACK response is not received for a data frame. If the CW reaches CWmax it remains at that value until it is reset. The CW is reset to CWmin after every successful MPDU transmit.

CW, CWmin, and CWmax may take values that are a power of 2 less 1. For the DCF, CWmin and CWmax are specified according to the particular PHY used. For 802.11a, 802.11g, and 802.11n PHYs, CWmin is 15 and CWmax is 1023. CW would thus start with the value 15 and when "doubled" take on the next higher power of 2 less 1 until it reaches 1023, i.e. 15, 31, 63, ..., 1023. The CW was defined this way for easy implementation: using binary notation, "doubling" is effectively a left shift operation with a lower order one inserted and the backoff value is obtained by using CW to mask a full word random number.

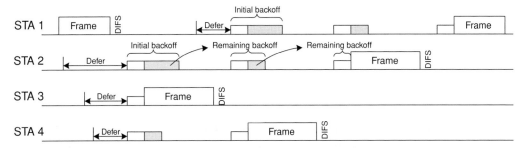

Figure 7.5 Backoff procedure.

7.3.1.6   Random backoff procedure

To begin the random backoff procedure, the station selects a random backoff count in the range [0, CW]. All backoff slots occur following a DIFS during which the medium is determined to be idle. During each backoff slot the station continues to monitor the medium. If the medium goes busy during a backoff slot then the backoff procedure is suspended. The backoff count is resumed when the medium goes idle again for a DIFS period.

The effect of this procedure is illustrated in Figure 7.5. When multiple stations are deferring and go into random backoff, then the station selecting the smallest backoff count (STA 3) will win the contention and transmit first. The remaining stations suspend their backoff and resume DIFS after the medium goes idle again. The station with the next smallest backoff count will win next (STA 4) and then eventually the station with the longest backoff count (STA 2). A station that begins a new access (STA 1 again) will select a random backoff from the full contention window and will thus tend to select a larger count than the remaining backoff for stations (such as STA 2) that have already suspended their backoff from a previous access attempt.

7.4   Data/ACK frame exchange

Transmission over a wireless medium is error prone. Data transfer benefits from a low latency, link level repeat mechanism that allows for the retransmission of frames that have not been successfully demodulated at the receiver. The basic mechanism by which this is achieved is to have the station that correctly receives a data frame addressed to it send an immediate, positive acknowledgement in the form of an ACK frame. If the station sending the data frame does not receive the ACK frame, it assumes the data frame was not received and may retransmit it.

Not all data frames can be acknowledged in this way. Broadcast and multicast data frames are directed to all or a subset of the stations in a WLAN and cannot be acknowledged this way. In an 802.11 WLAN, broadcast and multicast frames do not benefit from the additional reliability that the acknowledgement mechanism provides.

Data transfer using the Data/ACK exchange is illustrated in Figure 7.6. Here STA 1 is transferring data to STA 2. STA 1 accesses the medium after a contention period during

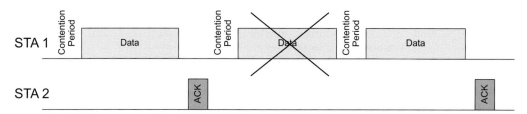

Figure 7.6 Basic data/ACK frame exchange sequence.

which it defers for DIFS followed by a random backoff period. If the medium remains idle, STA 1 transmits a data frame addressed to STA 2. If STA 2 detects and correctly demodulates the frame then it responds with an ACK. When STA 1 receives the ACK it knows that the frame was correctly received and begins channel access again in order to transmit the next frame. If, as with the second data frame in the figure, STA 2 fails to successfully demodulate the frame then STA 1 will not receive an ACK and will then begin channel access again to retransmit the data frame.

The number of retransmission attempts on a particular MSDU is limited. The transmitting station maintains a count of the number of retransmission attempts on an MSDU and when that count exceeds a configured retry limit the MSDU is discarded.

To enhance the reliability with which acknowledgement feedback is provided, the ACK frame is modulated robustly, i.e. it is sent using a lower PHY data rate than data frames sent to the same station. The additional overhead incurred with robust modulation is relatively small since the ACK frame itself is very short.

7.4.1   Fragmentation

Fragmentation is used to break up large MSDUs to improve the chance that the MSDU will be received correctly and to reduce the overhead of retransmission. At low data rates, an unfragmented MSDU can occupy a large amount of air time. For example, a 1500 byte data frame sent using the 1 Mbps 802.11b rate takes 12 ms to transmit, making it susceptible to changing channel conditions. A bit error in the frame would result in the entire frame being retransmitted. With fragmentation the MSDU would be broken into smaller sections and each section encapsulated in an MPDU. Each MPDU is sent in a separate PPDU with the preamble of each PPDU providing a new channel estimate. A bit error would result in only the MPDU carrying the errored segment being retransmitted.

The fragments making up an MSDU are sent as individual MPDUs. A station may send each fragment in a separate channel access, or the fragmented MSDU may be sent as a burst of data MPDUs following a single channel access, as illustrated in Figure 7.7.

An MSDU is fragmented when its length exceeds a threshold specified by the dot11FragmentationThreshold attribute. Each fragment contains an even number of bytes and all fragments are the same size, except the last fragment, which may be smaller. The fragments are delivered in sequence.

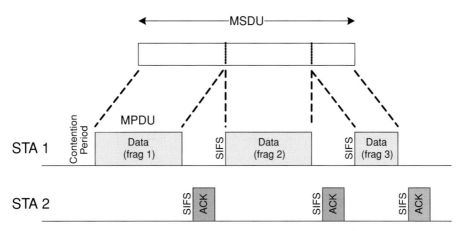

Figure 7.7 Fragment burst.

The transmitter of a fragmented MSDU maintains a timer. The attribute dot11MaxTransmitMSDULifetime specifies the maximum amount of time allowed to transmit a MSDU. The source station starts the timer on the first attempt to transmit the first fragment of the MSDU. If the timer exceeds dot11MaxTransmitMSDULifetime then all remaining fragments are discarded.

The receiver of a fragmented MSDU also maintains a timer. The attribute aMaxReceiveLifetime specifies the maximum amount of time allowed to receive an MSDU. The receive MSDU timer starts on reception of the first fragment of the MSDU. If the timer exceeds aMaxReceiveLifetime then all the fragments of the MSDU are discarded. Additional fragments which may be received later are also discarded.

7.4.2 Duplicate detection

With retransmission there is the possibility that a frame that was correctly received may be received again, for example if the transmitter retransmits a frame because the ACK response itself was not correctly demodulated. To detect duplicate frames the Data frame includes a Retry bit and a Sequence Control field consisting of a sequence number and a fragment number. The Retry bit is set on any frame that is retransmitted. The sequence number is generated as an incrementing sequence of integers assigned to MSDUs and management frames. If the MSDU or management frame is fragmented then each fragment receives the same sequence number with an incrementing fragment number.

To detect duplicate frames, the receiving station keeps track of the sequence number and fragment numbers of the last MSDU or management frame that it received from each station communicating with it. In other words, it maintains a cache of <transmit address, sequence number, fragment number> tuples for each fragment received for the last sequence number seen from each transmit address other than broadcast or multicast

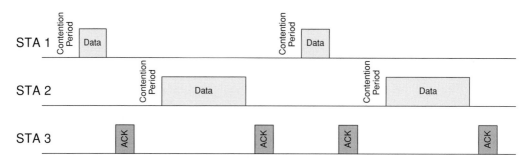

Figure 7.8 Two stations competing for access.

addresses. If the station receives an MPDU with the Retry bit set that matches an entry in the cache then it rejects the MPDU as a duplicate.

7.4.3   Data/ACK sequence overhead and fairness

The basic Data/ACK frame exchange has a fixed overhead associated with it. Since most data frames are successfully transmitted, this overhead includes the contention period during which the medium is essentially idle, the overhead associated with transmitting the data frame itself, the radio turnaround time at the receiver (SIFS), and the transmission of the ACK frame. While this overhead is essentially fixed, the duration of the data frame is not fixed since it depends on the modulation and coding scheme (data rate) used by the PHY. The higher the data rate the shorter the data frame duration and the greater the fixed overhead relative to the overall duration of the transfer.

The distributed channel access mechanism promotes fairness in the sense that all stations on the network with data to send will, on average, each send the same number of data frames. If they are all using the same packet size they will see the same throughput irrespective of their individual PHY data rates. So, for example, suppose there are two stations on the network, STA 1 and STA 2, both competing for access to send data to STA 3 (Figure 7.8). Suppose also that STA 1 is using a high data rate while STA 2 is using a low data rate. Each station competes for channel access to send one data frame and each station will on average get the same number of transmit opportunities. However, because STA 2 is using a lower data rate it will use proportionately more air time to send its data frames than STA 1.

7.5   Hidden node problem

The distributed nature of channel access in 802.11 WLANs makes the carrier sense mechanism critical for collision free operation. The physical carrier sense, which logically resides in the PHY, is responsible for detecting the transmissions of other stations. However, in some situations it may not be possible for the physical carrier sense to detect the transmissions of all stations. Consider the situation in Figure 7.9 where there

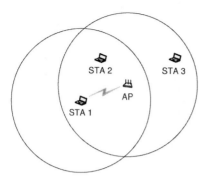

Figure 7.9 Hidden node problem.

is data transfer between STA 1 and the AP. Transmissions from STA 1 can be detected by the AP and STA 2. A distant node, STA 3, can detect transmissions from the AP but not from STA 1. STA 3 is a hidden node with respect to communication between STA 1 and the AP. When STA 1 transmits a frame to the AP there is a chance that STA 3 would still see the medium as idle and also begin a frame transmission.

7.5.1 Network allocation vector

One mechanism defined to overcome the hidden node problem is the network allocation vector (NAV). The NAV is a function that logically resides in the MAC and provides a virtual carrier sense to augment the physical carrier sense. Each MAC frame carries a Duration field that is used to update the NAV in any station other than the addressed station that successfully demodulates the frame. The Duration field holds a time value that indicates the duration for which the sending station expects the medium to be busy referenced from the end of the last symbol of the PPDU carrying the MAC frame.

All frames[2] include the Duration field and may set the NAV in neighboring stations. However, to do so the frame must be successfully demodulated by the neighboring stations. The NAV is most effectively set in neighboring stations using robustly modulated control frames, such as the RTS/CTS exchange, rather than data frames.

7.5.1.1 RTS/CTS frame exchange

To protect a station's transmissions from hidden nodes, a station may begin a sequence with an RTS/CTS exchange as illustrated in Figure 7.10. The RTS (request to send) is sent by the initiator (STA 1) and the station addressed by the RTS (STA 2) responds with a CTS (clear to send). The RTS frame occupies less air time than the data frame and is thus less susceptible to collision than the longer data frame transmitted alone. Also, loss of the RTS to collision would be quickly detected. The RTS and CTS are robustly modulating so that they are broadly received.

---

[2] Except the PS-Poll frame, which uses this field for other purposes.

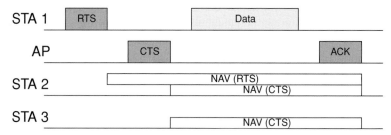

Figure 7.10 RTS/CTS exchange for hidden node protection.

The Duration field of the RTS frame carries a NAV setting to cover the CTS response plus the time needed for the subsequent frame exchange. The CTS response has its Duration field set to the Duration field value seen in the RTS less SIFS and the duration of the CTS response itself. In the diagram, the hidden node (STA 3) would receive the CTS frame and set its NAV to defer for the subsequent frame exchange. STA 2 sees both the RTS and CTS.

The RTS/CTS exchange is required when the length of a data or management frame exceeds the threshold set by the dot11RTSThreshold attribute. The dot11RTSThreshold is a local management attribute and may be set to 0 so that all MPDUs are delivered with an RTS/CTS exchange, to the maximum allowed MPDU length so that the RTS/CTS need not be used at all, or any value in between.

7.5.2   EIFS

Another mechanism used to protect against hidden nodes is the extended inter-frame space (EIFS). A station uses EIFS instead of DIFS to defer if a frame is detected but not correctly received, i.e. the MAC determines that the frame check sequence (FCS) is invalid. EIFS is defined as:

$$EIFS = aSIFSTime + ACKTxTime + DIFS \qquad (7.3)$$

where ACKTxTime is the time required to transmit an ACK frame at the lowest mandatory PHY data rate. EIFS is intended to prevent a station from transmitting over the ACK of a hidden node when a station is unable to demodulate the data frame and thus correctly set its NAV. If during the EIFS defer a valid frame is received (for example, the ACK) then a DIFS defer is used following the actual frame instead of continuing with EIFS. EIFS usage is illustrated in Figure 7.11.

7.6   Enhanced distributed channel access

Enhanced distributed channel access (EDCA) is an extension of the basic DCF introduced in the 802.11e amendment to support prioritized quality of service (QoS). The EDCA mechanism defines four access categories (ACs). Each AC is characterized by

Table 7.1 AC relative priorities and mapping from 802.1D user priorities

Priority	802.1D User priority	802.1D Designation	AC	Designation
Lowest	1	BK	AC_BK	Background
	2	–		
	0	BE	AC_BE	Best effort
	3	EE		
	4	CL	AC_VI	Video
	5	VI		
	6	VO	AC_VO	Voice
Highest	7	NC		

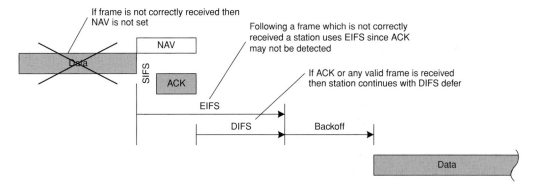

Figure 7.11 EIFS usage.

specific values for a set of access parameters that statistically prioritize channel access for one AC over another. The relative access priorities of the four ACs and the mapping of 802.1D (MAC bridging) user priorities to ACs are give in Table 7.1. An MSDU with a particular user priority is said to belong to a traffic category (TC) with that user priority.

Under EDCA, egress traffic (traffic leaving the system) is sorted logically into four queues, one for each AC (Figure 7.12). An instance of the EDCA access function operates for each queue contending for access with that AC's access parameters when the queue is non-empty. The EDCA access functions, like DCF, compete for access to the medium by deferring for a fixed period, the arbitration inter-frame space (AIFS), when the medium goes idle and then for a random backoff period. The parameters for EDCA access are similar to the parameters that are used for the DCF, but defined per AC. The AIFS value for each AC is referenced as AIFS[AC]. The contention window from which the random backoff count is selected is referenced as CW[AC].

The contention window for a particular AC, CW[AC], starts with the value CWmin[AC]. If a frame transmission for a particular AC is not successful CW[AC] is effectively doubled as described in Section 7.3.1.5. If CW[AC] reaches CWmax[AC]

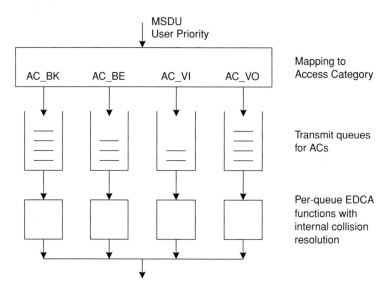

Figure 7.12 EDCA reference implementation. Reproduced with permission from IEEE (2007) © IEEE.

it remains at that value until it is reset. CW[AC] is reset to CWmin[AC] after a successful MPDU transmit in that AC.

If two (or more) instances of the EDCA access function gain access simultaneously, the internal collision is resolved by the highest priority AC gaining access and the other AC behaving as if an external collision occurred by doubling its contention window and re-arming for another access attempt.

7.6.1   Transmit opportunity

A key concept introduced in the 802.11e amendment is the transmit opportunity (TXOP). A TXOP is a bounded period during which a station may transfer data of a particular traffic class. Under EDCA, a TXOP is obtained by the station through the channel access procedure using access parameters for the particular class of traffic for which the TXOP will be used. Once the TXOP has been obtained, the station may continue to transmit data, control, and management frames and receive response frames, provided the frame sequence duration does not exceed the TXOP limit set for that AC. A TXOP limit of zero means that only one MSDU or management frame can be transmitted before competing again for access.

The TXOP concept promotes resource fairness rather than throughput fairness in that all stations accessing the network with traffic of the same class will on average receive the same amount of air time. Suppose two stations are competing for access, one with a high PHY data rate and the other with a low PHY data rate as shown in Figure 7.13. Both stations will on average receive the same amount of air time, but the station with the higher PHY data rate will see higher throughput than the station with the lower PHY

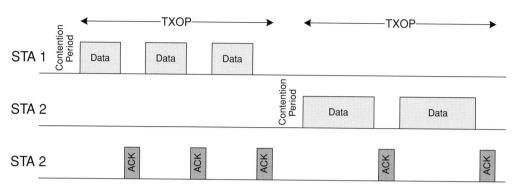

Figure 7.13 TXOP usage with different PHY data rates.

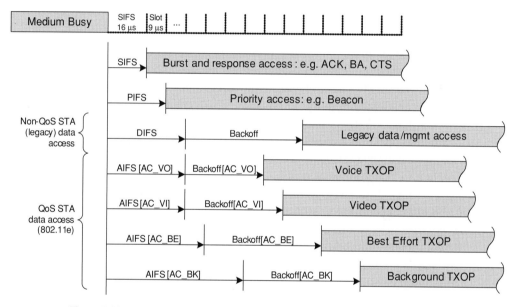

Figure 7.14 Channel access priorities including EDCA with associated timing.

data rate. Contrast this with the situation in Figure 7.8 where both stations see the same effective throughput but with unequal resource utilization.

7.6.2 Channel access timing with EDCA

Extending Figure 7.3 to include EDCA access timing we arrive at Figure 7.14. This figure shows the access priorities for the four ACs in relation to each other and to the DCF.

The arbitration inter-frame space (AIFS) for a particular AC is defined by the equation

$$AIFS[AC] = aSIFSTime + AIFSN[AC] \times aSlotTime \qquad (7.4)$$

where AIFSN[AC] is the slot count.

Table 7.2 Default EDCA access parameters for 802.11a, 802.11g, and 802.11n PHYs

AC	CWmin	CWmax	AIFSN	TXOP limit
AC_BK	31	1023	7	0
AC_BE	31	1023	3	0
AC_VI	15	31	2	3.008 ms
AC_VO	7	15	2	1.504 ms
legacy	15	1023	2	0

7.6.3   EDCA access parameters

The EDCA access parameters are provided in the EDCA Parameter Set information element that is present in Beacon and Probe Response frames. Stations on the BSS use the last seen version of the parameters and the AP may adjust the parameters over time, for example based on network load or the number of associated stations.

The default EDCA access parameters for the 802.11a, 802,11b, and 802.11n PHYs are given in Table 7.2. The default EDCA parameters are used if the AP does not broadcast a different set of parameters. While not an access category, the table also shows for comparison the equivalent parameters for the DCF (labeled legacy).

The EDCA access parameters determine the degree by which one AC is prioritized over another. The AIFSN parameter provides weak differentiation. ACs with lower values gain access more frequently than ACs with higher values, other parameters being equal. The CWmin parameter provides much stronger differentiation. The random backoff count is selected from the range [0, CW], where CW is typically equal to CWmin. Increasing the range from which the backoff count is selected has a bigger effect on the overall defer period than differences in AIFSN. The TXOP limit is also a strong differentiator. An AC with a large TXOP limit will receive more air time than an AC with a small TXOP limit assuming equal allocation of TXOPs.

7.6.4   EIFS revisited

Under DCF, a station must defer for EIFS instead of DIFS following a frame that is detected but not successfully demodulated. The intent is to protect a possible ACK response to the unsuccessfully demodulated frame from a more distant node that may not be detected by the station. To effect equivalent behavior under EDCA, a station must defer for EIFS − DIFS + AIFS[AC] following a frame that is detected but not successfully demodulated.

The EIFS and EIFS − DIFS + AIFS[AC] defer are a convoluted way of saying that following an unsuccessfully demodulated frame a station must defer for SIFS plus the duration of an ACK frame before performing the usual DIFS or AIFS defer.

7.6.5   Collision detect

When a station obtains a TXOP it may transmit for the duration of the TXOP, either as a single transmission or as a burst of back-to-back data frames. There is, however, a chance that two stations will obtain channel access simultaneously and the resulting colliding transmissions are likely to be unintelligible to the receiving peers.

To minimize the loss of air time due to these collisions, stations must perform a short frame exchange at the beginning of the TXOP to detect a collision. The short frame exchange may be either an RTS/CTS exchange or a short single Data/ACK exchange.

Collision detect is also necessary for the correct operation of the backoff algorithm, which must have its contention window size doubled should a collision occur. Doubling the contention window reduces the chance that a collision will occur the next time the two stations attempt channel access.

The short frame exchange performed at the beginning of the TXOP also allows the two stations involved in the frame exchange to set the NAV of their neighboring stations.

7.6.6   QoS Data frame

To support the QoS features and block acknowledgement (discussed below), the 802.11e amendment introduced a new data frame, the QoS Data frame. The QoS Data frame has the same fields as the regular Data frame, but includes an additional QoS Control field (see Section 11.1.5). The QoS Control field carries various subfields for managing QoS and other features introduced in the amendment. The TID or traffic identifier identifies the TC to which the frame belongs. Under EDCA, the TID field carries the user priority, which is mapped to the AC through Table 7.1.

The Ack Policy subfield determines how the data frame is acknowledged by the receiving peer and carries one of the following values:

- **Normal Ack** – if correctly received, the recipient responds to the QoS Data frame with an ACK response.
- **No Ack** – the recipient does not respond to the QoS Data frame. This may be useful for traffic that has a low tolerance for jitter or delay and does not benefit from retransmission.
- **No Explicit Ack** – there may be a response frame, but it is not an ACK. This policy is used when polling under centrally coordinated channel access (see Section 9.2).
- **Block Ack** – the recipient takes no action on the received frame except to record its reception. This policy is used under the block acknowledgement protocol.

7.7   Block acknowledgement

The block acknowledgement protocol, introduced with the 802.11e amendment, improves efficiency by allowing for the transfer of a block of data frames that are acknowledged with a single Block Acknowledgement (BA) frame instead of an ACK frame for each of the individual data frames. Unlike the normal acknowledgement mechanism, however, the block acknowledgement mechanism is session oriented and a station must establish a block acknowledgement session with its peer station for each traffic identifier (TID) for which block data transfer is to take place. A particular block acknowledgment session is thus identified by the <transmit address, receive address, TID> tupple.

The 802.11e amendment introduced two flavors of the block acknowledgement protocol: immediate block ack and delayed block ack. The two flavors differ in the manner in which the block acknowledgement control frames are exchanged. Under immediate block ack the Block Ack Request (BAR) frame solicits an immediate Block Ack (BA) frame response, i.e. the BA is returned within SIFS of receiving the BAR and thus within the same TXOP. With delayed block ack the BAR is sent in one TXOP and the BA response is returned in a separate, subsequent TXOP. Immediate block ack provides lower latency and improved performance over delayed block ack, which was primarily defined for ease of implementation.

The block acknowledgement protocol is described in more detail in Section 8.3, but a brief overview is provided here to illustrate the basic concept. Block acknowledgement is enabled in one direction for a particular TID with the exchange of an ADDBA Request and ADDBA Response. The station with data to send – the originator – sends an ADDBA Request to the station that will receive the data – the recipient. The recipient acknowledges the correct receipt of the ADDBA Request with an ACK and then responds some time later with an ADDBA Response to which the originator responds with an ACK. The ADDBA exchange allows the originator and responder to exchange parameters such as the responder reorder buffer size. To tear down a block ack session, the originator or the recipient sends a DELBA request which, if correctly received, is acknowledged with an ACK.

Block data transfer occurs as follows. The originator transmits one or more QoS Data frames addressed to the recipient and containing the TID of the block acknowledgement session. The Ack Policy field is set to Block Ack. The block of data frames need not be transmitted in sequence and may include retransmitted frames. The recipient is responsible for reordering the frames and delivering them in sequence to the higher layer and performs this function using a reorder buffer. The recipient will hold frames in the reorder buffer until gaps in the sequence number space are filled. The originator limits the sequence number range for which acknowledgements are outstanding so as not to overrun the recipient reorder buffer.

After sending a block of data frames, the originator sends a Block Ack Request (BAR) frame. The BAR frame performs two functions; it flushes the recipient's reorder buffer and it solicits a Block Ack (BA) frame. The recipient's reorder buffer may need to be flushed to advance passed holes in the sequence number space resulting from MSDUs that did not make it through after exhausting their retransmission count or lifetime limit. Flushing the reorder buffer releases MSDUs that may be held up behind these holes.

The BAR frame includes a Starting Sequence Control field that contains the sequence number of the oldest MSDU in the block for which an acknowledgement is expected.

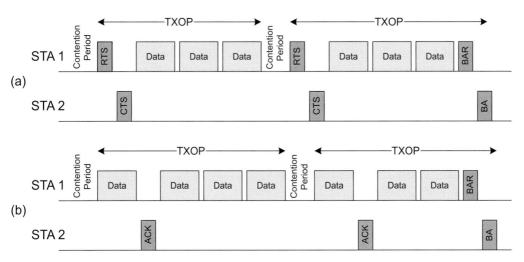

Figure 7.15 Block data frame exchange sequence.

MSDUs in the recipient's buffer with sequence numbers that precede the starting sequence number are either forwarded to the LLC layer (if complete) or discarded (if one or more fragments are missing). The solicited BA frame contains a bitmap that represents the acknowledgement state of the data frames received beginning with the starting sequence number from the BAR frame.

On receiving the BA frame the originator discards acknowledged data frames and re-queues data frames not acknowledged for retransmission. The originator may also discard data frames that have reached their retransmission count or lifetime limit.

With a block ack session in place, the originator may still solicit a regular ACK for QoS Data frames by setting the Ack Policy field to Normal Ack.

7.7.1 Block data frame exchange

The block data frame exchange using the immediate block ack protocol is illustrated in Figure 7.15(a) with STA 1 transferring data to STA 2. After a contention period, STA 1 gains a TXOP. As a collision detect mechanism and to set the NAV in neighboring stations, STA 1 performs a short frame exchange, in this case RTS/CTS. STA 1 then sends a back-to-back burst of data frames with SIFS separating the individual transmissions until the TXOP limit is reached. Since it has more data to send, STA 1 again accesses the wireless medium and gains a TXOP. An RTS/CTS exchange is again performed followed by the remaining frames in the block sent as a back-to-back burst. STA 1 then sends a BAR frame, which solicits a BA response from STA 2. The BA response indicates which of the data frames in the block were correctly received.

As an alternative to performing the RTS/CTS exchange, STA 1 may use a Data/ACK exchange with STA 2 to effect collision detect, as illustrated in Figure 7.15(b). Collision detect is necessary through one of these mechanisms as it reduces the degradation in network throughput that would otherwise occur if two stations collided for the full duration of the TXOP. The Data/ACK exchange provides more limited protection near the transmitter (due to the higher order modulation used for data frames), but is more efficient than the RTS/CTS exchange with which no information transfer takes place.

It should be noted that the block transfer is independent of the TXOP. The block transfer may occur over multiple TXOPs or it may be contained within a single TXOP.

References

IEEE (2004). *IEEE Std 802.1X™-2004, IEEE Standard for Local and Metropolitan Area Networks: Port-Based Network Access Control.*

IEEE (2007). *IEEE Std 802.11™-2007, IEEE Standard for Information Technology – Telecommunications and Information Exchange Between Systems – Local and Metropolitan Networks – Specific Requirements. Part 11: Wireless LAN Medium Access Control (MAC) and Physical Layer (PHY) Specifications.*

8 MAC throughput enhancements

Early on in the 802.11n standardization process it was recognized that even with significantly higher data rates in the PHY the fixed overhead in the MAC protocol was such that little of that gain would be experienced above the MAC. It was clear, as this chapter will show, that without throughput enhancements in the MAC the end user would benefit little from the improved PHY performance.

8.1   Reasons for change

Since the original 802.11 specification was completed, a number of amendments have introduced new PHY capabilities and with them enhanced performance. In addition, the 802.11e amendment which primarily added QoS features also enhanced MAC performance with the introduction of the TXOP concept and block acknowledgment. However, these MAC performance improvements were only slight, and with the potential for significantly higher PHY performance it was soon realized that the existing MAC protocol did not scale well with PHY data rate.

8.1.1   Throughput without MAC changes

The poor scaling of throughput above the MAC with PHY data rate is illustrated in Figure 8.1 where the theoretical throughput is given for unicast data sent from one station to another assuming a 3 ms TXOP limit, block ack protocol, and a 10% packet error rate (PER). As the PHY data rate is increased beyond the 54 Mbps peak data rate of 802.11a/g, throughput begins to level off. A 40 MHz 2 × 2 system with a 270 Mbps PHY data rate only achieves 92 Mbps above the MAC. Worse, a 40 MHz 4 × 4 system with a 540 Mbps PHY data rate achieves almost exactly the same throughput.

The drop in efficiency (Figure 8.2) with increasing data rate results from the fixed overhead in the preamble and inter-frame space. Not only does this overhead account for a greater percentage of the air time as the data payload gets shorter in duration, but the preamble itself needs to be longer to support the multiple spatial streams at the higher data rates introducing additional fixed overhead. The relative preamble overhead for a typical 1500 byte data frame is illustrated in Figure 8.3 for a select set of data rates.

As the payload gets shorter in duration and the preamble length increases in duration we see diminishing returns reducing the air time occupied by a single frame. Clearly

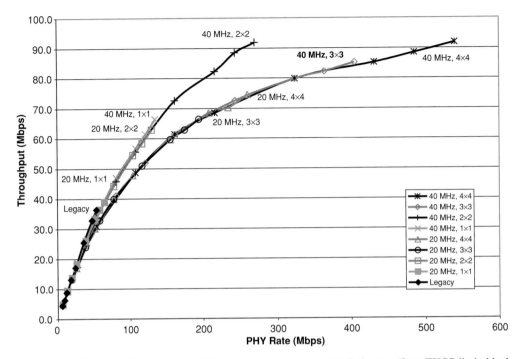

Figure 8.1 Throughput vs PHY data rate assuming no MAC changes (3 ms TXOP limit, block ack, 10% PER).

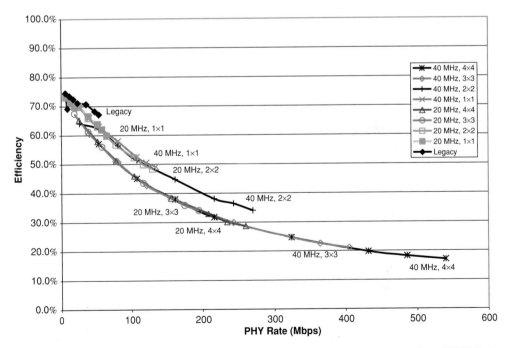

Figure 8.2 MAC efficiency vs PHY data rate assuming no MAC changes (3 ms TXOP limit, block ack, 10% PER).

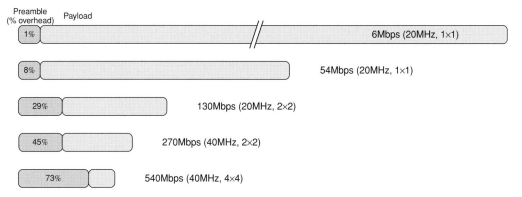

Figure 8.3 Relative preamble overhead for a 1500 byte frame at different PHY data rates.

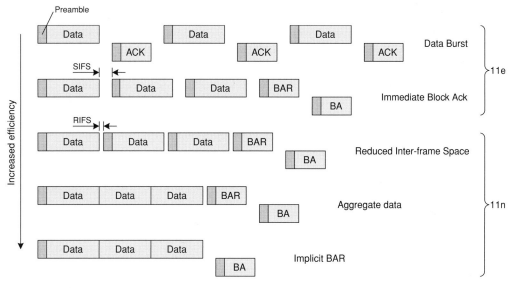

Figure 8.4 Basic throughput enhancements to the 802.11 MAC.

changes are needed to improve efficiency if applications are to see any significant gain in throughput.

8.1.2   MAC throughput enhancements

The 802.11n amendment developed a number of simple enhancements to the 802.11e MAC that significantly improved efficiency. The most important of these are summarized in Figure 8.4.

The first two sequences in Figure 8.4 show data bursting within a TXOP, features that are supported with the 802.11e amendment. The first of these sequences shows data bursting using normal ack and the second using immediate block ack. The block ack protocol allows data frames to be grouped together and is the key to further efficiency improvements introduced in 802.11n.

An easy enhancement under block ack is to reduce the inter-frame space for back-to-back transmissions during the data burst. Since the station remains in transmit mode for the duration of the burst, there is no need for the long SIFS between frames, a duration that was originally designed to accommodate receive to transmit switching. With back-to-back transmissions the inter-frame space need only be long enough for the receiver to re-arm for acquisition of the new signal.

Taking things a step further, one could eliminate the inter-frame space and preamble altogether and concatenate data frames in a single transmission. In 802.11n this is referred to as aggregation and is the key throughput enhancing feature introduced in the 802.11n MAC.

An additional enhancement that was considered was to concatenate the BAR frame with the data frames, improving efficiency slightly. This, however, reduces robustness since the BAR is transmitted at the end of the aggregate transmission and at the data rate rather than the typically more robust MCS used for control frames. Instead of this, it was recognized that one function of the BAR frame – soliciting a BA frame – could be performed with a single bit piggybacked on each of the data frames making up the aggregate. This change is both more efficient and more robust since it eliminates the single point of failure in the BAR frame being sent at the end of the aggregate transmission and at the data rate. As long as one of the data frames making up the aggregate gets through, the recipient will respond with a BA frame.

Recognizing that fragmentation has little benefit at high data rates, particularly when data aggregation is being performed, it is also possible to reduce the size of the BA frame. Compressing the BA frame so that it only acknowledges MSDUs and not MSDU fragments further enhances efficiency.

All these techniques and more have been adopted with the 802.11n amendment and will be described in detail in this and the following chapters.

8.1.3   Throughput with MAC efficiency enhancements

With these basic MAC efficiency enhancements, the 802.11n system performs as shown in Figure 8.5. Notice that throughput now scales near linearly with PHY data rate primarily as a result of aggregation, which allows for long data transmissions bounded by the TXOP limit. Implicit BAR and the use of the compressed BA frame format improve efficiency further.

Throughput of 100 Mbps at the top of the MAC, a target set in the 802.11n PAR (IEEE, 2006), is now easily reached with a PHY data rate of roughly 130 Mbps, achievable in 20 MHz bandwidth with two spatial streams or in 40 MHz with one spatial stream.

The improved efficiency is shown in Figure 8.6. MAC efficiencies of between 70% and 80% are maintained over the full PHY data rate range. In fact, efficiency is improved over 802.11e even at legacy data rates (54 Mbps and below). The legacy plot reflects 802.11a/g PHY data rates with the 802.11e protocol.

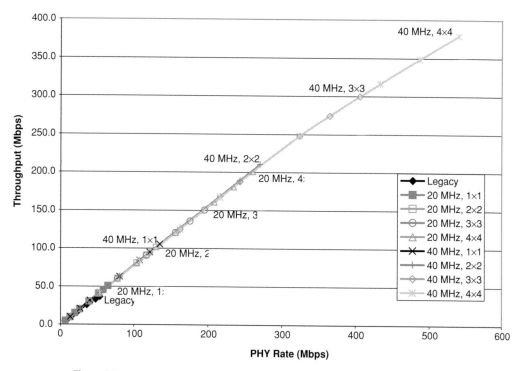

Figure 8.5 Throughput vs PHY data rate with basic MAC enhancements (3 ms TXOP limit, block ack, 10% PER).

The slight roll off in efficiency at very high PHY data rates is the result of hitting the block ack window limit of 64 MSDUs, which starts to limit the aggregate transmission length. While not reflected here, this can be largely overcome using a combination of A-MSDU and A-MPDU aggregation techniques as described later in this chapter.

8.2   Aggregation

Early on in the standardization process it was recognized that some form of aggregation would be required. Three techniques were proposed of which two were ultimately adopted in the standard. These two types of aggregation logically reside at the top and bottom of the MAC as illustrated in Figure 8.7.

The third aggregation technique, which was ultimately not adopted, logically resides at the top of the PHY. This is discussed briefly later in this section as it has some interesting characteristics.

At the top of the MAC is MSDU aggregation (or A-MSDU), which in the egress direction aggregates MSDUs as the first step in forming an MPDU. At the bottom of the MAC is MPDU aggregation (or A-MPDU), which in the egress direction aggregates multiple MPDUs to form the PSDU that is passed to the PHY to form the payload of

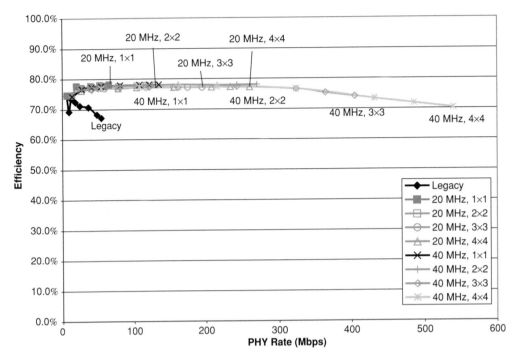

Figure 8.6 Efficiency vs PHY data rate with basic MAC enhancements (3 ms TXOP limit, block ack, 10% PER).

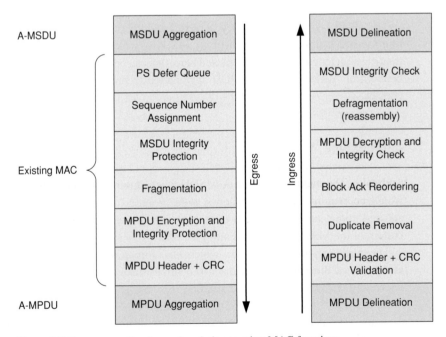

Figure 8.7 Two aggregation layers in relation to other MAC functions.

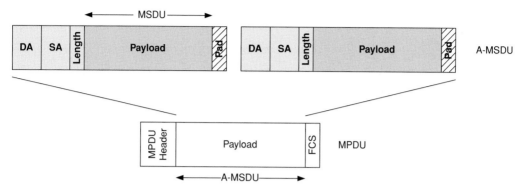

Figure 8.8 A-MSDU encapsulation.

a single transmission. The reverse functions, MPDU and MSDU delineation, reside in the same logical positions in the ingress direction.

8.2.1 Aggregate MSDU (A-MSDU)

With A-MSDU, MAC service data units (MSDUs) received from the LLC and destined for the same receiver and of the same service category (same traffic identifier or TID) may be accumulated and encapsulated in a single MAC protocol data unit (MPDU). The encapsulation is shown in Figure 8.8.

The MSDU as received from the LLC is prefixed with a 14 byte subframe header consisting of the destination address (DA), source address (SA), and a length field giving the length of the SDU in bytes. The header together with the SDU is padded with 0 to 3 bytes to round the subframe to a 32-bit word boundary. Multiple such subframes may be concatenated together to form the payload of the QoS Data frame, provided the total length of the data frame does not exceed the maximum MPDU size.

Support for A-MSDU is mandatory at the receiver under Normal Ack policy. Support is negotiated for use under Block Ack policy during the block ack establishment handshake. The maximum length A-MSDU that a station can receive is declared in its HT Capabilities information element as either 3839 bytes or 7935 bytes. These limits are derived from implementation considerations. Some early draft 2.0 systems were only capable of allocating a single 4 kB buffer per A-MSDU and, allowing for some implementation-specific use of the buffer (257 bytes), one arrives at the lower limit figure. The upper limit is arrived at by assuming an 8 kB receive buffer size and again allowing for some implementation specific use.

The channel access rules for a QoS Data MPDU carrying an A-MSDU are the same as a data MPDU carrying an MSDU of the same TID. The maximum lifetime of an A-MSDU is the lifetime of the longest lived constituent MSDU.

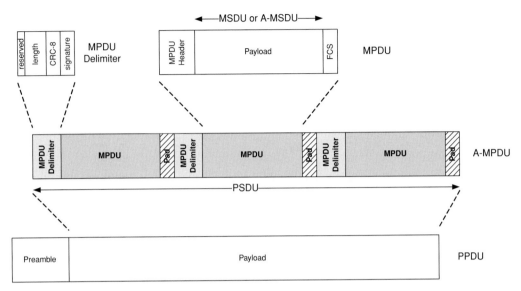

Figure 8.9 A-MPDU encapsulation.

8.2.2   Aggregate MPDU (A-MPDU)

With A-MPDU, fully formed MAC PDUs are logically aggregated at the bottom of the
MAC. A short MPDU delimiter is prepended to each MPDU and the aggregate presented
to the PHY as the PSDU for transmission in a single PPDU. A-MPDU encapsulation is
illustrated in Figure 8.9.

The MPDU delimiter is 32 bits in length and consists of a 4-bit reserved field, a
12-bit MPDU length field, an 8-bit CRC field, and an 8-bit signature field. The 8-bit
CRC covers the 4-bit reserved and 12-bit length fields and validates the integrity of the
header. The signature byte is set to the ASCII character "N" and was intended to aid
software implemented delineation. The MPDU is padded with 0–3 bytes to round it up
to a 32-bit word boundary.

A receiver implementation parses the A-MPDU framing structure by using the length
in each delimiter to extract the following MPDU. If a delimiter is corrupt (has an out of
range length value, or an invalid signature or CRC-8) then the receiver can scan forward
looking for the next valid delimiter on a 32-bit boundary. With high probability the
receiver will re-sync on an actual delimiter (as opposed to a random 32-bit word that
appears to be a delimiter) and be able to extract the MPDU and subsequent MPDUs.
In this sense the A-MPDU framing structure is robust since it is possible to recover
MPDUs following a corrupt delimiter.

All the MPDUs in an A-MPDU are addressed to the same receiver and all are of
the same service category (same TID). The Duration/ID field in the MAC header of all
MPDUs in an A-MPDU is set to the same value.

A-MSDU aggregation may be used together with A-MPDU aggregation if negotiated
through the block acknowledgement handshake. This typically offers little advantage

over straight A-MPDU aggregation; however, there are two circumstances under which it may be useful.

Under the anomalous situation in which a high volume of short packets are being transported, using A-MPDU alone may result in the block ack limit of 64 outstanding unfragmented MPDUs being reached in a single aggregate that does not fully utilize the TXOP. A similar situation may arise even with large packets at very high PHY rates. In both situations, using A-MSDU to aggregate two or more MSDUs into an MPDU and then aggregating the resulting larger MPDUs increases the amount of data in the aggregate transmission, thus improving efficiency.

8.2.2.1 A-MPDU contents

All MPDUs within an A-MPDU are addressed to the same receiver address. This restriction simplifies the frame sequences that would need to be supported but it also means that a power saving station can read the MAC header of the first MPDU in the aggregate and immediately sleep for the remainder of the PPDU if it is not addressed to it.

The Duration/ID fields in the MAC headers of all MPDUs in an A-MPDU carry the same value. There are also restrictions on the types of MPDUs that can be carried in an A-MPDU.

Under HT-immediate block ack, an A-MPDU may carry:

- a single BA as the first MPDU in the aggregate
- QoS Data MPDUs belonging to the same TID and subject to the constraints of the block ack protocol
- any management MPDU of subtype Action No Ack.

Under HT-delayed block ack, an A-MPDU may carry:

- One or more BA MPDUs with the BA Ack Policy field set to No Ack
- QoS Data MPDUs belonging to the same TID and subject to the constraints of the block ack protocol
- any management MPDU of subtype Action No Ack
- BAR MPDUs with the BA Ack Policy field set to No Ack.

Additional MPDU types are allowed with PSMP and these are detailed in a later chapter. The fragmentation of QoS Data MPDUs is not permitted in an A-MPDU.

8.2.2.2 A-MPDU length and MPDU spacing constraints

A station advertises the maximum A-MPDU length that it can receive in its HT Capabilities element. The advertised maximum length may be one of the following: 8191, 16 383, 32 767, or 65 535 octets. The sending station must not send an A-MPDU of greater length.

Some implementations have MPDU processing limitations and could be overwhelmed by sequences of short data frames. To prevent data loss through buffer overrun, a station may advertise a minimum MPDU start spacing such that MPDUs in an aggregate do not arrive at a rate faster than the implementation can process them. In its HT Capabilities

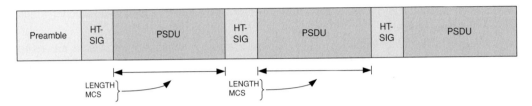

Figure 8.10 Early proposal for A-PSDU encapsulation

element, such a station would indicate the minimum start spacing between MPDUs (see Section 11.3.2.2).

A transmitting station ensures that when forming an aggregate it does not violate that spacing constraint. If MPDUs packed into an aggregate will violate the spacing constraint then the transmitter may send the MPDUs in separate transmissions, use A-MSDU to create larger MPDUs, or insert null MPDU delimiters to increase the spacing. A null MPDU delimiter contains a length field indicating a zero length MPDU.

8.2.3　Aggregate PSDU (A-PSDU)

During the proposal phase of the 802.11n development an aggregation scheme was proposed (Hansen and Edwards, 2004) that logically resides at the top of the PHY. This technique is illustrated conceptually in Figure 8.10. Essentially the proposal called for a single training sequence at the front of the PPDU followed by a framing structure that consists of the PHY signaling field (HT-SIG) delimiting one or more PSDUs. The LENGTH and MCS fields in the PHY signaling field give the duration of the PSDU that followed. A bit in the HT-SIG would indicate the last HT-SIG in the aggregate.

The A-PSDU concept had the appealing characteristic that the data rate could be changed for each PSDU making up the aggregate. This would support multi-rate aggregation allowing the aggregation scheme to be used to efficiently send data to multiple receivers in the same burst.

Ultimately this scheme was rejected, as the far more common use case where data is aggregated to a single station is more efficiently handled using A-MPDU aggregation. For this usage scenario the A-PSDU technique had higher overhead (each HT-SIG field occupied at least one 4 μs symbol) and was less robust since a single error in one of the HT-SIG fields would prevent the remainder of the aggregate from being demodulated.

8.3　Block acknowledgement

The block acknowledgement mechanism was introduced in the 802.11e amendment to improve efficiency by allowing for the transfer of a block of data frames that are acknowledged with a single Block Acknowledgement (BA) frame instead of an ACK frame for each of the individual data frames. The station with data to send is referred to as the originator and the receiver of the data as the recipient.

Two flavors of block ack were originally defined in the 802.11e amendment: immediate block ack and delayed block ack. Both flavors are enhanced in the 802.11n amendment to improve efficiency and take advantage of aggregation and the higher data rates. The enhanced mechanisms are referred to as HT-immediate block ack and HT-delayed block ack. All four flavors may be supported by HT stations, although the original mechanisms would only be used for interoperability with legacy stations.

In this section, the original 802.11e block ack mechanisms are described and then the HT variants. It should be noted that the original mechanisms were modified slightly to support HT stations, for example signaling A-MSDU support in the ADDBA Request and Response. However, these modifications were minor and the major changes were made in the HT variants.

8.3.1   Immediate and delayed block ack

While the normal acknowledgement mechanism is always in operation and in fact forms part of the underlying DCF, the block ack mechanism needs to be enabled by establishing a block ack session through the exchange of an ADDBA Request and Response. The block ack session is established between two stations for a particular traffic identifier (TID) and for data transfer in one direction, originator to responder.

Following a successful ADDBA exchange the data transfer phase is entered. The originator will send a block of data followed by a BAR, to which the responder will respond with a BA. The BA acknowledges correctly received data frames from the previous block. The originator requeues data frames which were not correctly received and may send them in the subsequent block.

The originator or recipient may tear down the block ack session by sending a DELBA Request which, if correctly received, is acknowledged with an ACK.

Immediate block ack and delayed block ack differ in the handling of the BAR and BA frames during the data transfer phase. With immediate block ack the BAR solicits an immediate BA response, while with delayed block ack correct reception of the BAR frame itself is acknowledged with an ACK and the BA is returned in a separate channel access and acknowledged with another ACK. Immediate and delayed block ack sessions are illustrated in Figure 8.11 and described in more detail below.

8.3.2   Block ack session initiation

Stations indicate their ability to support block ack by setting the Immediate Block Ack and/or Delayed Block Ack capability bits in the Capability Information field in their Beacon, Association/Reassociation Request, and Response frames. If a station advertises that it supports one or both flavors of block ack then a peer station may establish a compatible block ack session for a particular traffic class with that station.

The block ack session is initiated by the originator sending an ADDBA Request frame. In response to a correctly received ADDBA Request frame, the responder sends an ACK. After further processing the responder sends an ADDBA Response frame to which the originator responds with an ACK if correctly received. The originator and

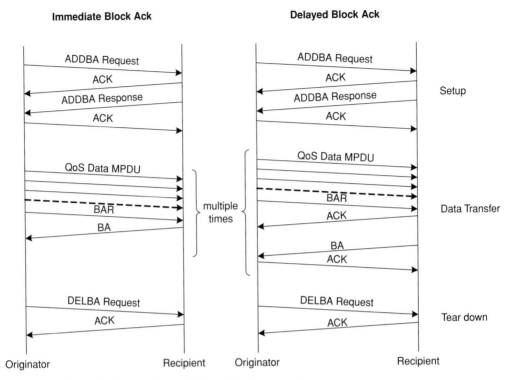

Figure 8.11 Immediate and delayed block ack sessions.

responder will retransmit the ADDBA Request/Response if the expected ACK is not received. An inactivity timeout at the originator will detect a failed session setup.

The ADDBA Request and Response frames include the following fields:

- **Block Ack Policy**. This field indicates whether the session is an immediate block ack session or delayed block ack session. If one of the stations is a non-HT station, then the value set by the originator in the ADDBA Request is advisory and the value returned by the responder indicates the type of session should the originator continue with the session. If both stations are HT stations then the block ack policy established by the originator must either be accepted or rejected by the recipient.
- **TID**. This field gives the identifier for the traffic class or traffic stream for the session.
- **Buffer Size**. This field indicates the number of frame buffers the responder has available for reordering frames. The value, if set by the originator, is the desired value; the value set by the responder is binding. The originator must not have more than this number of MPDUs outstanding before soliciting a block ack.
- **A-MSDU Supported**. This field is set in the ADDBA Request to indicate that the originator may send A-MSDUs under this session and is set in the ADDBA Response if the recipient is capable of receiving A-MSDUs in this session. If the field is not set in the response then the originator will not send A-MSDUs.
- **Block Ack Timeout Value**. This field gives the duration after which the block ack session is terminated when there are no frame exchanges in this session.

- **Start Sequence Number (SSN).** This is the sequence number of the first data frame from the originator.

The responder may reject a block ack session from an originator by sending a DELBA frame to the initiator after acknowledging receipt of the ADDBA Request.

8.3.3   Block ack session data transfer

During the data transfer phase, the originator may transmit a block of QoS Data frames, either as a burst, separated by SIFS or RIFS, or as part of an A-MPDU. Each QoS Data frame in the block has its Ack Policy field set to Block Acknowledgement. The recipient maintains a scoreboard to track which MPDUs have been received correctly. The data block may be wholly contained within a single TXOP or it may straddle multiple TXOPs. The data block and TXOP are not coupled in any way.

After transferring the data block, the originator sends a BAR frame. This frame includes a starting sequence number (SSN), which is the sequence number of the oldest MSDU in the block for which an acknowledgement is needed. On receiving the BAR, the recipient performs two functions. First, it prepares a BA response using the scoreboard for that session. The scoreboard is converted into a bitmap where the first bit represents the MPDU with the same sequence number as the SSN from the BAR frame and subsequent bits indicate successive sequence numbers. The bitmap thus forms an array indexed by sequence number with the SSN as starting reference.

Second, it examines its reorder buffer for MPDUs with sequence numbers that precede the SSN value. These MPDUs are either reassembled into complete MSDUs and forwarded to the higher layers or discarded if complete MSDUs cannot be created.

The primary difference between immediate and delayed block ack is in the timing with which the recipient responds to the BAR. Under immediate block ack, the recipient responds to the BAR with a BA frame after SIFS. Under delayed block ack, the recipient responds to the BAR with an ACK. Later, in a separate channel access, the recipient generates a BA frame and sends it to the originator. The originator responds to the delayed BA with an ACK.

Immediate block ack provides better performance while delayed block ack was defined for ease of implementation. With delayed block ack, the recipient has more time to process the BAR and is suited to implementations where the bulk of the BA processing is performed in software on the host system.

On receiving the BA, the originator releases MPDUs that are acknowledged and requeues MPDUs that were not acknowledged for retransmission provided their time to live has not been exceeded.

8.3.4   Block ack session tear down

When the originator has no additional data to send and the final block ack exchange has completed, it may disable the block ack session by sending a DELBA frame to the recipient. The recipient sends an ACK in response and releases any resources allocated for the block ack session.

The block ack session may also be torn down by the originator or recipient if either does not receive a BA, BAR, or QoS Data frame belonging to that session within the duration of the block ack timeout value.

8.3.5   Normal ack policy in a non-aggregate

Small efficiency gains are possible using normal ack in a block ack session. Many traffic patterns are bursty and frequently have short periods where only a single frame needs to be sent. If the last block transfer has completed and all frames have been acknowledged to that point then it is more efficient to send the data frame using the normal ack procedure than to perform a BAR/BA exchange. In this case the QoS Data frame has its Ack Policy field set to Normal Ack and is sent in a non-aggregate PHY transmission. If correctly received the responder responds with an ACK. The frame is marked as correctly received in the block ack session scoreboard.

8.3.6   Reorder buffer operation

When the recipient receives a QoS Data frame for which a block ack session is in place, the recipient will buffer the MPDU. If the arriving MPDU completes the MSDU at the head of the reorder buffer, then the recipient forwards the complete MSDU and subsequent complete MSDUs in the reorder buffer in sequence to the higher layers until an incomplete MSDU forming a hole in the sequence space is encountered. If, when the MPDU arrives, there are incomplete preceding MSDUs in the reorder buffer then the MSDU is held until those preceding MSDUs are complete.

If an MPDU arrives and the reorder buffer is full then the first MSDU in the reorder buffer is discarded (since it is incomplete) to make room. This may also result in the release of complete subsequent MSDUs to the higher layers.

If a BAR frame is received, all complete MSDUs with a lower sequence number than the starting sequence number of the BAR are forwarded to the higher layers and all incomplete MSDUs with a lower sequence number are discarded. The BAR frame thus has a dual role. In addition to soliciting a block ack response, it provides the originator with a mechanism for flushing the recipient's reorder buffer of incomplete MSDUs or holes representing MSDUs whose retransmit lifetime has expired. If the originator discards one or more MPDUs due to lifetime expiry it must send a BAR to flush the recipient reorder buffer so that subsequent MSDUs are not needlessly held up waiting for the sequence to be completed.

Figure 8.12 gives an example of reorder buffer behavior. A block of QoS Data frames composed of fragmented MSDUs is sent. In the diagram, the QoS Data MPDUs are numbered with the number before the decimal point being the MSDU sequence number and the number after the decimal point being the MSDU fragment number. MSDU 1 is received completely, reassembled and forwarded to the higher layers. The second fragment of MSDU 2 is lost and thus the received fragment is stored until MSDU 2 can be completed. This holds up subsequent MSDUs even if they are complete.

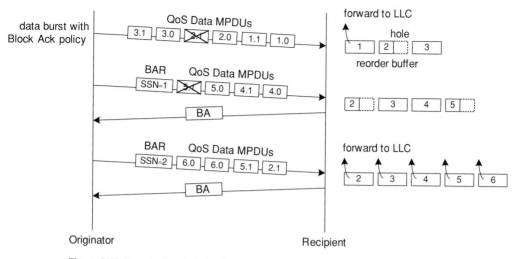

Figure 8.12 Reorder buffer behavior with fragmented MSDUs.

After a BAR/BA exchange, the originator learns of the lost fragments and retransmits them together with additional MSDUs that are available and that will fit in the reorder buffer. All MSDUs are now complete and forwarded in sequence to the higher layers. The originator learns that all MSDUs have been successfully transferred with another BAR/BA exchange.

8.4 HT-immediate block ack

HT-immediate block ack is a significant modification to the immediate block ack protocol and stands as a separate protocol for the purposes of backward compatibility with legacy devices. An HT station that wishes to establish a block ack session with a non-HT station must use the original immediate or delayed block ack protocol. An HT station that establishes a block ack session with another HT station uses the HT-immediate or HT-delayed block ack protocol. There is a lot of commonality between the HT variants and the original protocol, which eases implementation.

All HT stations are required to support HT-immediate block ack as a recipient.

8.4.1 Normal Ack policy in an aggregate

The block ack protocol was introduced in the 802.11e amendment prior to the introduction of aggregation in the 802.11n amendment. Since an aggregate is a single PHY transmission containing multiple MPDUs it was thought that a mechanism analogous to the original Data/ACK mechanism might be possible. The mechanism introduced in the 802.11n amendment to support this changes the meaning of the Normal Ack policy when present in QoS Data frames in an aggregate transmission. If one or more of the QoS Data MPDUs in an aggregate have their Ack Policy field set to Normal Ack then the

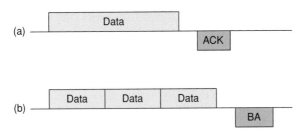

Figure 8.13 Normal Ack policy in (a) non-aggregate and (b) aggregate.

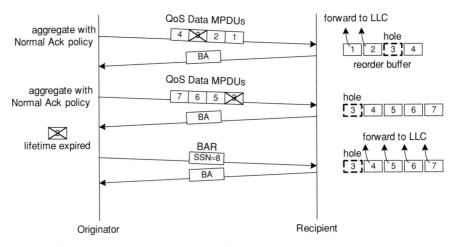

Figure 8.14 Using BAR to flush the reorder buffer.

responder will return a BA in response to the aggregate. The two response mechanisms to Normal Ack policy are illustrated in Figure 8.13.

The use of Normal Ack policy to solicit a BA does not eliminate the need for the BAR. Recall that the BAR frame performs two functions: it solicits a BA response and it flushes the MSDUs in the reorder buffer that are held up due to an earlier incomplete MSDU. If the originator does not receive an acknowledgement for an MSDU whose lifetime has expired then the originator must send a BAR to flush the recipient's reorder buffer of complete MSDUs subsequent to the MSDU that will never make it across. This is illustrated in the example given in Figure 8.14.

In this example, MSDU 3 is not successfully received by the recipient after a number of retries. On lifetime expiry the originator discards MSDU 3. To eliminate the hole in the recipient's reorder buffer, the originator must send a BAR with a SSN that is greater than the sequence number of the discarded MSDU. In practice, the SSN is set to the sequence number of the next MSDU to be transmitted (although that MSDU may not be available for transmission yet) which in this case is 8 since all MSDUs with lower sequence numbers have been acknowledged and discarded by the originator.

8.4.2   Compressed block ack

At the higher HT rates, fragmentation does not provide much benefit. The original BA frame was defined with a 1024-bit scoreboard (128 octets) to support 64 MSDUs, each of which can be fragmented with up to 16 fragments. 802.11n introduces a compressed BA variant (see Section 11.2.1.6) that does away with the 16 bits per MSDU for fragmentation, resulting in a 64-bit scoreboard (8 octets). This reduces both the on-air overhead and the memory requirements in the recipient.

8.4.3   Full state and partial state block ack

The block ack mechanism defined in the 802.11e amendment is referred to as full state block ack to distinguish it from partial state block ack, introduced in the 802.11n amendment. Partial state block ack is backward compatible with full state block ack in the sense that an originator using partial state rules will operate correctly with a recipient implementing full state operation. The distinction between full state and partial state operation is discussed in more detail in the next section.

8.4.3.1   Full state block ack operation

Under full state block ack, the recipient maintains an ack state scoreboard for each block ack session. The scoreboard records the ack state of up to 64 MSDUs. When fragmentation is used each MSDU may be fragmented in up to 16 fragments, thus the scoreboard is an up to 64 entry by 16-bit array. A recipient implementation with limited memory may constrain the extent of the array by setting the BufferSize parameter in the ADDBA Response.

The MSDU sequence number is a 12-bit value, thus the scoreboard represents a window in the sequence number space of 4096 values. The scoreboard widow is defined by a beginning sequence number WinStart, an ending sequence number WinEnd, and an extent WinSize. With the establishment of the block ack session the scoreboard is initialized with WinStart set to the starting sequence number provided in the ADDBA request.

When a QoS Data frame arrives, if the sequence number falls within the space represented by the scoreboard then the recipient will index the scoreboard using the data frame's sequence number (SN) and record its correct receipt. If the SN is outside the space represented by the scoreboard but within the range WinEnd to WinStart $+2^{11}$ (half the sequence number space) then the recipient will shift the scoreboard to the right until it includes the new sequence number on the rightmost edge of its window.

When a BAR arrives, the scoreboard window is shifted to the right so that WinStart is equal to the SSN provided in the BAR frame and a BA response is returned with the contents of the scoreboard.

8.4.3.2   Motivation for partial state block ack

With the original block ack mechanisms, it is required that the scoreboard state persist for the duration of the block ack session. This burdens the recipient implementation

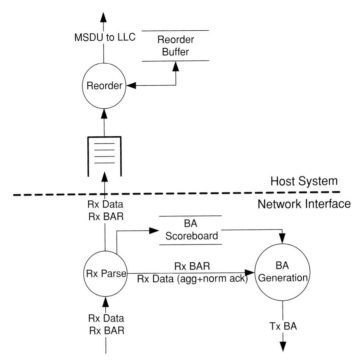

Figure 8.15 Typical functional split for immediate block ack implementation in recipient station.

with the need to maintain state for all active block ack sessions and, in practice, with the low latency required to produce a BA in response to a BAR, this means using expensive on-chip memory. With the 802.11n amendment the rules are relaxed so that on-chip memory reserved for block ack state may be reused by different block ack sessions. The state memory effectively serves as a cache, storing the state of the most recently active block ack session. The newer rules are referred to as partial state block ack and are fully backward compatible with the original full state block ack rules.

To understand the motivation for the change consider a typical implementation illustrated in Figure 8.15. Under immediate BA, when the recipient receives a BAR or an aggregate frame containing QoS Data frames with Normal Ack policy, the recipient must transmit a BA Response frame SIFS after receiving the BAR or aggregate QoS Data frame. With decode latencies on the receive path and encode latencies on the transmit path, there is little time available to locate the appropriate state information and form the BA Response. This largely necessitates on-chip storage of the block ack scoreboard that will be returned in the BA Response.

The other major function in the block ack mechanism is the reassembly and reorder function, which reassembles complete MSDUs and forwards them in order to the higher layers. This function is not time critical and requires a large buffer for storing packets during the reorder and reassembly process and as such is typically implemented in the system hosting the network interface.

The new partial state rules do not affect the reorder and reassembly process, but reduce the resource requirements in the network interface for storing the block ack scoreboard. A reorder buffer is still required for each block ack session, but since this is typically stored in the host memory it is relatively cheap. Much less memory is required on-chip to store the block ack scoreboard since the same memory can be reused for multiple block ack sessions.

8.4.3.3   Partial state block ack operation

On receiving a QoS Data frame with sequence number SN, the recipient checks to see if it has a record of the block ack scoreboard for that block ack session, where the session is identified by the transmit address (TA) and TID. If not, then it creates a scoreboard for that session with WinEnd = SN and WinStart = WinEnd − WinStart + 1, perhaps reusing memory from another session. The correct reception of the data frame is recorded by setting a 1 in the position representing SN, i.e. WinEnd.

For each data frame thereafter:

- If SN is within the current scoreboard window, i.e. WinStart $\leq$ SN $\leq$ WinEnd, then the scoreboard records receipt at the offset represented by SN.
- If SN is outside the current scoreboard window, but within half sequence space range, i.e. WinEnd $<$ SN $<$ WinStart $+ 2^{11}$, then the scoreboard is shifted right to accommodate SN.
- If SN is more than half the sequence space beyond the window, i.e. WinStart $+2^{11} <$ SN $<$ WinStart, then no change is made.

The scoreboard operation is illustrated in Figure 8.16. Here a QoS Data frame with sequence number 102 is received and the recipient creates a new scoreboard with a mark for 102 at its rightmost edge. With the next aggregate, QoS Data frames 103 and 105 are received correctly and the scoreboard is shifted right to accommodate the new entries. QoS Data frame 100 falls within the scoreboard sequence number range and is simply marked up. Note that in this sequence the QoS Data frames have their Ack Policy field set to Block Ack.

When the BAR is received, the recipient shifts the scoreboard to the right so that WinStart = SSN from the BAR (100 in this case) and returns a BA frame with the contents of the scoreboard.

Note that the sequence numbers shown are for illustration only. In practice, the sequence number is a 16-bit value composed of a 12-bit MSDU number and a 4-bit fragment number.

The major difference between partial state and full state block ack operation is the transient nature of the scoreboard retained by the recipient. Under partial state block ack the originator should ensure that it retrieves the ack state with high probability before another station has a chance to send data to the recipient and potentially erase the session's block ack scoreboard. In practice this means that the originator should attempt to retrieve the block ack scoreboard before the end of each TXOP. If occasionally the block ack scoreboard is not retrieved during the TXOP (perhaps the BA frame is received in error) then there is still a good chance that a subsequent immediate channel access

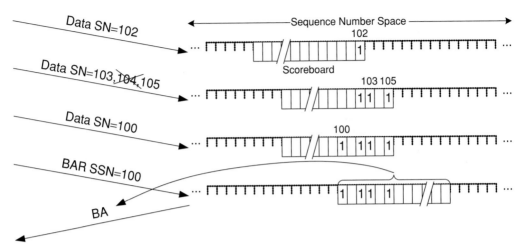

Figure 8.16 Scoreboard operation.

by the same station will occur before data belonging to another block ack session is received by the recipient, erasing the session's block ack scoreboard. The originator can thus simply use a BAR frame in a subsequent channel access to retrieve the ack state. In the unlikely event that the block ack scoreboard is no longer available, the BA will show all zeros and the originator will need to retransmit the MSDUs. Alternatively, if the originator sent a single aggregate consisting of QoS Data frame with Normal Ack policy and no BA was received, then the originator may assume that none of the QoS Data frames made it through and simply retransmit the QoS Data frames.

While more complex behavior is permitted, most implementations will conform to partial state rules by sending a single aggregate in each TXOP, soliciting a BA by setting the ack policy of the QoS Data frames making up the aggregate to Normal Ack. On the rare occasion when an MSDU lifetime expires before it is acknowledged, the originator will send a BAR frame to flush the holes in the reorder buffer. An originator implementing this behavior would work with both a full state and partial state implementation in the recipient.

8.4.4   HT-immediate block ack TXOP sequences

Typical TXOP sequences under HT-immediate block ack are shown in Figure 8.17. The TXOP begins with a short frame exchange as a collision detect mechanism. In these sequences an RTS/CTS exchange, which also provides enhanced protection, is used, but the more efficient Data/ACK exchange could also be used. In sequence (a) a single aggregate data transmission is made with the ack policy set to Normal Ack in the QoS Data MPDUs making up the aggregate. This solicits an immediate BA response.

Sequence (b) is similar except that two aggregate transmissions are sent back-to-back as a burst. This might be done to improve robustness, for example if a single transmission was too long and possibly subject to rapidly changing channel conditions for which a

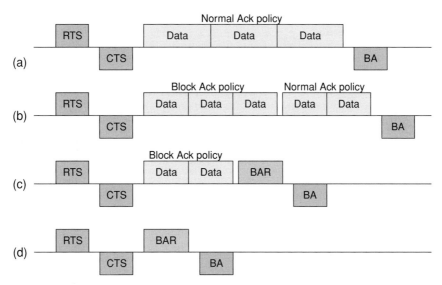

Figure 8.17 Typical HT-immediate TXOP sequences.

new channel estimate would be beneficial. The first aggregate would be sent with Ack Policy set to Block Ack and the second aggregate with Ack Policy set to Normal Ack in order to solicit a BA.

Sequences (c) and (d) may be used when an MSDU is discarded due to lifetime expiry, when the BAR is required to release potentially stalled complete MSDUs in the recipient's reorder buffer. In sequence (c) data frames are available and would typically be sent ahead of the BAR. The aggregate is sent with Ack Policy set to Block Ack since the BAR will solicit the BA response. Note that it is permitted to aggregate the BAR with the data frames; however, the BAR would typically be sent as a separate more robustly modulated frame. Since the BAR is sent infrequently, sequences involving BAR do not need to be optimized as the overall gain is small.

8.5   HT-delayed block ack

HT-delayed block ack is an extension to the delayed block ack protocol and differs from it in the manner in which BAR and BA frames are acknowledged. Support for HT-delayed block ack is optional and a station advertises support by setting the HT-delayed Block Ack capability bit in the HT Capabilities element (see Section 11.3.2.2). A peer station may establish a HT-delayed block ack session with a station that advertises itself as HT-delayed block ack capable.

Under HT-delayed block ack the BAR and BA frames carry a BAR Ack Policy and BA Ack Policy field, respectively. If set to 1 this indicates that the receiver of the frame should not return an ACK response.

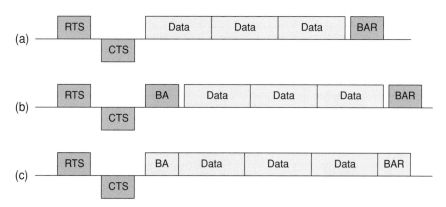

Figure 8.18 Typical HT-delayed TXOP sequences.

8.5.1   HT-delayed block ack TXOP sequences

Typical TXOP sequences under HT-delayed block ack are shown in Figure 8.18. The TXOP begins with a short frame exchange which could be RTS/CTS (as shown) or Data/ACK. HT-delayed block allows the TXOP to be fully utilized by the originator since no immediate response is required from the responder. Sequence (a) shows the originator sending an aggregate with ack policy set to Block Ack followed by a BAR. When data flows both ways, the BA may be combined with data in the reverse direction. This is shown in sequence (b).

It is possible, with reduced robustness, to aggregate the BA and BAR frames together with the data. This is shown in sequence (c).

References

Hansen, C. and Edwards, B. (2004). *WWiSE Proposal: High Throughput Extension to the 802.11 Standard*, IEEE 802.11-04/0886r6.

IEEE (2006). IEEE 802.11n Project authorization request, 26 May 2006, available at: http://standards.ieee.org/board/nes/projects/802-11n.pdf.

9 Advanced channel access techniques

This chapter introduces some of the advanced channel access techniques in the 802.11 standard. In addition to, and built upon, the distributed channel access techniques described in Chapter 7, the 802.11 standard includes two centrally coordinated channel access techniques. The PCF was introduced in the original 802.11 specification and HCCA was introduced in the 802.11e amendment to support parameterized QoS and to fix some of the deficiencies in the PCF. The chapter then introduces new channel access techniques in the 802.11n amendment.

A very simple technique called the reverse direction protocol was introduced with one bit of additional signaling and some simple changes to the rules for operating a TXOP. This technique is particularly effective under EDCA for improving throughput for certain traffic patterns.

During the development of the 802.11n amendment, a strong interest emerged among many participants for improving the power efficiency of the MAC protocol. While outside the scope of the 802.11n PAR this resulted in the power-save multi-poll (PSMP).

9.1 PCF

Infrastructure network configurations may optionally include the point coordination function (PCF). With PCF, the point coordinator (PC), which resides in the AP, establishes a periodic contention free period (CFP) during which contention free access to the wireless medium is coordinated by the PC. During the CFP the NAV of all nearby stations is set to the maximum expected duration of the CFP. In addition, all frame transfers during the CFP use an inter-frame spacing that is less than that used to access the medium under DCF, preventing stations from gaining access to the medium using contention-based mechanisms. At the end of the CFP, the PC resets the NAV of all stations and regular contention-based access proceeds.

9.1.1 Establishing the CFP

The PC periodically establishes a CFP using a Beacon frame as illustrated in Figure 9.1. The CFP repetition interval is a fixed multiple of the DTIM (delivery traffic indication message) period where the DTIM or delivery traffic indication message is an element that

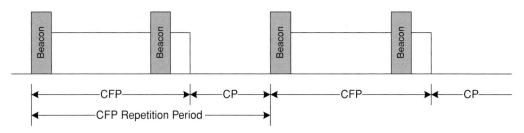

Figure 9.1 Contention free and contention period alternation.

appears periodically in the Beacon frame. The DTIM period is a multiple of the beacon period or the same as the beacon period if every beacon contains a DTIM element.

The length of the CFP is controlled by the PC, with the maximum duration specified by the CFPMaxDuration parameter in the CF Parameter Set element of the Beacon frame. If the CFP duration is longer than the beacon interval, the PC will transmit beacons at the appropriate times during the CFP. The actual length of the CFP depends on the traffic exchanged during the CFP but the PC ensures that it is never longer than CFPMaxDuration. At or before CFPMaxDuration, the PC terminates the CFP by transmitting a CF-End frame. The CF-End resets the NAV of nearby stations allowing contention-based access.

To send the Beacon frame that begins a CFP the PC, operating under a DCF (distributed coordination function), must first ensure that the medium is idle. If the medium is busy at the desired CFP start time, the PC must wait until the current access completes and this will introduce some jitter in the CFP repetition interval. It may also be necessary to foreshorten the CFP duration by the delay introduced since the CFPMaxDuration applies from the ideal start time of the CFP not the actual start time.

9.1.2   NAV during the CFP

Each station, except the AP containing the PC, sets its NAV to the CFPMaxDuration value obtained from the CF Parameter Set element in the Beacon frame that starts the CFP. Stations update their NAV from the CFPDurRemaining value in the CF Parameter Set element in Beacons that are transmitted during the CFP. On receiving a CF-End frame a station resets its NAV. Stations set, update, or reset their NAV irrespective of the BSS from which the Beacon or CF-End frame is received. This prevents the stations, even in neighboring BSSs, from taking ownership of the medium during the CFP.

9.1.3   Data transfer during the CFP

The contention free transfer protocol is based on a polling scheme controlled by the PC in the AP of the BSS. Frame transfers consist of frames alternately sent from the PC and sent to the PC.

An example of data transfer during the CFP is illustrated in Figure 9.2. During the CFP, the AP may send unicast data to individual stations. The AP should also poll the stations to see if they have data to transfer. Polling the station is done using a CF-Poll,

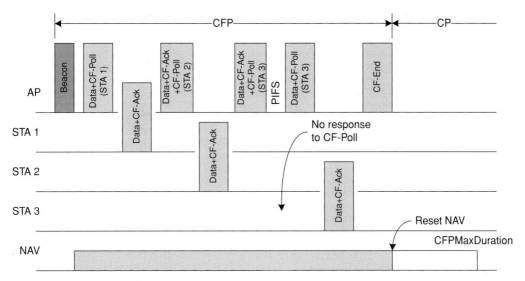

Figure 9.2 Data transfer during the CFP.

which may be sent as an individual CF-Poll frame or, for increased efficiency, may be piggybacked on a data frame.

9.1.3.1 Contention free acknowledgement

PCF permits the piggybacking of data frame acknowledgements on data frames. The piggybacked CF-Ack acknowledges the correct reception of the previously sent data frame irrespective of the station to which the current data frame is addressed. In the example in Figure 9.2, the data frame sent by the PC together with the CF-Poll to STA 1 is acknowledged with a CF-Ack piggybacked on the data frame returned in response to the CF-Poll. That data frame is in turn acknowledged by a CF-Ack piggybacked on a data frame sent to STA 2. This mechanism requires that the frame carrying the piggybacked ack be receivable by both the sender of the data frame being acknowledged and the receiver of the data frame itself, necessitating careful selection of the data rate at which the data frames are sent.

9.1.4 PCF limitations

The PCF suffers from a number of limitations. One major limitation is the potential delay it introduces for both PCF and DCF traffic. Since contention free periods only occur as frequently as the DTIM period, a station operating under PCF with traffic to send at the end of one CFP may only be polled again in the next CFP. Also, traffic to be sent under DCF, but arriving during a CFP, must wait until the end of the CFP before channel access can be gained. This can severely impact delay sensitive traffic.

Another limitation of the PCF is that it does not work well if used on neighboring BSSs sharing the same channel. If a CFP is in progress on one BSS a neighboring BSS would need to wait for that CFP to end before beginning its own CFP. Also, if the total duration of the two CFPs is longer than the CFP repetition interval on either BSS then the service guarantees will not be met.

The PCF was never widely implemented and remains in very limited use today, if it is used at all. This is partly the result of the above limitations, but also because the generally unpredictable nature of the unlicensed wireless spectrum makes it difficult to design schedulers that provide service guarantees under all operating conditions. It turns out that the much simpler distributed contention based access mechanisms are more reliable and robust in many ways.

9.2   HCCA

Hybrid coordination function (HCF) controlled channel access (HCCA) is a channel access technique that is centrally coordinated by the hybrid coordinator (HC) residing in the AP. Under HCCA, stations do not compete for access to the wireless medium, but rely on the AP to poll them regularly and thus gain channel access. In this sense it is similar to the PCF; however, HCCA improves on PCF in two important respects:

(1) The HC may poll the station during both the CP and CFP. This reduces latency since a station that relies on being polled may be polled more frequently than if it were polled during the CFP only.
(2) When polled, a station is granted a TXOP during which the station may transmit multiple frames subject to the limit on the TXOP duration. Under PCF, the PC polls the station for each frame individually.

9.2.1   Traffic streams

HCCA is designed to support parameterized QoS. With parameterized QoS, a station registers QoS requirements for a traffic stream (TS) with the HC and, if admitted, the HC establishes a polling schedule to meet those requirements.

A TS is a specific set of MSDUs traveling in one direction that need to be delivered subject to specific QoS constraints. A TS is identified in the uplink (toward the AP) by its traffic stream identifier (TSID) and direction, and in the downlink (toward the station) by its TSID, direction, and the station address. It is the station that establishes the TS regardless of the direction. The AP may support up to eight uplink and eight downlink TSs per associated station.

The QoS requirements for a TS are provided by the station in a traffic specification (TSPEC), which may include, for example, a maximum service interval that specifies the maximum amount of time that may elapse between polls and a minimum data rate that specifies the lowest expected average data rate. The purpose of the TSPEC is to reserve resources within the HC and modify the HC's scheduling behavior. It also allows other parameters to be specified that are associated with the TS, such as the traffic classifier and acknowledgement policy.

In the downlink, a TS may have one or more traffic classification (TCLAS) elements associated with it. A TCLAS element contains parameters that identify packets belonging to the TS. The AP uses the TCLAS elements to filter the downstream MSDUs and assign them to the appropriate TS. For example, a TCLAS element associated with a particular TS may identify packets carrying a UDP payload with a specific UDP port number as belonging to that TS. In the uplink, MSDUs belonging to a TS are classified by the station internally.

TSPECs, while described here in the context of HCCA for which TSPECs are required, may also be used under EDCA, for example to support admission control. Traffic admitted in the context of the TSPEC can be sent using EDCA or HCCA or a hybrid of the two called HEMM (HCCA, EDCA mixed mode).

9.2.1.1   TS setup and maintenance

A TS is set up by the station sending an Add Traffic Stream (ADDTS) Request to the HC. The ADDTS Request includes the TSPEC parameters that define the QoS requirements of the TS. On receiving the ADDTS Request, the HC decides whether to admit the TSPEC as specified, refuse the TSPEC, or not admit but suggest an alternative TSPEC.

In response to the ADDTS Request, the HC transmits an ADDTS Response. The response frame includes a result code that indicates success, invalid parameters, rejected with suggested changes or rejected for delay period. If the result code indicates success then the TS enters the active state. Otherwise the whole process can be repeated using the same TSID and a modified TSPEC until the station decides that the granted TSPEC is adequate or inadequate and cannot be improved.

The parameters that are set for the TS can be renegotiated in a similar manner, i.e. the station sends a new ADDTS Request with an updated TSPEC and the HC responds with an ADDTS Response.

9.2.1.2   Data transfer

Once a TSPEC has been set up, MSDUs are classified above the MAC and the MAC delivers the MSDUs based on a schedule using QoS Data frames. In the case of a station the MSDUs are transmitted when the station is polled by the HC.

9.2.1.3   TS deletion

The TS may be deleted by either the station or the HC sending a DELTS frame and the receiver acknowledging correct receipt with an ACK frame response. The TS may be deleted for numerous reasons, including:

- The station deciding that the TS is no longer required.
- The HC or station determining that there is no activity on that TS.
- Traffic is received for an unknown TSID, i.e. a TSID for which the station or HC does not have TS state.
- A timeout waiting for an ADDTS Response.

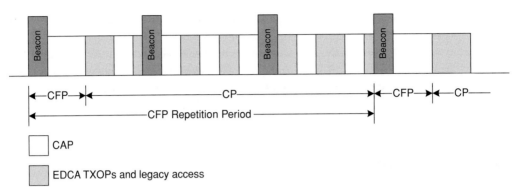

Figure 9.3 Controlled access phases. Reproduced with permission from IEEE (2007a) © IEEE.

TSPECs that have been set up are also deleted on disassociation or reassociation. Reassociation will cause the station and AP to clear their state and the station will have to reinitiate the TS setup.

9.2.2 Controlled access phases

The HC gains control of the wireless medium as needed for a controlled access phase (CAP) by waiting a shorter time following another transmission than stations using EDCA procedures. During a CAP, response frames follow HC initiated transmissions by SIFS. If the HC does not detect a response frame after SIFS it may reclaim the channel PIFS after the last frame if the channel remains idle. A CAP ends when the HC does not reclaim the channel PIFS after the last transmission.

The HC may also use the PCF mechanism for establishing a contention free period (CFP) following a Beacon frame. The use of controlled access phases is illustrated in Figure 9.3.

9.2.3 Polled TXOP

The HC polls stations with a CF-Poll that is carried by a QoS CF-Poll frame, but could also be piggybacked on another QoS Data frame and is thus generally referred to as a QoS (+)CF-Poll frame to indicate that it is any QoS Data frame carrying the CF-Poll operation (see Section 11.1.1.2). When polled by the HC with a QoS (+)CF-Poll frame, the station is granted a TXOP with a duration limit specified in the TXOP Limit subfield of the QoS Control field. The TXOP is protected by the NAV set by the Duration field of the frame that contained the CF-Poll function.

When granted a TXOP, the station may transmit multiple frame exchange sequences within the given polled TXOP subject to the limit on TXOP duration. The station will not initiate transmission of a frame unless the transmission and any immediate response expected from the MAC peer are able to complete prior to the end of the TXOP. If the TXOP Limit subfield in the QoS Control field of the QoS Data frame that includes

CF-Poll is set to 0 then the station will respond with either one MPDU or one QoS Null frame.

The QoS (+)CF-Poll frame that grants the TXOP also contains a TID subfield in the QoS Control field that indicates the TS for which the poll is intended. The station typically responds with data frames for that TS; however, it is non-binding and the station may respond with any frames, i.e. QoS Data frames belonging to any TID as well as management frames.

9.2.4   TXOP requests

A station makes a request for a subsequent TXOP by setting the TXOP Duration Requested or Queue Size subfield in a QoS Data frame sent to the AP. These subfields are mutually exclusive and the station sets one or the other (see Section 11.1.5). By setting the TXOP Duration Requested field the station is requesting a certain amount of additional airtime in the form of a single TXOP in order to transmit data frames in its queue. By setting the Queue Size subfield the station is providing the AP with an indication of the amount of traffic awaiting transmission of the station's queue.

The station may send the request in a polled TXOP or an EDCA TXOP. Both types of request provide the AP with information that it can consider together with similar information from other stations and the TSPECs for the various types of traffic streams involved and allocate TXOPs accordingly.

9.2.5   Use of RTS/CTS

Stations may send an RTS frame as the first frame of any exchange sequence when improved NAV protection is desired, during either the CP or CFP.

If NAV protection is desired for a transmission to the AP in response to a QoS (+)CF-Poll, then the full RTS/CTS exchange may not be needed. Instead, the station can send a CTS-to-self as a robust mechanism for setting the NAV of stations in its vicinity.

9.2.6   HCCA limitations

Although HCCA significantly improves PCF, it too has not been widely implemented. This is due to a number of factors.

Contention-based techniques are remarkably effective at fairly apportioning bandwidth based on demand. The prioritized access techniques introduced with EDCA have also been remarkably effective at protecting fixed rate or near fixed rate traffic such as VoIP and Video from data traffic. As a result there has been little market demand for HCCA.

To ensure that throughput and delay guarantees are met for traffic streams, the HC establishes a schedule and based on that schedule gains regular priority access to the medium using the PIFS defer. However, this mechanism breaks down in the presence of neighboring QoS BSSs. If the HCs of both BSSs are regularly attempting to gain channel access then there may be frequent collisions. The collision problem can be

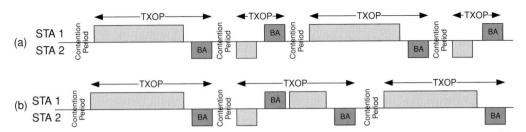

Figure 9.4 TXOP utilization (a) without and (b) with the reverse direction protocol.

mitigated to some extent using random backoff; however, the protocol now devolves into something very similar to contention-based access without prioritization for different traffic categories.

9.3   Reverse direction protocol

Many traffic patterns are highly asymmetric with a large volume of traffic in one direction and very little in the reverse direction. Consider the transfer of a large file using the file transfer protocol (FTP) or hypertext transfer protocol (HTTP), both of which operate over TCP. During the actual file transfer, the direction from the source to the destination carries large TCP data packets with the file content while the reverse direction carries short TCP ack packets.

In practice this means that while TXOPs in the forward direction are fully utilized, TXOPs in the reverse direction are often under utilized as illustrated in Figure 9.4(a). The reverse direction protocol allows the owner of an underutilized TXOP to sublease the remainder of the TXOP to its peer. The result with asymmetric traffic is that all TXOPs are fully utilized as illustrated in Figure 9.4(b). The overhead associated with the contention period and the short frame exchange at the beginning of the TXOP (not shown in the diagram) are amortized over a fully utilized TXOP rather than a partially utilized TXOP improving efficiency. Simulation results (Mujtaba *et al.*, 2005) showed network throughput improvement ranging from 8% to 43% with TCP traffic, depending on the usage scenario.

The reverse direction protocol is an optional feature. A station indicates that it is able to act as an RD responder by setting the RD Responder capability bit in the HT Capabilities element (see Section 11.3.2.2). An RD initiating station may grant the remainder of an otherwise underutilized TXOP to a station that indicates that it is capable of acting as an RD responder. Note that because the RD protocol makes use of the HT Control (HTC) field, the responder must also indicate HTC Support in the HT Capabilities element.

9.3.1   Reverse direction frame exchange

The reverse direction protocol makes use of the RDG/More PPDU bit in the HT Control field of a QoS Data, BAR, or BA frame. The HT Control field must thus be present in frames carrying reverse direction signaling.

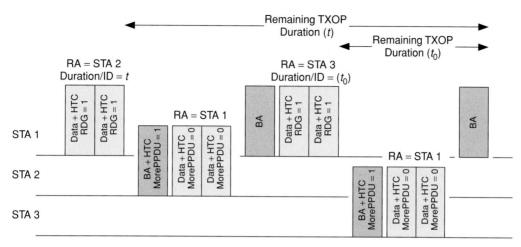

RDG = Reverse Direction Grant

Figure 9.5 Example of an RD exchange.

An example of an RD frame exchange sequence is shown in Figure 9.5. The RD initiator grants the remainder of a TXOP to the RD responder by setting the Reverse Direction Grant (RDG) bit in an MPDU, which is typically a QoS Data or BAR MPDU. The time remaining in the TXOP and the time granted to the responder is carried in the Duration/ID field of the MPDU. The RDG bit is typically set in a frame which would otherwise solicit a response anyway, often the QoS Data MPDUs in an aggregate carrying an implicit BAR or a BAR MPDU itself.

The RD responder responds with an ACK or BA. If the responder intends to continue with a burst, it sets the More PPDU bit in the response frame indicating that more PPDUs are to follow. The last PPDU in the burst has the More PPDU bit set to zero in the MPDU(s) it is carrying. The RD responder may solicit a response from the initiator in the last PPDU of the burst.

The RD initiator takes control of the TXOP again when it sees the More PPDU bit set to zero or through error recovery. A reverse direction frame that does not contain the HTC field is interpreted as having the More PPDU bit set to zero. Once it has gained control of the TXOP again, the RD initiator may make further grants to the same responder or to other stations subject to the TXOP constraints.

9.3.2   Reverse direction rules

In an EDCA TXOP, all data frames belong to the same AC. That includes data from the initiator as well as responder(s). The AC is indicated by the TID in the MPDU(s) carrying the reverse direction grant or, if a management frame is carrying the grant, then the access category is AC_VO.

In an HCCA TXOP, the RD initiator determines whether or not the responder is constrained on which access category data frames it should belong to. The RD initiator sets the AC Constraint bit to zero if the RD responder is allowed to send traffic with any TID and sets the AC Constraint bit to one if the RD responder may only send traffic of the designated TID in its response burst.

The RD responder addresses all MPDUs in the response burst to the RD initiator and may not use the subleased TXOP to send frames to other stations.

9.3.3   Error recovery

After granting the remaining TXOP, the RD initiator may take control of the TXOP again if it senses the medium idle for PIFS and there is sufficient time remaining in the TXOP to transmit another frame. This recovery mechanism allows the RD initiator to regain control of the TXOP should the responder fail to successfully demodulate the PPDU containing the reverse direction grant.

If the RD initiator receives a PPDU response which does not include an MPDU carrying the More PPDU bit, it does not transmit again until sensing the medium idle for PIFS. This assures that the RD initiator does not transmit over a burst from the RD responder should the MPDU carrying the More PPDU bit not be demodulated correctly.

9.4   PSMP

Power-save multi-poll (PSMP) is a scheduling technique introduced in 802.11n to optimize channel access for devices that receive and transmit small amounts of data periodically and would like their communications interface to remain inactive most of the time to conserve power, i.e. when they are not actively transmitting or receiving frames.

PSMP provides a small improvement in channel utilization and power saving capability over alternative techniques such as HCCA and scheduled or unscheduled automatic power save delivery (APSD). PSMP uses a single PSMP frame to schedule multiple stations instead of the directed QoS (+)CF-Poll frame used in HCCA, which may be slightly more efficient in some usage scenarios. Channel access is optimized by grouping downlink transmissions and scheduling uplink transmissions to follow each other immediately. Downlink transmissions can thus be sent as a burst and uplink transmissions scheduled to follow each other closely.

PSMP optimizes for power consumption by providing the downlink and uplink schedule at the start of the PSMP phase so that individual stations can shut down their receivers until needed in the downlink phase and transmit when scheduled during the uplink phase without performing CCA.

A basic PSMP sequence is illustrated in Figure 9.6. The sequence begins with a PSMP frame, which contains the schedule for the subsequent downlink and uplink transmissions. After receiving the PSMP frame, a station need only be awake for the PSMP downlink transmission time (PSMP-DTT) containing broadcast frames (if any), during its allotted PSMP-DTT and during its PSMP uplink transmission time (PSMP-UTT). During the downlink phase, PPDUs are sent as a continuous burst with individual PPDUs separated by SIFS or RIFS. The use of RIFS may be restricted as described in

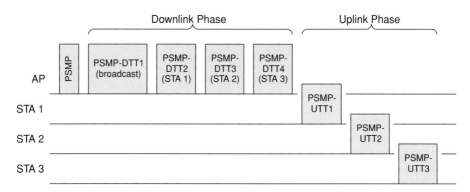

Figure 9.6 PSMP sequence.

Section 10.5.2.1. A particular PSMP-DTT may cover one or more PPDUs where the MAC frames carried by the PPDUs are addressed to a specific station (have the same RA). If there are any broadcast or multicast frames these are grouped together in the first PSMP-DTT.

A station that has one or more frames to send will start transmitting at the start of its PSMP-UTT without performing CCA and disregarding any NAV setting. The station must complete its transmission within the allocated PSMP-UTT.

The PSMP-UTTs are separated by either SIFS or aIUStime, where aIUStime is shorter than SIFS and only used if RIFS is permitted. PPDUs transmitted within a PSMP-UTT are separated by SIFS or RIFS where RIFS, again, is only used when permitted.

During a PSMP sequence, the AP and station may only exchange the following frame types:

- PSMP management frame
- QoS Data
- Multi-TID BAR
- BAR (HT-delayed session only)
- Multi-TID BA
- BA (HT-delayed session only)
- An MPDU that does not require an immediate response (e.g. Action No Ack management frame).

9.4.1   PSMP recovery

A station may not detect or successfully demodulate the PSMP frame and thus miss its PSMP-UTT. The AP will detect that this has occurred if no transmission is detected PIFS after the start of the PSMP-UTT. If the PSMP-UTT is sufficiently long to warrant it, the AP may attempt recovery by transmitting a PSMP recovery frame with a new schedule for this station.

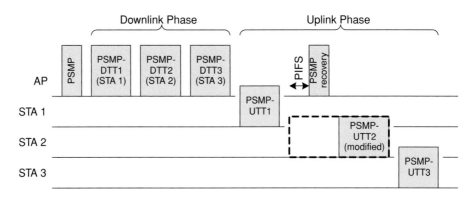

Figure 9.7 PSMP recovery.

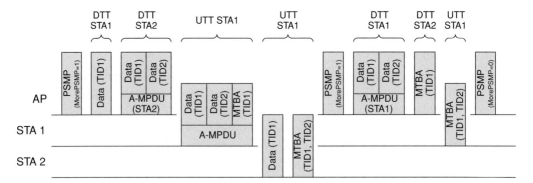

Figure 9.8 PSMP burst. Reproduced with permission from IEEE (2007) © IEEE.

This is illustrated in Figure 9.7 where STA 2 has not transmitted a frame at the start of PSMP-UTT2. The AP detects this condition and transmits a PSMP recovery frame PIFS after the start of the UTT. The PSMP recovery frame contains an uplink schedule for STA 2 using the remaining time of the original PSMP-UTT2 and referenced from the end of the PSMP recovery frame.

9.4.2   PSMP burst

The AP may perform a burst or series of PSMP exchanges as illustrated in Figure 9.8. After an initial data exchange in the first PSMP sequence, additional resource allocation can be made and retransmissions scheduled in subsequent PSMP exchanges. The PSMP frame in the first exchange has its MorePSMP bit set to 1 to indicate that another PSMP exchange will follow. The MorePSMP bit is set in the PSMP frame of each subsequent exchange as long as the burst continues. The burst is terminated with a PSMP frame that has the MorePSMP bit set to 0.

9.4.3   Resource allocation

A station uses the TSPEC mechanism defined for HCCA to reserve resources. The AP establishes a PSMP schedule based on the TSPECs received from one or more stations such that the QoS requirements of all stations are met.

To accommodate variability in traffic loads and retransmission, a station may adjust the resource allocation long term by updating its TSPEC or, more dynamically, using the Queue Size or TXOP Duration Request fields in the QoS Data frame (see Section 11.1.5). If a station cannot transmit all its queued data within the allocated PSMP-UTT, the station will transmit as much data as possible and piggyback a resource request for an additional PSMP-UTT. If the AP receives a resource request it will attempt to satisfy that request in the next PSMP sequence.

9.4.4   Block ack usage under PSMP

PSMP does not permit the use of ACK for data exchanges, thus data transfer that requires acknowledgment must occur under an established block ack session. Data may be transferred without a block ack session, however, it can only do so using No Ack ack policy.

PSMP requires that the block ack session be HT-immediate or HT-delayed. Under HT-immediate, use of the Multi-TID BAR and Multi-TID BA is required, while under HT-delayed the BAR and BA frames should have their BAR/BA ack policy set to No Ack.

References

IEEE (2007). *IEEE P802.11n*™*/D3.00, Draft Amendment to STANDARD for Information Technology – Telecommunications and Information Exchange Between Systems – Local and Metropolitan Networks – Specific Requirements. Part 11: Wireless LAN Medium Access Control (MAC) and Physical Layer (PHY) Specifications: Amendment 4: Enhancements for Higher Throughput.*

Mujtaba, S. A., Stephens, A. P., Myles, A., *et al.* (2005). *TGnSync Proposal MAC Results*, IEEE 802.11–04/0892r7, July.

10 Interoperability and coexistence

The 802.11n amendment, more so than any previous 802.11 amendment, introduces many optional features geared toward specific market segments that will likely only be deployed in specific classes of devices. Many of the optional features are complex and, given time to market concerns and cost constraints, many implementations will only adopt a subset of the available features, perhaps phasing in features with time or in higher end products. Some features appear in multiple flavors. This is due to the many unknowns present at the time the features where being discussed during the standard development process. Very often there was no single clear direction to take that would clearly suit all situations and so variations were introduced.

The popularity of 802.11 also means that there are a large number of legacy 802.11 devices deployed making interoperability and coexistence with those devices essential.

This chapter discusses various features that help ensure interoperability and coexistence between 802.11n compliant devices as well as legacy 802.11 devices.

10.1 Station and BSS capabilities

With the large number of optional features in the 802.11n amendment a fair bit of signaling is required to establish device capabilities and ensure interoperability. Also, care must be taken to ensure that a feature used by one station does not adversely affect the operation of a neighboring station that is not directly involved in the data frame exchange.

HT station capabilities are signaled using the HT Capabilities element. This element is present in Beacon, Association Request, Association Response, Reassociation Request, Reassociation Response, Probe Request, and Probe Response management frames sent by an HT AP or HT station. The optional capabilities signaled by the HT Capabilities element are summarized in the following sections, with the details on the HT Capabilities element itself provided in Section 11.3.2.2.

10.1.1 HT station PHY capabilities

The HT station PHY capabilities include the following:

- **Supported channel width set**. Indicates whether the station is capable of 20 MHz operation only or both 20 MHz and 40 MHz operation.

- **Greenfield format PPDUs**. Indicates whether or not the station is capable of receiving HT Greenfield format PPDUs.
- **LDPC coding**. Indicates whether or not the station is capable of receiving LDPC coded PPDUs.
- **Short GI in 20 MHz PPDU formats**. Indicates whether or not the station is capable of receiving 20 MHz frames using the short guard interval.
- **Short GI in 40 MHz PPDU formats**. Indicates whether or not the station is capable of receiving 40 MHz frames using the short guard interval.
- **Transmitting STBC formats**. Indicates whether or not the station is capable of transmitting space-time block coded frames.
- **Receiving STBC formats**. Indicates whether or not the station is capable of receiving space-time block coded frames and if so, whether that capability is limited to one spatial stream, one or two spatial streams or one, two, and three spatial streams.
- **DSSS/CCK in 40 MHz**. Indicates whether or not the station is capable of transmitting and receiving DSSS/CCK format frames (802.11b) while operating with 40 MHz channel width.
- **40 MHz intolerant**. Indicates whether or not the station requires the AP to prevent 40 MHz transmissions.

10.1.2   HT station MAC capabilities

The HT station MAC capabilities include the following:

- **Spatial multiplexing power save mode**. Indicates the level of support for Spatial Multiplexing (SM) Power Save: static, dynamic or disabled.
- **HT delayed block ack**. Indicates whether or not the station supports HT-delayed block ack.
- **PSMP support**. Indicates whether or not the station supports PSMP operation.
- **L-SIG TXOP protection**. Indicates whether or not the station supports the L-SIG TXOP protection mechanism.
- **+HTC field support**. Indicates whether or not the station can parse received frames containing the HT Control (HTC) field.
- **Reverse direction protocol**. Indicates whether or not the station can act as a reverse direction responder.
- **Maximum A-MSDU length**. Indicates the maximum A-MSDU length that the station can receive: 3839 or 7935 octets.
- **A-MPDU parameters**. Indicates the processing limitations with respect to A-MPDU delineation. The station may be limited in the length of A-MPDU it can receive as well as the rate at which MPDUs within the A-MPDU can be de-aggregated.

10.1.3   BSS capabilities

Some of the HT Capability fields have additional meaning that applies to the BSS when transmitted by the AP in Beacon, Probe Response, Association Response, and Reassociation Response frames. These fields are:

- **PCO Support**. This indicates the BSS supports phased coexistence operation (PCO).
- **PSMP Support**. This indicates that the BSS supports PSMP operation.
- **DSSS/CCK Mode in 40 MHz**. This indicates whether or not member stations of the BSS are permitted to transmit DSSS/CCK format PPDUs during 20/40 MHz BSS operation.

10.1.4   Advanced capabilities

Some advanced capabilities for which support is indicated in the HT Capabilities element are:

- MCS feedback
- transmit beamforming support
- antenna selection.

These capabilities are discussed in more detail in Chapter 12.

10.2   Controlling station behavior

To control the operation of stations on a BSS, the AP uses the HT Information element in Beacon, Probe Response, Association Response, and Reassociation Response frames. Details on the HT Information element can be found in Section 11.3.2.3.

The AP controls 20 MHz and 20/40 MHz operation of the BSS through the Supported Channel Width Set field in the HT Capabilities element and the following fields in the Information element:

- **Primary Channel**. This field gives the channel number of the primary channel for the BSS.
- **Secondary Channel Offset**. This field gives the offset of the secondary channel relative to the primary channel.
- **STA Channel Width**. Defines the channel widths with which stations may transmit to the AP.

Twenty MHz and 20/40 MHz BSS operation is described in more detail in later sections.

The AP also provides information on the capabilities of stations that are members of the BSS or, optionally, detected operating in the same channel(s). This enables protection for certain HT sequences a station may use but for which nearby station may not correctly defer. The fields in the HT Information element that provide information on nearby stations for the purpose of enabling protection are the following:

- **HT Protection**. This indicates the types of stations that are present as members or non-members of the BSS and thus the nature of the protection required. Four categories are distinguished as listed in Table 10.1.

Table 10.1 HT Protection field encoding

Encoding	Condition
0 (HT members with same channel width, HT non-members)	One of the following is true: • This is a 20/40 MHz BSS with only 20/40 MHz HT station members. Any non-member stations detected in the primary or secondary channel are also HT, i.e. no non-HT stations are detected • This is a 20 MHz BSS with 20 MHz HT station members. Any non-member stations detected in the channel are also HT, i.e. no non-HT stations are detected
1 (HT members with different channel widths, non-HT non-members)	One of the following is true: • This is a 20/40 MHz BSS with either 20/40 MHz HT station or 20 MHz HT station members and at least one non-HT station non-member has been detected in the primary or secondary channel • This is a 20 MHz BSS with 20 MHz HT station members and at least one non-HT station non-member has been detected in the primary channel
2 (HT members with different channel widths, HT non-members)	This is a 20/40 MHz BSS possibly with 20/40 MHz HT station members, but with at least one 20 MHz HT station member and only HT station non-members, if any, have been detected in the primary and secondary channels
3 (HT and non-HT members)	Otherwise, i.e. 20/40 MHz or 20 MHz BSS with both HT and non-HT members

- **Non-Greenfield HT STAs Present**. This indicates that there are HT stations associated with the BSS or in neighboring BSSs that are unable to receive HT Greenfield format PPDUs. When set, stations must protect HT Greenfield format transmissions.
- **L-SIG TXOP Protection Full Support**. This indicates whether or not all stations on the BSS support L-SIG TXOP Protection. A station may use this to determine whether or not to use L-SIG TXOP Protection.
- **OBSS Non-HT STAs Present**. This indicates that there may be non-HT stations present that are not members of the BSS. A station may optionally use this to determine if protection is necessary for HT sequences.
- **RIFS Mode**. The AP can directly control whether or not RIFS bursting is permitted through the RIFS Mode bit. Stations may only use RIFS bursting when this bit is set to 1.

The HT Information element also provides information for controlling PSMP operation (see Section 9.4), PCO operation (see Section 10.4) and the use of Dual Beacons to extend range.

Finally, the Basic MCS Set field of the HT Information element provides the set of MCS values that are supported by all HT stations in the BSS.

10.3   20 MHz and 20/40 MHz operation

802.11n introduces optional 40 MHz channel width operation but must maintain interoperability with legacy and HT 20 MHz stations. There are thus three basic classes of device: 20 MHz legacy or non-HT stations, 20 MHz HT stations, and 20/40 MHz HT stations. Twenty MHz HT stations are capable of sending and receiving 20 MHz transmissions, while 20/40 MHz stations are capable of sending and receiving both 20 MHz and 40 MHz transmissions. A 20/40 MHz HT station is also capable of operating as a 20 MHz HT station, i.e. a 20/40 MHz station may choose to advertise itself as a 20 MHz HT station and send and receive only 20 MHz transmissions. A legacy or non-HT station will not include the HT Capabilities element in any management frames.

A 20 MHz HT station sets the Supported Channel Width Set subfield in the HT Capabilities Info field of the HT Capabilities element to 0, indicating that it only supports 20 MHz operation. A 20/40 MHz HT station sets the Supported Channel Width Set subfield in the HT Capabilities Info field of the HT Capabilities element to 1, indicating that it supports both 20 MHz and 40 MHz operation.

In addition to, and because of, the various device classes there are two basic modes of BSS operation:

- **20 MHz BSS operation**. If 40 MHz operation is not feasible, either because the AP is not capable of supporting a 20/40 MHz BSS or because the environment does not permit 40 MHz operation then the AP operates a 20 MHz BSS.
- **20/40 MHz BSS operation**. If 40 MHz operation is feasible then the AP may operate a 20/40 MHz BSS. For 20/40 MHz BSS operation two adjacent 20 MHz channels are used, one of which is designated the primary channel and the other the secondary channel. A 20/40 MHz capable HT station may associate with the AP and operate with 40 MHz channel width. 20 MHz only HT stations and legacy stations may also associate with the AP and operate only on the primary 20 MHz channel.

10.3.1   Beacon transmission

The Beacon is generally transmitted using a format that is most broadly supported for a particular frequency band. The Beacon is always transmitted in a 20 MHz non-HT format on the primary channel (which is the only channel in the case of 20 MHz operation). In the case of the 5 GHz bands, a non-HT OFDM format is used since all 802.11 devices using these bands are expected to be backward compatible with 802.11a. In the case of the 2.4 GHz band, a DSSS/CCK format is typically used since most devices are backward compatible with 802.11b. A non-HT OFDM format may be used in the 2.4 GHz band if the AP is only capable of, or only wishes to support, OFDM transmissions.

An AP may also transmit a secondary Beacon using an STBC PPDU to extend the range over which the Beacon may be detected. This Beacon would be transmitted at half beacon period shift relative to the primary Beacon and has the STBC Beacon bit set in the HT Information element.

10.3.2   20 MHz BSS operation

Any HT AP (20 MHz or 20/40 MHz capable) may operate a 20 MHz BSS. A 20/40 MHz capable AP may choose to operate a 20 MHz BSS as a configuration option or if it is unable to find a suitable pair of 20 MHz channels on which to operate a 20/40 MHz BSS.

The HT AP indicates 20 MHz BSS operation by setting the Secondary Channel Offset subfield in the HT Information element to 0 to indicate that there is no secondary channel. The Primary Channel subfield in the HT Information element is set to indicate the 20 MHz channel number of the BSS (the channel on which the Beacon and all other frames are transmitted).

An HT station, whether it is 20 MHz only or 20/40 MHz capable, that associates with a 20 MHz BSS will send and receive 20 MHz transmissions.

In other respects, the 20 MHz BSS operation is relatively straightforward and similar to legacy 20 MHz BSS operation.

10.3.3   20/40 MHz BSS operation

Any HT AP (20 MHz or 20/40 MHz capable) may operate a 20/40 MHz BSS. The HT AP indicates 20/40 MHz operation by setting the Secondary Channel Offset subfield in the HT Information element to a non-zero value indicating the offset of the secondary channel relative to the primary channel, i.e. +1 or −1. The Primary Channel subfield indicates the channel number of the primary channel.

An HT station may associate with the AP as either a 20 MHz capable HT station or as a 20/40 MHz capable HT station. A 20 MHz HT station would send and receive only 20 MHz PPDUs and would only use the primary 20 MHz channel. A 20/40 MHz station would be able to send and receive either 20 MHz or 40 MHz PPDUs. All 20 MHz transmissions would only use the primary channel. The secondary channel should remain clear except for 40 MHz use and possible neighboring BSS use, although this should be avoided through careful channel selection.

If the HT AP is only 20 MHz capable the AP will not be able to send or receive 40 MHz transmissions. In this case, 40 MHz transmissions will only be exchanged between 20/40 MHz capable member stations of the BSS using direct links.

Coexistence with neighboring BSSs and among the different device classes associated with the BSS during 20/40 MHz operation is managed through a number of mechanisms, including the following:

- Overlapping BSS scanning and careful channel selection on initial BSS setup to avoid channels already in use by other BSSs.
- Changing channels or operating width after BSS setup if a new BSS is detected operating on the secondary channel.
- Independent CCA on both the primary and secondary channels.
- Protection mechanisms such as CTS-to-self sent in legacy compatible format on both primary and secondary channels.
- BSS Operating modes such as phased coexistence operation.

Also, because of the very different nature of the 2.4 GHz and 5 GHz bands, different rules apply in each case.

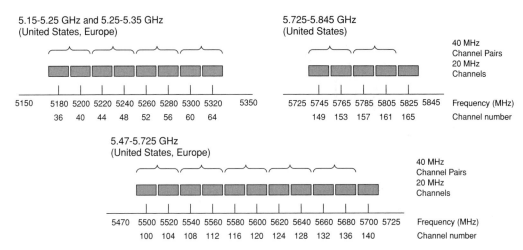

Figure 10.1 20 MHz channel pairings for 40 MHz operation in the 5 GHz bands.

10.3.3.1   20/40 MHz operation in the 5 GHz bands

Forty MHz operation in the 5 GHz bands is accomplished by utilizing two adjacent 20 MHz channels. One of the 20 MHz channels is designated the primary channel and the other the secondary channel. The permissible channel options for various regulatory domains are listed in Appendix 5.1 and summarized in Figure 10.1.

Compared to the 2.4 GHz band, the 5 GHz bands have far more bandwidth and easily accommodate multiple 40 MHz channels. Also, unlike the 2.4 GHz band, there are no partial channel overlaps. The 5 GHz band is thus preferred for 40 MHz operation, particularly in environments where multiple APs are deployed, such as a large enterprise.

10.3.3.2   20/40 MHz operation in the 2.4 GHz band

Forty MHz operation in the 2.4 GHz band is complicated by the limited spectrum available, overlapping channels and the ubiquitous deployment of legacy 802.11b and 802.11g devices operating in this band. During the development of the standard, arguments were made to disallow 40 MHz operation in the 2.4 GHz band because of the difficult coexistence issues. However, enough members in the task group believed that a solution was possible, however imperfect, and the argument was made that since 40 MHz operation provided a significant performance gain over 20 MHz operation at little additional product cost, vendors would develop 40 MHz devices whether or not the standard permitted 40 MHz operation. It was felt, although perhaps by only a slim majority of the members, that standardized operation would provide better coexistence in the end than attempting to disallow it, which would effectively lead to non-standardized 40 MHz operation.

The 2.4 GHz band is composed of channels spaced 5 MHz apart. The vast majority of legacy deployments, however, make use of channels 1, 6, and 11, which is clearly shown in an informal survey of access points taken in the San Francisco Bay area, USA (1088 APs) and in the Netherlands, Belgium, and Italy (1722 APs) shown in

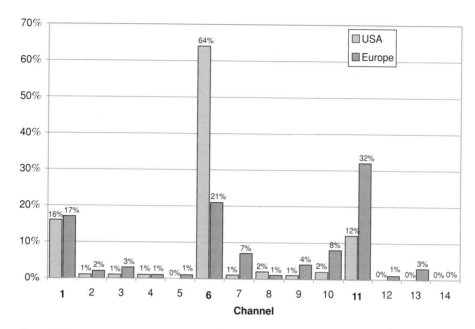

Figure 10.2 Percentage of legacy APs using a given channel from an informal survey of 1088 APs in San Francisco Bay area (representing the USA) and 1722 APs in the Netherlands, Belgium, and Italy (representing Europe). Reproduced from van Nee *et al.* (2006).

Figure 10.2. These are the so called non-overlapping channels and are spaced 25 MHz apart and referred to here as the widely used channels.

A major issue with supporting 40 MHz operation in the 2.4 GHz band is coexistence with neighboring BSSs on one or more of these widely used channels. To highlight the issue consider Figure 10.3. This shows two 40 MHz channel options coexisting with possible neighboring 20 MHz BSSs on channels 1, 6, and 11. If neighboring BSSs exist on only two of the widely used channels then it is possible to select adjacent primary and secondary channels such that the primary channel coincides with one of the used channels and the secondary channel occupies spectrum overlapping only unutilized channels.

If all three of the widely used channels are occupied then channel selection is more problematic. If one of the widely used channels is selected for the primary channel then the secondary channel will partially overlap another of the widely used channels.

Another issue with 40 MHz operation in the 2.4 GHz band is the limited bandwidth available. Two 40 MHz BSSs can be established in regulatory domains, such as Europe, where 13 (or more) channels are available. However, in the USA where only 11 are available, only one non-overlapping 40 MHz BSS can be established.

10.3.3.3 A brief history of 40 MHz in the 2.4 GHz band

The first draft of the 802.11n amendment (IEEE, 2007b) paid scant attention to the coexistence issues in the 2.4 GHz band. CCA was not required on the secondary channel and this meant that a 20/40 MHz station could disregard any traffic on channels overlapping

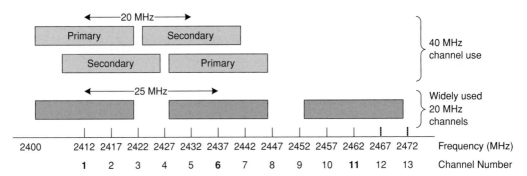

Figure 10.3 40 MHz and 20 MHz channel use in the 2.4 GHz ISM band. Reproduced with permission from van Nee *et al.* (2006) © IEEE.

the secondary channel when transmitting. When selecting a channel for a 20/40 MHz BSS, it was recommended that the AP scan for neighboring BSSs prior to the establishment of the 40 MHz BSS, but no recommendations were made on how to behave if a legacy BSS started operating nearby after a channel had been selected.

Leading up to the second draft, simulation results were presented (Kandala *et al.*, 2006) that showed that a heavily loaded 20/40 MHz BSS would significantly impair the operation of a neighboring 20 MHz BSS on or overlapping with the secondary channel. The results also showed that simply performing an energy detect CCA on the secondary channel and deferring for any detected transmissions significantly improved the situation. Combining CCA on the secondary channel with CCA on the primary channel and performing the full random backoff procedure provided a moderate further improvement for traffic on the neighboring BSS. Based on these results the group adopted the requirement that CCA be performed on the secondary channel and that the 20/40 MHz station defer for traffic on the secondary channel. It was decided, however, that the full random backoff based on CCA on the secondary channel was not required.

Scanning for neighboring BSSs prior to establishing a 20/40 MHz BSS was made mandatory. Rules were adopted that would prevent the operation of a 20/40 MHz BSS when existing BSSs are detected on or overlapping the secondary channel.

Concerns were also expressed that non-802.11 use of the 2.4 GHz band might be interfered with by 40 MHz operation. The example given was an IEEE 802.15.1 (Bluetooth) wireless personal area network (WPAN) that can operate on a reduced channel set in the presence of 20 MHz 802.11 transmissions but would fail if the number of available channels fell below a certain threshold, as may happen in the presence of 40 MHz 802.11 transmissions. As a result, the 40 MHz Intolerant information bit was adopted which may be set by any station to prevent 40 MHz operation. A station that has knowledge of incompatibilities with 40 MHz operation (perhaps a system with both 802.11 and 802.15.1 radios) could set this bit and thus prevent an AP from operating a 40 MHz BSS. Somewhat controversially, there are no restrictions on the use of this bit and any station can set this bit for any reason.

10.3.4   Clear channel assessment in 20 MHz

An HT station with an operating channel width of 20 MHz must provide clear channel assessment (CCA) as follows:

- CCA must indicate busy with a sensitivity of at least −82 dBm to a valid 20 MHz PPDU transmission. This is referred to as signal detect CCA.
- CCA must indicate busy with a sensitivity of at least −62 dBm to any signal in the 20 MHz channel. This is referred to as energy detect CCA.
- A station that does not support the reception of Greenfield format PPDUs must still indicate CCA busy with a sensitivity of at least −72 dBm for any valid Greenfield format PPDU.

The first two requirements are identical to the 802.11a and 802.11g requirements. The last requirement ensures that even a station that is unable to receive the optional Greenfield format PPDUs will defer for Greenfield format PPDUs at sensitivity levels better than the normal energy detect levels.

10.3.5   Clear channel assessment in 40 MHz

An HT station with an operating channel width of 40 MHz must provide CCA on both the primary and secondary channels. CCA goes busy under the following conditions:

- When the secondary channel is idle, CCA on the primary channel must go busy with a sensitivity of at least −82 dBm for a valid 20 MHz PPDU transmission.
- When the secondary channel is idle, CCA for the primary channel must go busy with a sensitivity of at least −62 dBm to any signal on the primary channel.
- A station that does not support the reception of Greenfield format PPDUs must still indicate CCA busy with a sensitivity of at least −72 dBm for any valid 20 MHz Greenfield format PPDU on the primary channel when the secondary channel is idle.
- CCA for both the primary and secondary channel must go busy with a sensitivity of at least −79 dBm to a valid 40 MHz PPDU transmission.
- CCA for both the primary and secondary channel must go busy with a sensitivity of at least −59 dBm to any signal present on both channels.
- A station that does not support reception of Greenfield format PPDUs must still indicate CCA busy for both the primary and secondary channel with a sensitivity of at least −69 dBm for any valid 40 MHz Greenfield format PPDU.
- When the primary channel is idle, CCA for the secondary channel must go busy with a sensitivity of at least −62 dBm to any signal on the secondary channel.

The 40 MHz CCA requirements are thus a simple extension of the 20 MHz requirements and essentially allow for expected 20 MHz traffic on the primary channel and 40 MHz traffic on the two bonded 20 MHz channels. Only energy detect is performed on the secondary 20 MHz channel.

10.3.6   Channel access for a 40 MHz transmission

A 20/40 MHz station gains an EDCA TXOP for transmission on slot boundaries based on the 20 MHz primary channel CCA. The station shall, however, only transmit a 40 MHz PPDU if the secondary channel has been idle for at least a PIFS interval prior to the expiry of the backoff counter. If the secondary channel has not been idle for a PIFS interval then the station must restart its backoff count by selecting a random number in the same contention window as the previous backoff count.

A TXOP obtained for the transmission of 40 MHz PPDUs may also be used for 20 MHz PPDUs. Note that this means that during the TXOP the secondary channel would be idle during the 20 MHz transmission and a 20 MHz station on the secondary channel may begin transmitting, possibly interfering with the 40 MHz transmissions.

10.3.7   NAV assertion in a 20/40 MHz BSS

An HT station updates its NAV using the Duration/ID field value in any frame received in a 20 MHz PPDU in the primary channel or received in a 40 MHz PPDU. NAV is set if the RA of the received frame does not match the MAC address of the receiving station. A station does not need to set its NAV in response to 20 MHz frames received on the secondary channel even if it is capable of receiving those frames.

10.3.8   OBSS scanning requirements

Before an AP establishes a 20/40 MHz BSS it must scan for existing BSSs in the channels in the frequency range affected by the new BSS. This type of scanning is referred to as overlapping BSS (OBSS) scanning. It is expected that the scanning will allow the AP, either autonomously or through a network operator, to select the most appropriate channels for 20/40 MHz BSS operation from the channels available in the regulatory domain. The 802.11n amendment does not dictate rules for selecting the most appropriate pair of channels, but does provide rules that prevent 20/40 MHz operation in a manner that would interfere with existing 20 MHz or 20/40 MHz BSSs.

As with legacy 20 MHz BSSs, scanning to improving coexistence is not required for 20 MHz HT BSSs. However, it is recommended that a 20 MHz BSS not be established in the secondary channel of an existing 20/40 MHz BSS.

10.3.8.1   Establishing a 20/40 MHz BSS in the 5 GHz bands

If an AP chooses to start a 20/40 MHz BSS in the 5 GHz bands that occupies the same two channels as an existing 20/40 MHz BSS then it must ensure that the primary and secondary channels of the new BSS match the primary and secondary channels of the existing BSS. If two or more 20/40 MHz BSSs are present with non-matching primary and secondary channels then the AP is free to match primary and secondary channels with any of the existing BSSs. This situation may arise if existing BSSs cannot detect each other as shown in Figure 10.4.

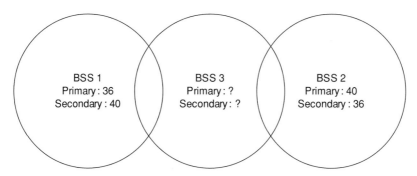

Figure 10.4 Establishing a 20/40 MHz BSS between two existing 20/40 MHz BSSs with non-matching primary and secondary channels. Reproduced with permission from Chan *et al.* (2007) © IEEE.

It is recommended that the AP not establish a 20/40 MHz BSS where the secondary channel is occupied by a 20 MHz BSS. Conversely, it is recommended that an AP not establish a 20 MHz BSS in the secondary channel of an existing 20/40 MHz BSS.

Unlike the 2.4 GHz band, there are no overlapping 20 MHz channels in the 5 GHz band. Once the 20/40 MHz BSS has been established the AP may periodically scan the secondary channel for 20 MHz OBSSs.

10.3.8.2 Establishing a 20/40 MHz BSS in the 2.4 GHz band

Because of the limited bandwidth and overlapping 20 MHz channels, the rules for establishing a 20/40 MHz BSS in the 2.4 GHz band are more restrictive than for the 5 GHz bands. Before an AP can establish a 20/40 MHz BSS in the 2.4 GHz band, it must determine that the following conditions are true over the frequency range affected by the new BSS:

1. The primary and secondary channels of the new BSS are the same as the primary and secondary channels of all existing 20/40 MHz BSSs.
2. The primary channel is the same as the operating channel of all existing 20 MHz BSSs.

The frequency range affected includes all channels with center frequencies that are within 25 MHz of the center frequency of the 40 MHz channel, which could be up to 11 channels. Figure 10.5 shows an example of a permitted 20/40 MHz BSS under these rules.

In the USA where there are 11 channels in the 2.4 GHz band, only one non-overlapping 20/40 MHz BSS can be established. Thus the first rule implies that if there is an existing 20/40 MHz BSS in the 2.4 GHz band the AP must ensure that the primary and secondary channels of the new BSS be the same as the primary and secondary channels of the existing BSS.

In Europe where there are 13 channels available in the 2.4 GHz band, it is possible to establish two non-overlapping 20/40 MHz BSSs, one using the channel pair (1, 5) and another using the channel pair (9, 13). However, if an existing 20/40 MHz BSS is not

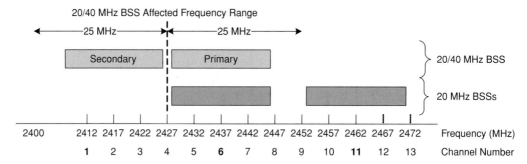

Figure 10.5 An example of permitted 20/40 MHz BSS operation.

using one of these two channel pairs, then the new BSS must use the same primary and secondary channel as the existing BSS.

There are some instances where establishing a 20/40 MHz BSS may not be possible at all. For example, if there are existing 20 MHz BSSs on each of the widely used channels 1, 6, and 11 or even when there are just two 20 MHz BSSs, for example using channels 4 and 11.

In a situation similar to that in Figure 10.4, where two existing 20/40 MHz are hidden from each other and using non-matching primary and secondary channels, it is not permitted to establish a 20/40 MHz BSS.

10.3.8.3   OBSS scanning during 20/40 MHz BSS operation

During the operation of a 20/40 MHz BSS in the 2.4 GHz band, active 20/40 MHz HT stations associated with the BSS are required to monitor the channels affected by the frequency range of the BSS to ensure that the conditions that allowed the establishment of the BSS do not change. The AP itself is not required to perform any monitoring, although it may still do so.

Monitoring stations keep a local record of the channels that are in use by other BSSs. Monitoring stations that receive Beacons determine the primary channel on which the BSS is operating by examining the DS Parameter Set element in the Beacon frame rather than from the channel number of the channel on which the Beacon was received. This is because it is possible for a station to receive a DSSS format Beacon with a 5 MHz center frequency offset, i.e. from an adjacent channel in the 2.4 GHz band, thus the center frequency of the receiver is not a reliable indicator of the channel. The existence of the secondary channel is determined by examining the Secondary Channel Offset field in the HT Information element, if present.

If a change occurs to the local record that would prohibit the operation of the 20/40 MHz BSS, or if the channels in use by neighboring BSSs change, then the station sends a 20/40 Coexistence Management frame to the AP. If the 20/40 Coexistence Management frame is sent because conditions prohibit the operation of the 20/40 MHz BSS, then the 20 MHz BSS Width Request field is set to 1. On receiving a 20/40 MHz Coexistence Management frame with the 20 MHz BSS Width Request field set to 1, the AP must immediately switch the BSS to 20 MHz operation.

With the 20/40 Coexistence Management frame, the monitoring station also reports the list of occupied channels. The channel list would include channel numbers for the channels, other than the primary channel of the BSS to which the station is associated, on which at least one 20 MHz BSS has been detected. The list would also include the channel number of the primary channel of any 20/40 MHz BSS with a primary channel and secondary channel that does not match the primary and secondary channel of the BSS to which the station is associated.

If the channels in use by neighboring BSSs change, the monitoring station reports the change in the channel list in the 20/40 Coexistence Management frame. The AP uses the in-use channel information reported by all monitoring stations as well as its own information to determine if conditions permit the operation of a 20/40 MHz BSS. If such conditions persist for dot11BSSWidthChannelTransitionDelayFactor × dot11BSS-WidthTriggerScanInterval seconds (see MIB variables for 20/40 MHz scanning in Table 10.1), the AP may resume 20/40 MHz BSS operation.

An AP switches from 20/40 MHz BSS operation to 20 MHz BSS operation by setting the Secondary Channel Offset field to 0 in the next Beacon transmission. The AP will also set the STA Channel Width to 0 indicating that it can only receive 20 MHz transmissions. Similarly, the AP may switch from 20 MHz BSS operation to 20/40 MHz BSS operation by setting the Secondary Channel Offset field to +1 or −1 and updating its STA Channel Width field. An AP may also use the Channel Switch Announcement mechanism to change the BSS operating width or switch channels.

10.3.8.4 Scanning requirements for 20/40 MHz stations

An OBSS scan is a passive or active scan of a set of channels that may be affected by 20/40 MHz BSS operation. In the 5 GHz bands the affected channels are simply the two channels the 20/40 MHz BSS occupies or will occupy. In the 2.4 GHz band the affected channels would be any channels in the regulatory domain that are within 25 MHz of the center frequency of the paired 20 MHz channels the 20/40 MHz BSS will occupy or occupies.

OBSS scans are performed by an AP before it establishes a 20/40 MHz BSS and by active 20/40 MHz HT stations associated with a 20/40 MHz BSS. The parameters for OBSS scanning are listed in Table 10.2 together with their default values. The AP may override the default values in HT stations associated with the BSS by transmitting an Overlapping BSS Scan Parameters element (see Section 11.3.2.4).

An active 20/40 MHz HT station must perform at least 2 scans of each channel every dot11BSSWidthTriggerScanInterval. The scan of each channel may be an active scan or a passive scan. Passive scanning involves simply listening for Beacons while active scanning involves sending Probe Request frames and processing any received Probe Response frames. Active scanning is prohibited in some regulatory domains, for example in some radar bands, necessitating passive scanning.

Each time the station scans a particular channel it must examine that channel for a minimum duration of dot11OBSSScanPassiveDwell for a passive scan and dot11OBSSScanActiveDwell for an active scan. Typically the station will not stay in the scanned channel for much longer since by doing so it may miss traffic on the

Table 10.2 Parameters for 20/40 MHz scanning

Parameter	Default Value
dot11BSSWidthTriggerScanInterval	300 seconds
dot11OBSSScanPassiveDwell	20 TU (where TU = 1024 μs)
dot11OBSSScanActiveDwell	10 TU
dot11OBSSScanPassiveTotalPerChannel	200 TU
dot11OBSSScanActiveTotalPerChannel	20 TU
dot11BSSWidthChannelTransitionDelayFactor	5
dot11OBSSScanActivityThreshold	25 × 1/100 percent (= 0.25%)

BSS channels. Thus the station will revisit the channel periodically over the scan interval. The station must scan each channel for a minimum total time of dot11OBSSScanPassiveTotalPerChannel or dot11OBSSScanActiveTotalPerChannel respectively for a passive or active scan respectively.

Not all stations are required to scan, only the AP prior to establishing a 20/40 MHz BSS and active stations. An active station is a station for which the fraction of time spent transmitting MSDUs and receiving unicast MSDUs directed to that station over the prior dot11BSSWidthChannelTransitionDelayFactor scan intervals exceeds the threshold dot11OBSSScanActivityThreshold/100. In other words, if

$$
\begin{aligned}
&\text{STA activity} \\
&= \frac{\text{time spent transmitting MSDUs} + \text{time spent receiving unicast MSDUs}}{\text{activity period}}
\end{aligned}
\tag{10.1}
$$

where

$$
\begin{aligned}
\text{activity period} = \; &\text{dot11BSSWidthChannelTransitionDelayFactor} \\
&\times \text{dot11BSSWidthTriggerScanInterval}
\end{aligned}
\tag{10.2}
$$

and

$$
\text{STA activity} > \frac{\text{dot11OBSSActivityThreshold}}{100}
\tag{10.3}
$$

then the station is considered active and must perform OBSS scanning over the next scan interval.

During the development of the 802.11n amendment there was some controversy surrounding the requirement that stations perform scanning. Companies with an interest in handhelds wanted those devices to be exempt from the scanning requirements citing concern regarding power consumption. However, an analysis of the additional power consumption likely from the scanning requirements (Perahia, 2007) shows that the burden is minimal, even for a handheld device in standby (waiting for a call). The compromise reached was to set an activity threshold below which a station would not be required to scan. The default activity threshold of 0.25% would require devices with an active voice over IP (VoIP) call to scan (during a VoIP call a station typically receives or sends a short frame every 20 ms), but not if the station was in standby where there is only a small amount of data transfer to maintain connectivity or where short text messages may be sent and received. Stations can also always re-associate with the AP as 20 MHz stations (more than adequate for VoIP) to avoid the scanning requirement.

10.3.9   Signaling 40 MHz intolerance

A station operating in the 2.4 GHz band may set the Forty MHz Intolerant bit in the HT Capabilities Information element to 1 to indicate to the AP with which it is associating that it may not operate a 20/40 MHz BSS. A station may also periodically broadcast a 20/40 BSS Coexistence Management frame containing a 20/40 BSS Coexistence element with the Forty MHz Intolerant bit set to indicate to any neighboring BSSs that they may not operate a 20/40 MHz BSS. The concept of 40 MHz intolerance does not apply to stations operating in the 5 GHz bands and such stations must always set this bit to 0.

An AP with a member station that has the Forty MHz Intolerant bit set in its HT Capabilities element will switch to 20 MHz BSS operation if it is operating a 20/40 MHz BSS at the time of the association. The AP will continue with 20 MHz BSS operation as long as the station remains associated.

An AP that receives a 20/40 BSS Coexistence element with the Forty MHz Intolerant bit set must also switch to 20 MHz BSS operation. The prohibition on 20/40 MHz BSS operation continues for dot11BSSWidthChannelTransitionTime × dot11WidthTriggerScanInterval seconds after the Forty MHz Intolerant bit was last detected and an AP may resume 20/40 MHz operation if it no longer receives a Forty MHz Intolerant indication after that period.

The net effect of these rules is that a station that cannot tolerate 40 MHz operation is able to disable 40 MHz operation in the BSS to which it is associated and any neighboring BSSs that detect its periodic 20/40 BSS Coexistence Management frame transmission.

10.3.10   Channel management at the AP

The 802.11h amendment (Spectrum and Transmit Power Management Enhancements in the 5 GHz band in Europe) introduced a mechanism for moving a BSS to a different channel. A Channel Switch Announcement element and a Channel Switch Announcement frame were defined along with a procedure for performing the channel switch.

The ability to switch channels dynamically is considered an important mechanism for supporting 40 MHz coexistence with neighboring 20 MHz BSSs. In addition, it was thought that the mechanism could be used to switch channel operating widths. An AP that detects significant traffic from neighboring BSSs on the secondary channel, or significant traffic on the primary channel for that matter, could move the BSS to a channel pair with less traffic and/or narrow the operating channel width. However, the original Channel Switch Announcement element, designed for 20 MHz channels, could not handle the channel pairing used to create 40 MHz channels.

As it turned out, another task group working on the 802.11y amendment (3650-3700 MHz Operation in the USA (IEEE, 2007b)) encountered a similar limitation in that the Channel Switch Announcement element did not support switching to a new regulatory domain. To overcome these limitation and maintain backward compatibility, a new Extended Channel Switch Announcement element and Extended Channel Switch Announcement frame are defined. The new element and frame are flexible enough to support the requirements of both task groups. The Extended Channel Switch Announcement element includes a New Regulatory Class field to indicate the regulatory class of the new channel to which the BSS is switching. For 802.11n, the regulatory class codes (see Appendix 5.1) have been defined so that the channel width (20 MHz or 40 MHz) can be determined from it and, if 40 MHz wide, which 20 MHz channel is the primary and which the secondary.

The basic procedure for switching channels is as follows. The decision to switch channels is made by the AP and the AP should select a new channel that is supported by all associated stations. The AP informs associated stations that the BSS is moving to a new channel and/or changing operating channel width at some point in the future by including the Extended Channel Switch Announcement element in Beacon frames and Probe Response frames. It may also send out one or more Extended Channel Switch Announcement frames that include the Extended Channel Switch Announcement element.

The AP will attempt to schedule the channel switch so that all stations in the BSS, including stations in power save mode, have an opportunity to receive at least one Extended Channel Switch Announcement element before the switch. A scheduled channel switch occurs just before a target beacon transmission time (TBTT). The Channel Switch Count field indicates the number of TBTTs until the switch, including the TBTT just before which the switch occurs. A value of 1 indicates that the switch occurs just before the next TBTT.

An AP may indicate that the switch is to occur immediately (or anytime in the near future) by setting the Channel Switch Count value to 0. The AP may force stations in the BSS to stop transmissions until the channel switch occurs by setting the Channel Switch Mode field to 1.

10.4   A summary of fields controlling 40 MHz operation

A number of fields in various management frames control the use of 40 MHz transmissions. These fields are as follows:

- The Supported Channel Width Set field in the HT Capabilities element.
- The Secondary Channel Offset field in the HT Information element.
- The STA Channel Width field in the HT Information element.
- The STA Channel Width field of the Notify Channel Width action frame.
- The Extended Channel Switch Announcement element.

The Supported Channel Width Set field is used to indicate whether the station is capable of transmitting and receiving 40 MHz PPDUs. Once a station has associated with the BSS it cannot change this value without first disassociating. The AP may change this value to indicate a new channel width for the BSS.

The Secondary Channel Offset field is used by the AP to indicate whether or not the BSS is occupying a pair of 20 MHz channels, and if it is, whether the secondary channel is above or below the primary channel in frequency. This field will only have a non-zero value, indicating the presence of a secondary channel, if the Supported Channel Width field is set to 1.

The STA Channel Width field is used to indicate a station's current operating width (both the AP and BSS member stations). A value of 0 indicates that the station is not capable of receiving or transmitting a 40 MHz PPDU. A station can only set this field to 1 if the Supported Channel Width Set field is set to 1.

The Extended Channel Width Switch Announcement element can be used to indicate a change in the operating width of the BSS. Following the change in operating width, subsequent management frames will reflect the new operating width in the Supported Channel Width Set and Secondary Channel Offset fields.

A station that is associated with a BSS or the AP itself can only send a 40 MHz PPDU if the following conditions are met:

- The Supported Channel Width Set field of both stations is set to 1.
- The Secondary Channel Offset field of the most recently received HT Information element is set to a non-zero value.
- The AP has not executed a channel switch to 20 MHz channel width. Specifically, the most recently received Extended Channel Switch Announcement element transmitted by the AP did not have the Channel Switch Count field set to 0 and a New Regulatory Class value that does not indicate a 40 MHz channel.

A station may also dynamically change its channel width from 40 MHz to 20 MHz (or vice versa). So a station will also not transmit a 40 MHz PPDU if either of the following occurred:

- The most recently received Notify Channel Width Action frame from the peer station had the STA Channel Width field set to 0.
- The most recently sent Notify Channel Width Action frame to the peer station had the STA Channel Width field set to 0.

10.5   Phased coexistence operation (PCO)

Phased coexistence operation (PCO) is an optional BSS mode with alternating 20 MHz and 40 MHz phases of operation controlled by the AP. PCO allows a 20/40 MHz BSS to operate when there are neighboring 20 MHz BSSs on both the primary and secondary 20 MHz channels making up the 40 MHz channel. In practice, because of the overhead and potential real-time disruption associated with PCO, it is likely to only be used if

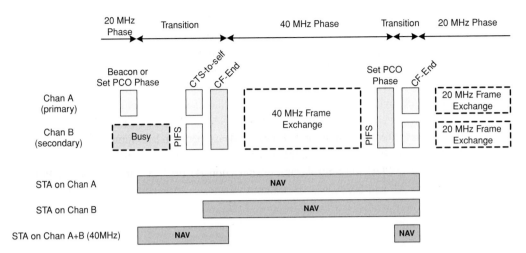

Figure 10.6 PCO 20 and 40 MHz phase transitions. Reproduced with permission from IEEE (2007a) © IEEE.

it is not possible to find a 40 MHz channel without neighboring traffic on both of the 20 MHz channels. Real-time disruption occurs with PCO because the 20 MHz stations are prevented from transmitting at all during the relatively long 40 MHz phases of operation. Also, if the neighboring 20 MHz BSSs are heavily loaded, then it may be more efficient from an overall network throughput perspective to operate in 20 MHz only.

A PCO capable AP advertises its BSS as 20/40 MHz BSS Channel Width and PCO Capable. A non-PCO capable station may associate with the BSS as it would any 20/40 MHz BSS.

A PCO capable station sets the PCO field in the HT Extended Capabilities information element to 1.

10.5.1   Basic operation

The PCO AP periodically transitions between 20 MHz and 40 MHz phases of operation using the sequences shown in Figure 10.6. To transition to 40MHz operation, the AP sends a Beacon or Set PCO Phase management frame on the primary channel and uses it to set the NAV of 20MHz stations operating in that channel as well as 20/40MHz stations. The AP then waits for the secondary channel to go idle. If the secondary channel does not go idle for an excessively long time, the AP may cancel the transition to 40 MHz operation by sending another Set PCO Phase management frame.

If the secondary channel goes idle then the AP transmits a CTS-to-self on both the primary and secondary channels using the non-HT duplicate mode. The CTS-to-self sets the NAV of stations on both 20 MHz channels to cover the expected 40MHz phase of operation. Finally, the AP sends a CF-End in a 40 MHz HT PPDU to reset the NAV of 40 MHz capable stations. The 40 MHz stations then begin contention for channel access.

The AP may extend the 40 MHz phase by transmitting a Set PCO Phase management frame in a non-HT duplicate PPDU followed by a CF-End sent in a 40 MHz HT PPDU. The Set PCO Phase management frame will set the NAV in all stations and the CF-End will reset the NAV in the 40 MHz stations so that they can compete for access again.

At some point the PCO AP begins a transition to 20 MHz operation by gaining priority channel access to send a Set PCO Phase management frame in a 40 MHz HT PPDU. It follows this with a CTS-to-self sent in a non-HT duplicate mode PPDU, which resets the NAV of 20 MHz stations and allows 20 MHz operation to resume.

During the 40 MHz phase, a 20/40 MHz PCO station transmits data frames using a 40 MHz HT PPDU and control frames using a non-HT duplicate or a 40 MHz HT PPDU, except for any CF-End frame, which can only be sent using a 40 MHz HT PPDU.

10.5.2   Minimizing real-time disruption

To minimize the access delay 20 MHz stations experience because of the 40 MHz phase of operation, the AP has a management variable dot11PCO40MaxDuration, which defines the maximum duration of the 40 MHz phase. Similarly, dot11PCO20MaxDuration defines the maximum duration of the 20 MHz phase.

Also, in order for the PCO AP to provide sufficient opportunity for stations to send frames, a minimum duration for the 40 MHz and 20 MHz phases is also defined in dot11PCO40MinDuration and dot11PCO20MinDuration, respectively.

10.6   Protection

Mechanisms are necessary to protect HT transmissions and certain HT sequences from stations that may not recognize these formats and thus not defer correctly. However, protection mechanisms add significant overhead and an attempt should be made to only use these mechanisms when needed. As with previous 802.11 amendments, it is expected that early networks will consist of a heterogeneous mix of legacy devices and 802.11n conformant devices. To complicate things, the 802.11n amendment includes many optional features and it is expected that early implementations will include varying support for many of the 802.11n optional features. Over time it is expected that implementations will become more consistent in the features supported as devices become more fully featured. It is expected that while early networks will be heterogeneous and require protection mechanisms, these mechanism may be required less and less as networks become more fully featured and homogeneous.

The protection mechanisms described here essentially ensure that a potential interferer defers transmission for a known period allowing an HT station to complete its HT frame sequence. Multiple mechanisms are available to perform this function, some of which depend on the capabilities of the devices directly involved in the exchange, the capabilities of third party devices, and the devices against which protection is sought.

10.6.1   Protection with 802.11b stations present

Protecting new frame formats and sequences is not new to 802.11. The 802.11g amendment, which introduced the enhanced rate PHY (ERP) for OFDM operation in the 2.4 GHz ISM band, required a protection mechanism against the widely deployed 802.11b (DSSS/CCK) devices present in the band at the time. Two mechanisms were available to protect OFDM transmissions: the RTS/CTS exchange and the lower overhead CTS-to-Self transmission. Both mechanisms use legacy compatible frames to set the NAV in the legacy station(s) to cover the OFDM transmission sequence.

Because of the inherent overhead associated with both these mechanisms, it is beneficial to only require protection in the presence of legacy 802.11b stations. The mechanism used to enable protection is for the AP to set the Use_Protection bit in the ERP Information element in the Beacon and Probe Response frames. The AP sets this bit if at least one 802.11b station is associated with BSS. The AP may, but is not required to, also set this bit if it detects the presence of 802.11b stations in a neighboring BSS. A station associated with the BSS, seeing this bit set, knows to begin a sequence containing an OFDM frame with a DSSS/CCK format RTS/CTS exchange or CTS-to-Self transmission.

This protection mechanism is carried forward to HT BSSs. If an HT station sees that the Use_Protection bit is set to 1 in the ERP Information element, then it must protect non-HT and HT OFDM sequences with a DSSS/CCK format RTS/CTS exchange or CTS-to-Self transmission. In the case of a 40 MHz HT sequence the RTS/CTS exhange or CTS-to-Self transmission occurs on the primary channel.

The RTS/CTS and CTS-to-Self mechanisms are described in more detail below.

10.6.2   Protection with 802.11g or 802.11a stations present

HT transmissions are inherently protected against 802.11g and 802.11a (non-HT) stations through the use of the HT mixed format frame. This frame format includes a legacy compatible preamble, which allows non-HT stations to defer correctly for the duration of the frame.

However, some HT transmissions may not be interpreted correctly by non-HT stations and even some HT stations. In particular, non-HT stations and some HT stations may not be able to defer correctly for HT Greenfield format PPDUs. Also, certain non-HT stations may not defer correctly for burst sequence where the shorter RIFS spacing is used between frames. For these situations, additional mechanisms are needed to protect these sequences.

A broader set of protection mechanisms are available for protecting HT Greenfield format and RIFS burst transmissions when only 802.11g or 802.11a non-HT and HT stations are present, than is available for protection when 802.11b stations are present. These protection mechanisms are:

- RTS/CTS
- CTS-to-Self

- a non-HT or HT mixed format PPDU soliciting a non-HT response PPDU
- L-SIG TXOP protection.

These mechanisms offer protection with varying degrees of robustness and overhead and are discussed in more detail later in this section. To begin with though we consider first the conditions under which RIFS bursts and Greenfield format transmissions require protection.

10.6.3   Protection for OBSS legacy stations

The AP may set the OBSS Non-HT STAs Present bit in the HT Information element if it detects non-HT stations on the primary or secondary channel that are not members of the BSS. On seeing this bit set, member stations of the BSS may optionally use protection mechanisms to protect sequences for which a non-HT station may not defer correctly.

10.6.4   RIFS burst protection

Burst sequences that use RIFS may not be receivable by some legacy stations. A legacy station, designed expecting frames to be spaced at least SIFS apart, may not re-arm its acquisition circuit in time following a received frame to detect a second frame if the second frame arrives less than SIFS after the first frame. If the legacy station fails to detect the frame then it may not defer correctly for the duration frame and may begin transmitting before the frame completes.

   An AP can prevent RIFS from being used in a sequence by any station associated with the BSS by setting the RIFS Mode subfield in the HT Information element to 0. A station may only use RIFS bursting if the RIFS Mode subfield is set to 1.

   If RIFS bursting is permitted then the station may, but is not required to, protect RIFS sequences if there are legacy stations present. Given the overhead associated with protection it is likely that implementations would use SIFS rather than provide protection unless protection was needed for other reasons. A station knows that legacy (non-HT) stations may be present when the HT Protection subfield in the HT Information field is set to 1 (HT BSS members, non-HT non-members) or 3 (Non-HT BSS members present).

10.6.5   Greenfield format protection

The Greenfield format preamble is shorter and thus more efficient than the mixed format preamble. However, the Greenfield format preamble is not compatible with legacy stations and, support being optional, is also not receivable by some HT stations. An HT station indicates that it is not able to receive the Greenfield format by setting the Greenfield bit in the HT Capabilities element to 0.

   If there are HT stations associated with the AP that are not able to receive Greenfield format PPDUs then the AP will set the Non-Greenfield HT STAs Present bit to 1 in the

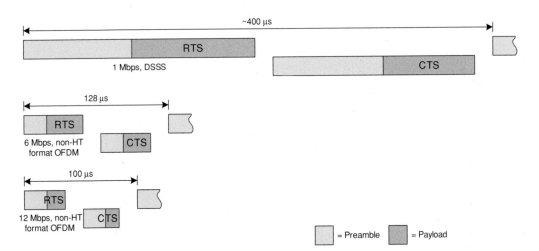

Figure 10.7 RTS/CTS overhead with DSSS and non-HT format PPDUs.

HT Information element. If there are non-HT stations associated with the BSS the AP will set the HT Protection field to 3. When the Non-Greenfield HT STAs Present bit is set to 1 or the HT Protection field is set to 3 in the HT Information element then stations associated with the BSS must protect Greenfield format PPDUs.

There was some opposition in the development of the 802.11n amendment with the requirement that protection is needed when non-Greenfield HT stations are present since these stations are still required to defer for Greenfield format transmissions at signal levels lower than energy detect threshold (see Sections 10.3.4 and 10.3.5).

10.6.6   RTS/CTS protection

An RTS/CTS exchange at the beginning of a TXOP sets the NAV of stations in the vicinity of both the initiator and recipient. The RTS and CTS frames are transmitted at a basic rate and are thus widely received. The robust modulation used and the fact that the widely received frames are transmitted from both ends of the link makes the RTS/CTS exchange the most robust mechanism for establishing protection.

The initiator begins a TXOP by transmitting a RTS frame, setting the Duration field in the RTS to the expected duration of the TXOP less the duration of the RTS frame itself. The responder sends a CTS frame setting the Duration field to the value seen in the RTS frame less SIFS less the duration of the CTS frame. The CTS frame is transmitted with the same modulation and coding as the RTS frame. Stations that successfully demodulate either the RTS or CTS frame or both frames have their NAV set for the duration of the TXOP and will not transmit during that time.

Figure 10.7 shows the relative overhead of an RTS/CTS exchange using 1 Mbps DSSS format, 6 Mbps non-HT OFDM format and 12 Mbps non-HT OFDM format, PPDUs. The DSSS format PPDUs may be used in the presence of 802.11b stations. However,

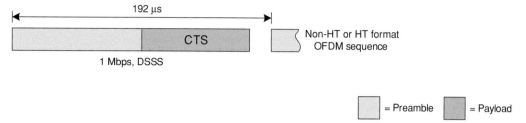

Figure 10.8 CTS-to-Self overhead with 1 Mbps DSSS (short preamble) format PDDU.

because of the high overhead of this exchange, it is more likely that the CTS-to-Self mechanism will be used since it is rare that protection needs to be established from both ends of the link, especially at DSSS rates. The non-HT format exchange may be used to protect HT Greenfield format and RIFS burst sequences in the presence of 802.11a or 802.11g stations. The 12 Mbps rate offers slightly reduced robustness over the 6 Mbps rate, but with lower overhead.

10.6.7   CTS-to-Self protection

CTS-to-Self (Figure 10.8) is a MAC level mechanism for protecting frame sequences from stations in the vicinity of the initiator. This was originally introduced with the 802.11g amendment to protect OFDM transmissions in the 2.4 GHz ISM band from 802.11b stations which were only capable of detecting DSSS/CCK transmissions. In most networks all stations are able to detect transmissions of all other stations and hidden nodes are rare. For such situations, the CTS-to-Self mechanism offered significantly reduced overhead compared with a full RTS/CTS exchange.

At the beginning of a sequence, the initiator sends a CTS frame with the RA field set to its own MAC address and the Duration field set to the expected duration of the sequence less the duration of the CTS frame itself. The CTS frame is transmitted using a basic rate PPDU compatible with the legacy stations present and against which protection is desired.

In practice, CTS-to-Self may only be used to protect a Data/ACK or MMPDU/ACK sequence since TXOP rules require that burst sequences and A-MPDU sequences have a short frame exchange performed at the start of the TXOP as a collision detect mechanism.

CTS-to-Self sent in a non-HT OFDM format PPDU may be used by HT stations that wish to protect Greenfield data/ACK sequences in the presence of 802.11g or 802.11a stations, however, the mandatory existence of the HT mixed format preamble and its lower overhead obviate this use. Thus it is likely that CTS-to-Self will continue to only be used when 802.11b stations are present.

10.6.8   Protection using a non-HT or HT mixed PPDU with non-HT response

The NAV of nearby stations is set by any frame that is correctly received for which the RA is not the address of the station receiving the frame. Thus a sequence that begins

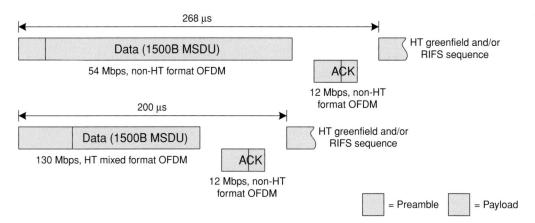

Figure 10.9 A non-HT or HT mixed format PPDU soliciting a non-HT PPDU response.

with a legacy compatible frame, for example a data frame, and that solicits a response frame, an ACK frame in this example, could be used to set NAV in nearby stations. In the example sequence in Figure 10.9, the data frame is likely transmitted using a high order MCS, however, the ACK frame is robustly modulated using a non-HT format PPDU and should be widely received. The ACK frame could thus be used to carry the NAV setting for nearby stations.

The initiator would set the Duration field of the data frame to the expected duration of the TXOP less the duration of the data frame itself. The ACK response would have its Duration field set to the value of the Duration field in the data frame less SIFS less the duration of the ACK frame itself. With NAV established in the surrounding stations, the remainder of the TXOP could include HT Greenfield format PPDUs and/or RIFS burst sequences.

10.6.9   Non-HT station deferral with HT mixed format PPDU

The HT mixed format preamble is designed to be compatible with legacy 802.11g and 802.11a stations in the sense that a legacy station receiving the frame will defer for the duration indicated by the length (L_LENGTH) and rate (L_DATARATE) given in the signaling field (L-SIG) of the legacy compatible portion of the preamble. This is illustrated in Figure 10.10.

The L_DATARATE field is set to indicate a rate of 6 Mbps, which is the lowest OFDM PHY data rate. The L_LENGTH field is set to indicate the desired duration. A legacy station would interpret the L_DATARATE and L_LENGTH fields as follows. At 6 Mbps each 4 μs symbol carries 3 octets (24 bits). At this rate, the overhead in the payload in the 16-bit SERVICE field (which carries the scrambler initialization) and 6-bit Tail (used to flush the decoder) account for one symbol.[1] The payload duration would thus

---

[1] Actually 22 bits, but since the PSDU is an integral number of octets the rounding works.

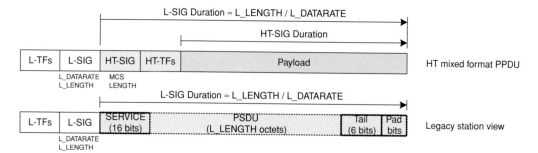

Figure 10.10 Non-HT station defer with HT mixed format PPDU.

be:

$$\text{L-SIG duration} = \left(1 + \left\lceil \frac{\text{L\_LENGTH}}{3} \right\rceil \right) \times 4\,\mu s \qquad (10.4)$$

where $\lceil\,\rceil$ indicates a rounding up to the next integer value. Since the maximum L_LENGTH value is 4095, an HT station could indicate an L-SIG duration of up to 5.464 ms on 4 µs boundaries. An HT station must set the L_LENGTH and L_DATARATE to a value that indicates an L-SIG duration of at least the duration of the frame.

It is worth noting that some legacy devices may not defer correctly for L_LENGTH field values greater than 2340. The legacy signaling field is weakly protected by a single parity bit. Many implementations perform additional checks to ensure that the frame has been correctly detected and the signaling field successfully demodulated, including checking that the length field is less than 2340 (the maximum MPDU size prior to 802.11n). These implementations would interpret a large length value as a false detect and not assert CCA if the signal strength is below the energy detect threshold.

10.6.10   L-SIG TXOP protection

For an HT station receiving an HT mixed format PPDU, the L_LENGTH and L_DATARATE in the legacy signaling field are not needed to demodulate the HT portion of the frame since the actual PSDU length and MCS used are specified in the HT signaling field (HT-SIG). The L_DATARATE and L_LENGTH may thus be used to signal a duration that is longer that the actual frame duration and in so doing protect a sequence rather than just a single frame. This is the basis for L-SIG protection and is illustrated in Figure 10.11.

An HT station sets the L-SIG duration to at least cover the duration of the HT PPDU. If the HT station is sending the PPDU to an HT station that supports L-SIG TXOP protection, the station may set the L-SIG duration to cover a sequence of frames. Since a legacy station is not able to receive another frame for the full L-SIG duration, none of the frames in the protected sequence should be addressed to a legacy station.

An HT station that supports L-SIG TXOP protection and that receives an HT mixed format PPDU will set its NAV at the end of the PPDU to the remaining L-SIG duration

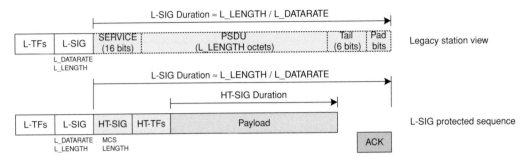

Figure 10.11 L-SIG protection.

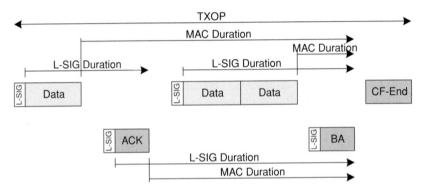

Figure 10.12 Example of an L-SIG TXOP protected sequence. Reproduced with permission from IEEE (2007a) © IEEE.

if the HT station could not successfully extract an MPDU from the PPDU with which it would normally update its NAV using the MPDU's Duration/ID field.

An example of an L-SIG protected sequence is given in Figure 10.12. An L-SIG protected sequence begins with an initial handshake, which is the exchange of two short frames each of which are in HT mixed format PPDUs. Any initial frame sequence may be used that is valid for the start of a TXOP provided the duration of the response frame is known beforehand by the initiator.

The L-SIG duration of the initial PPDU is set to cover the duration of the initial PPDU less the legacy compatible preamble plus SIFS plus the duration of the response frame. The Duration field in the MAC frame carried by the PPDU is set to the expected duration of the remainder of the sequence.

The L-SIG duration in the response PPDU will be set to the duration indicated in the Duration field from the MAC frame less SIFS less the duration of the legacy compatible portion of the preamble.

Following the successful completion of the initial handshake the sequence proceeds similarly using HT mixed format PPDUs. The L-SIG Duration of the subsequent PPDU from the initiator has its L-SIG set to cover the remainder of the sequence. The Duration fields in the MAC frames carried by the PPDU also indicate a value that extends to the end of the sequence. The response frames use the Duration field in the MAC frames to set the L-SIG duration.

A drawback to the use of L-SIG TXOP protection is that legacy stations do not successful demodulate any of the frames and will use an EIFS defer for channel access at the end of the protected period. This would put them at a disadvantage in gaining channel access as compared to HT stations which may successfully demodulate the final HT mixed format PPDU. To prevent this, the initiator should follow the protected sequence with a CF-End frame using a basic rate non-HT PPDU.

References

Chan, D., Hart, B., and Qian, Q. (2007). *20/40 MHz Coexistence for 5 GHz: Issues and Proposed Solution Overview*, IEEE 802.11-07/2564r1.

IEEE (2007a). *IEEE P802.11n™/D3.00, Draft Amendment to STANDARD for Information Technology – Telecommunications and Information Exchange Between Systems – Local and Metropolitan Networks – Specific Requirements – Part 11: Wireless LAN Medium Access Control (MAC) and Physical Layer (PHY) Specifications: Amendment 4: Enhancements for Higher Throughput.*

IEEE (2007b). *IEEE P802.11y™/D5.0, Draft Amendment to STANDARD for Information Technology – Telecommunications and Information Exchange Between Systems – Local and Metropolitan Networks – Specific Requirements – Part 11: Wireless LAN Medium Access Control (MAC) and Physical Layer (PHY) Specifications: Amendment 3: 3650-3700 MHz Operation in USA.*

Kandala, S., Jones, V. K., Raissinnia, A., and de Vegt, R. (2006). *Extension Channel CCA Proposed Solutions*, IEEE 802.11-06/0608r2.

Perahia, E. (2007). *MIB Attributes for 40 MHz Scanning in 2.4 GHz*, IEEE 802.11-07/2446r0.

Van Nee, R., Jones, V. K., Awater, G., and de Vegt, R., (2006). *Transmitter CCA Issues in 2.4 GHz*, IEEE 802.11-06/0869r1.

11 MAC frame formats

This section provides details on the MAC frame formats. The information provided here is sufficiently detailed to act as a reference for the topics discussed in this book, but it does not provide an exhaustive list of all field elements, particularly in the management frames. For a detailed treatment of the frame formats refer to the actual specification (IEEE 2007a, 2007b).

11.1 General frame format

Each MAC frame consists of the following:

- a MAC header
- a variable length frame body that contains information specific to the frame type or subtype
- a frame check sequence or FCS that contains a 32-bit CRC.

This frame format consists of a set of fields that occur in a fixed order as illustrated in Figure 11.1. Not all fields are present in all frame types.

11.1.1 Frame Control field

The Frame Control field is shown in Figure 11.2 and is composed of a number of subfields described below.

11.1.1.1 Protocol Version field

This field is 2 bits in length and is set to 0. The protocol version will only be changed when a fundamental incompatibility exists between a new revision and the prior edition of the standard, which to date has not happened.

11.1.1.2 Type and Subtype fields

The Type and Subtype fields together identify the function of the frame. There are three frame types defined: control, data, and management. Each frame type has several subtypes defined and the combinations are listed in Table 11.1.

The Data frame, Null, CF-Ack, CF-Poll, and variants (subtypes 0000 through 0111) were introduced in the original 802.11 specification. The QoS Data frame subtype

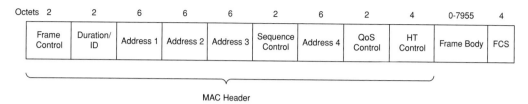

Octets	2	2	6	6	6	2	6	2	4	0-7955	4
	Frame Control	Duration/ ID	Address 1	Address 2	Address 3	Sequence Control	Address 4	QoS Control	HT Control	Frame Body	FCS

MAC Header

Figure 11.1 MAC frame format. Reproduced with permission from IEEE (2007b) © IEEE.

B0	B1 B2	B3 B4	B7	B8	B9	B10	B11	B12	B13	B14	B15
Protocol Version	Type	Subtype		To DS	From DS	More Frag	Retry	Pwr Mgt	More Data	Protected Frame	Order

Figure 11.2 Frame Control field. Reproduced with permission from IEEE (2007a) © IEEE.

and variants (1000 through 1111) were introduced with the 802.11e amendment (QoS Enhancements).

The Control Wrapper frame was introduced in 802.11n and may be used in place of any other control frame. It replicates the fields of the control frame it replaces and adds an HT Control field. It is called a wrapper frame since it effectively wraps the original control frame in a new frame and adds an HT Control field.

The Action management frame was introduced with the 802.11h amendment (Spectrum and Transmit Power Management Extensions). Since the numbering space for management frames is almost exhausted, the Action frame subtype has been used for many of the new management frames subsequent to that amendment. The Action No Ack management frame was introduced in the 802.11n amendment and serves a similar purpose, but is not acknowledged by the receiving peer.

11.1.1.3   To DS and From DS fields
The meanings of the To DS and From DS combinations are given in Table 11.2.

11.1.1.4   More Fragments field
This field is set to 1 in all data or management frames that have another fragment of the current MSDU, A-MSDU, or MMPDU to follow. It is set to 0 in MPDUs that contain a complete MSDU or A-MSDU and MPDUs that contain the last fragment of an MSDU or A-MSDU.

11.1.1.5   Retry field
The Retry field is set to 1 in any data or management frame that is a retransmission of an earlier frame. It is set to 0 in all other frames. A receiving station uses this indication to aid in eliminating duplicate frames.

Table 11.1 Valid type and subtype combinations (IEEE, 2007b)

Type	Type description	Subtype	Subtype description
00	Management	0000	Association Request
		0001	Association Response
		0010	Reassociation Request
		0011	Reassociation Response
		0100	Probe Request
		0101	Probe Response
		0110–0111	Reserved
		1000	Beacon
		1001	ATIM
		1010	Disassociation
		1011	Authentication
		1100	Deauthentication
		1101	Action
		1110	Action No Ack
		1111	Reserved
01	Control	0000–0110	Reserved
		0111	Control Wrapper
		1000	Block Ack Request
		1001	Block Ack
		1010	PS-Poll
		1011	RTS
		1100	CTS
		1101	ACK
		1110	CF-End
		1111	CF-End + CF-Ack
10	Data	0000	Data
		0001	Data + CF-Ack
		0010	Data + CF-Poll
		0011	Data + CF-Ack + CF-Poll
		0100	Null (no data)
		0101	CF-Ack (no data)
		0110	CF-Poll (no data)
		0111	CF-Ack + CF-Poll (no data)
		1000	QoS Data
		1001	QoS Data + CF-Ack
		1010	QoS Data + CF-Poll
		1011	QoS Data + CF-Ack + CF-Poll
		1100	QoS Null (no data)
		1101	Reserved
		1110	QoS CF-Poll (no data)
		1111	QoS CF-Ack + CF-Poll (no data)
11	Reserved	0000–1111	Reserved

Table 11.2 To/From DS field value combinations (IEEE, 2007a)

From DS	To DS	Meaning
0	0	Indicates: • a data frame direct from one station to another within the same IBSS • a data frame direct from one non-AP station to another non-AP station within the same BSS • all management and control frames
0	1	A data frame destined for the distribution system (DS) or being sent by a station associated with an AP to the Port Access Entity in that AP
1	0	A data frame exiting the DS or being sent by the Port Access Entity in an AP
1	1	A data frame using the four-address format (not defined in the standard)

11.1.1.6 Power Management field

The Power Management field indicates the power management mode of the station after completing a frame sequence. A value of 0 indicates that the station will be in active mode, while a value of 1 indicates that the station will be in Power Save (PS) mode. This field is always set to 0 by the AP.

11.1.1.7 More Data field

The More Data field is used to indicate to a station in PS mode that more MSDUs or MMPDUs are buffered for that station at the AP.

The More Data field is set to 1 in directed data or management frames transmitted by an AP to a station in PS mode.

The More Data field is set to 1 in broadcast/multicast frames transmitted by the AP when additional broadcast/multicast MSDUs or MMPDUs remain to be transmitted by the AP during this beacon interval.

The AP may set the More Data field to 1 in ACK frames to a QoS capable station with APSD enabled to indicate that the AP has a pending transmission for that station.

11.1.1.8 Protected Frame field

The Protected Frame field, when set to 1, indicates that the Frame Body field has been encrypted. The Protected Frame field may only be set to 1 for data frames and Authentication management frames.

11.1.1.9 Order field

An AP may change the delivery order of broadcast and multicast MSDUs relative to unicast MSDUs to a power saving station. For example, broadcast or multicast traffic may be sent following the beacon when all power save stations are awake, while unicast traffic, perhaps received earlier than the broadcast/multicast traffic, may be delayed and delivered to a specific power save station later. If a higher layer protocol cannot tolerate this reordering then the optional StrictlyOrdered service class should be used. Frames sent using the StrictlyOrdered service class have their Order bit set to 1.

Figure 11.3 Sequence Control field. Reproduced with permission from IEEE (2007a) © IEEE.

As may be appreciated, this field was never widely used. It is reused in 802.11n to indicate the presence of HT Control field in QoS Data frames. Prior to 802.11n, this bit was reserved in QoS Data frames which always had it set to 0.

11.1.2   Duration/ID field

When the value of the Duration/ID field is less than 32 768 (high order bit not set), then the value is interpreted as duration in microseconds and used to update the network allocation vector (NAV). If the two high order bits are set in a PS-Poll frame then the low order 14 bits are interpreted as the association identifier (AID).

11.1.3   Address fields

There are four address fields in the MAC frame format, although not all fields are present in all frames. Address 1 contains the receive address (RA) and is present in all frames. Address 2 contains the transmit address (TA) and is present in all frames except ACK and CTS.

Address 3 is present in data and management frames. In a data frame, the address carried by the Address 3 field is dependent on the To DS and From BS bit settings and whether the frame is carrying a MSDU or A-MSDU (see Table 11.11). In management frames, Address 3 contains the BSSID. Address 4 is only present in data frames and only when both the To DS and From DS bits are set (Table 11.11). The 802.11 specification does not define a usage scenario for this case.

11.1.4   Sequence Control field

The Sequence Control field consists of a 4-bit Fragment Number and a 12-bit Sequence Number as shown in Figure 11.3.

The Sequence Number gives the sequence number of the MSDU or MMPDU. Each MSDU or MMPDU transmitted by a station is assigned a sequence number.

Non-QoS stations assign sequence numbers to MSDUs and MMPDUs from a single module-4096 counter, starting at 0 and incrementing by 1 for each MSDU or MMPDU. QoS stations, in addition to a single counter for legacy data frames and management frames, maintain a modulo-4096 counter for each receive address and TID for which they transmit data.

Table 11.3 QoS Control field (IEEE, 2007b)

Applicable frame subtypes	B0–B3	B4	B5–B6	B7	B8–B15
QoS (+)CF-Poll sent by AP		EOSP			TXOP Limit
QoS Data, QoS Data+CF-Ack, QoS Null sent by AP	TID	EOSP	Ack Policy	A-MSDU Present	AP PS Buffer State
QoS Data frames sent by non-AP station		0			TXOP Duration Requested
		1			Queue Size

Each fragment of an MSDU or MMPDU contains a copy of the sequence number of that MSDU or MMPDU. The fragments are numbered in sequence, starting with 0 for the first fragment.

11.1.5   QoS Control field

The QoS Control field identifies the traffic class (TC) or traffic stream (TS) to which the frame belongs and various other QoS related information about the frame. The QoS Control field is present in QoS Data frames, i.e. data frames of the QoS subtype. The usages of the various subfields in the QoS Control field are given in Table 11.3.

The Traffic Identifier (TID) subfield identifies the TC or TS to which the corresponding MSDU or MSDU fragment in the frame body belongs. The TID subfield also identifies the TC or TS for which a TXOP is being requested through the setting of the TXOP Duration Requested or Queue Size subfields.

The EOSP or End of Service Period subfield is used by the AP to indicate the end of the current service period. The AP sets the EOSP subfield to 1 in the frame which ends the service period and sets it to 0 otherwise.

The Ack Policy subfield determines the acknowledgement policy that is followed by the recipient of the QoS Data frame. The policies are enumerated in Table 11.4.

The A-MSDU Present subfield indicates the presence of an A-MSDU in the body of the frame. This field is reserved (always 0) in QoS CF-Poll, QoS CF-Ack+CF-Poll and QoS Null frames since these frames carry no payload.

11.1.5.1   TXOP Limit subfield

The TXOP Limit subfield is present in QoS (+)CF-Poll frames and specifies the time limit in 32 μs units that the AP has granted for the subsequent TXOP. The TXOP begins a SIFS period after the QoS (+)CF-Poll frame. A TXOP limit value of 0 implies that one MPDU or one QoS Null frame is to be transmitted immediately following the QoS (+)CF-Poll frame.

11.1.5.2   Queue Size subfield

The Queue Size subfield indicates the amount of buffered traffic for a given TID at the non-AP station sending this frame. It is present in QoS Data frames sent with bit 4 of the QoS Control field set to 1. The AP may use the information to determine the

Table 11.4 Ack Policy subfield in QoS Control field (IEEE, 2007b)

B5	B6	Meaning
0	0	Normal Ack or Implicit Block Ack Request QoS Data in non A-MPDU: The addressed recipient returns an ACK or QoS +CF-Ack frame QoS Data in A-MPDU: The addressed recipient returns a BA
1	0	No Ack The addressed recipient does not return an ACK. This is used when the sender does not require acknowledgement for the unicast frame sent or when the frame sent is a broadcast or multicast frame
0	1	No Explicit Acknowledgement or Scheduled Ack under PSMP There may be a response frame to the frame that is received but it is neither an ACK nor any data frame of subtype +CF-Ack. QoS CF-Poll and QoS CF-Ack+CF-Poll frames always use this value Under PSMP, this value indicates a scheduled acknowledgement in the next PSMP-DTT or PSMP-UTT
1	1	Block Acknowledgement The addressed recipient takes no action upon receipt of the frame except for recording the state. The recipient can expect a BAR frame or A-MPDU containing QoS Data frame(s) with Normal Ack ack policy in the future

TXOP duration it will grant the station taking into account competing requests from other stations.

The Queue Size value is the total size of the station's queue for the specified TID rounded up to the nearest multiple of 256 and expressed in units of 256 octets. A Queue Size of 0 indicates the absence of any buffered traffic for that TID. A Queue Size of 254 is used for all sizes greater than 64 768 octets. A Queue Size value of 255 is used to indicate an unspecified or unknown size.

11.1.5.3   TXOP Duration Requested subfield

As an alternative to the Queue Size, the non-AP station may request a desired TXOP Duration. The TXOP Duration Requested subfield is present in QoS Data frames sent with bit 4 of the QoS Control field set to 0.

The TXOP Duration Requested is specified in units of 32 μs. A TXOP Duration of 0 indicates that no TXOP is requested for the specified TID. The TXOP Durations requested are not cumulative; a TXOP Duration request for a particular TID supersedes any prior TXOP Duration request for that TID.

11.1.5.4   AP PS Buffer State subfield

The AP PS Buffer State subfield indicates the power save buffer state at the AP for a particular non-AP station. The AP PS Buffer State subfield is further subdivided into the fields Buffer State Indicated, Highest-Priority Buffered AC, and AP Buffered Load:

- **Buffered State Indicated**. This subfield is set to 1 when the AP Buffered State is specified. It is set to 0 otherwise.

Table 11.5 Link Adaptation Control subfields (IEEE, 2007b)

Bits	Subfield	Definition
B0	Reserved	
B1	TRQ	Sounding Request. Set to 1 to request a sounding PPDU from the responder
B2–B5	MAI	MRQ or ASEL indication. Set to 14 to indicate that the MFB/ASELC field should be interpreted as the ASELC field. Otherwise this field is interpreted as shown in Table 11.6
B6–B8	MFSI	MFB Sequence Identifier. Set to the received value of the MSI to which the MFB information refers. Set to 7 for unsolicited MFB
B9–B15	MFB/ASELC	When the MAI field is set to 14 this field is interpreted as the Antenna Selection Control (ASLEC) field. Otherwise, this field contains the recommended MCS feedback. A value of 127 indicates that no feedback is present

Table 11.6 MAI subfields (IEEE, 2007b)

Bits	Subfield	Definition
B2	MRQ	MCS Request. When set to 1, MCS feedback is requested. When set to 0, no MCS feedback is requested
B3–B5	MSI	MCS Request Sequence Identifier. When MRQ is set to 1, the MSI field contains the sequence number in the range 0 to 6 that identifies the request

B0 — B15	B16 — B17	B18 — B19	B20	B21 — B22 — B23	B24	B25 — B29	B30	B31
Link Adaptation Control	Calibration Position	Calibration Sequence	Reserved	CSI/Steering	NDP Announcement	Reserved	AC Constraint	RDG/More PPDU

Figure 11.4 HT Control field. Reproduced with permission from IEEE (2007b) © IEEE.

- **Highest Priority Buffered AC**. This subfield indicates the AC of the highest priority traffic remaining that is buffered at the AP, excluding the MSDU of the present frame.
- **AP Buffered Load**. This subfield indicates the total buffer size, rounded up to the nearest 4096 octets and expressed in units of 4096 octets, of all MSDUs buffered at the AP excluding the present QoS data frame. A value of 0 indicates the absence of any buffered traffic for the indicated AC.

11.1.6   HT Control field

The HT Control field is a new field introduced with the 802.11n amendment and is illustrated in Figure 11.4. This field is always present in Control Wrapper frames and is present in QoS Data frames when the Order bit in the Frame Control field is set to 1.

The subfields in Link Adaptation Control field are listed in Table 11.5.

The MAI subfields are listed in Table 11.6.

Table 11.7 ASELC subfield: ASEL Command and ASEL Data values (IEEE, 2007b)

ASEL Command B9–B11	Interpretation of ASEL Command	ASEL Data B12–B15
0	Transmit Antenna Selection Sounding Indication (TXASSI)	Number of remaining sounding PPDUs to be transmitted (0–15)
1	Transmit Antenna Selection Request (TXASSR) or Transmit Antenna Selection Sounding Resumption	0 when the command is Transmit Antenna Selection Sounding Request. A number in the range of values of 1 through 15, the number being the number of the first sounding PPDU to be transmitted when the command is Transmit Antenna Selection Sounding Resumption, where 0 corresponds to the first sounding PPDU in the original ASEL training sequence
2	Receive Antenna Selection Indication (RXASSI)	Number of remaining sounding PPDUs to be received (0 to 15)
3	Receive Antenna Selection Sounding Request (RXASSR)	Number of sounding PPDUs required (0 to 15)
4	Sounding Label	Sequence number of the sounding PPDU corresponding to a CSI frame in ASEL feedback (0 to 15)
5	No feedback due to ASEL training failure or stale feedback	A number in the range of values of 0 through 15, the number being the number of the first sounding PPDU that was not received properly, where 0 corresponds to the first sounding PPDU in the ASEL training sequence, or 0 if no sounding PPDUs were received properly, or 0 if this is a request for a full retraining sequence
6	Transmit Antenna Selection Sounding Indication (TXASSI) request feedback explicit CSI	Number of remaining sounding PPDUs to be transmitted (0 to 15)
7	Reserved	

The ASELC subfield contains the ASEL Command and ASEL Data subfields, the contents of which are detailed in Table 11.7.

The Calibration Position and Calibration Sequence fields are used during the calibration exchange for implicit feedback beamforming. These fields are defined in Table 11.8.

The CSI/Steering field indicates the type of beamforming feedback requested and takes the values listed in Table 11.9.

The NDP Announcement subfield indicates that an NDP will be transmitted after the frame. It is set to 1 to indicate that an NDP will follow, otherwise it is set to 0.

The AC Constraint field indicates whether the TID of reverse direction data is constrained to a single TID or not. A value of 0 indicates that the response to a reverse direction grant may contain data frames of any TID. A value of 1 indicates that the

Table 11.8 Calibration Control subfields (IEEE, 2007b)

Field	Meaning	Definition
Calibration Position	Position in calibration exchange sequence	0 = not a calibration frame 1 = Calibration Start 2 = Sounding Response 3 = Sounding Complete
Calibration Sequence	Calibration sequence identifier	This field is included in each frame in the calibration exchange and its value remains unchanged for that sequence

Table 11.9 CSI/Steering values (IEEE, 2007b)

Value	Definition
0	No feedback required
1	CSI
2	Non-compressed beamforming
3	Compressed beamforming

Table 11.10 RDG/More PPDU values (IEEE, 2007b)

Value	Role of transmitting STA	Interpretation of value
0	RD initiator	No reverse grant
	RD responder	The PPDU carrying the frame is the last PPDU from the RD responder
1	RD initiator	A reverse direction grant is present for the duration given in the Duration/ID field
	RD responder	The PPDU carrying the frame is followed by another PPDU

response to a reverse direction grant may contain data frames only from the same AC as the last data frame received from the reverse direction initiator.

The RDG/More PPDU field is interpreted as defined in Table 11.10.

11.1.7 Frame Body field

The Frame Body field contains information specific to the individual frame types and subtypes.

11.1.8 FCS field

The Frame Check Sequence (FCS) field contains a 32-bit CRC calculated over all the fields in the MAC header and the frame body. The FCS is used to validate the integrity of the MPDU.

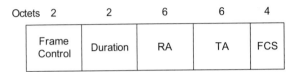

Figure 11.5 RTS frame. Reproduced with permission from IEEE (2007) © IEEE.

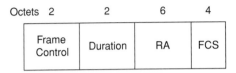

Figure 11.6 CTS frame. Reproduced with permission from IEEE (2007a) © IEEE.

11.2   Format of individual frame types

11.2.1   Control frames

11.2.1.1   RTS

The format of the RTS frame is shown in Figure 11.5. The RA field is the address of the station that is the intended recipient of the pending directed data or management frame. The TA field is the address of the station transmitting the RTS frame.

The Duration field is set to the duration in microseconds of the frame sequence consisting of the expected CTS (clear to send) response and subsequent frame exchanges.

11.2.1.2   CTS

The format of the CTS frame is shown in Figure 11.6. When a CTS frame follows an RTS frame the RA field is copied from the TA field of the RTS frame that solicited the CTS. The Duration field in this case is the value obtained from the RTS frame less the duration of a SIFS less the duration of the CTS frame.

When the CTS is the first frame in a sequence (providing NAV protection to the subsequent frames in the sequence) then the RA field is the transmit address of the sending station. This is called CTS-to-Self. The Duration field in this case is set to the duration of the subsequent frame exchange in microseconds.

11.2.1.3   ACK

The format of the ACK frame is shown in Figure 11.7. The RA field in the ACK frame is copied from the Address 2 field of the immediately preceding data, management, BAR, BA, or PS-Poll frame.

11.2.1.4   BAR

The format of the Block Ack Request (BAR) frame is shown in Figure 11.8. A variant of the BAR frame, called the Multi-TID BAR, is used under PSMP and is described below.

Octets 2   2   6   4

| Frame Control | Duration | RA | FCS |

Figure 11.7 ACK frame. Reproduced with permission from IEEE (2007a) © IEEE.

Octets 2   2   6   6   2   2   4

| Frame Control | Duration | RA | TA | BAR Control | Starting Sequence Control | FCS |

Figure 11.8 BAR frame. Reproduced with permission from IEEE (2007a) © IEEE.

B0   B1   B2   B3   B11 B12   B15

| BAR Ack Policy | Multi-TID | Compressed Bitmap | Reserved | TID/ NumTIDs |

Figure 11.9 BAR Control field. Reproduced with permission from IEEE (2007b) © IEEE.

The RA field is set to the address of the recipient station. The TA field is set to the address of the originator station. The Duration field is set to cover at least the responding ACK or BA frame plus SIFS.

The BAR Control field is shown in Figure 11.9. The individual subfields have the following meaning:

- **BAR Ack Policy**. This field is used under HT-delayed block ack. When set to 0 (Normal Acknowledgement) on an HT-delayed block ack session, the BAR frame will solicit an ACK frame if correctly received by the recipient. This is the same behavior as under delayed block ack. If set to 1 (No Acknowledgement) on an HT-delayed block ack session the BAR will not solicit an ACK response.
- **Multi-TID**. This field is always set to 0 in the basic BAR frame. If set to 1 then this is a Multi-TID BAR and the format differs from the basic BAR as described in the next section.
- **Compressed Bitmap**. If set to 1, the BAR frame solicits a BA with a compressed bitmap (see Section 11.2.1.6).
- **TID/NumTIDs**. If the Multi-TID field is not set then this field carries the TID of the block ack session. If the Multi-TID field is set then this field carries the number of TID fields in the Multi-TID BAR frame.

The Starting Sequence Control (SSC) field is shown in Figure 11.10. The Starting Sequence Number subfield is the sequence number of the first MSDU for which this BAR is sent.

Figure 11.10 Starting Sequence Control field. Reproduced with permission from IEEE (2007a) © IEEE.

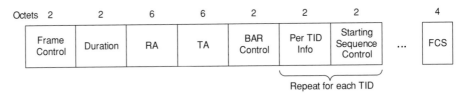

Figure 11.11 Multi-TID BAR frame.

Figure 11.12 Per TID Info field. Reproduced with permission from IEEE (2007b) © IEEE.

11.2.1.5   Multi-TID BAR

The Multi-TID BAR is a variant of the BAR frame and is used under PSMP. As the name suggests, this variant is intended to support multiple block ack sessions and solicits a Multi-TID BA. The Multi-TID BAR is not a separate frame type from the BAR, although for descriptive purposes we treat it as such here. The format of the Multi-TID BAR frame is shown in Figure 11.11.

The Multi-TID BAR frame is identified by it being a control frame of subtype BAR and having the Multi-TID and Compressed Bitmap fields set in the BAR Control field. The TID/NumTIDs field in the BAR Control field is set to indicate the number of TIDs for which this Multi-TID BAR applies. For each TID there is a repeated Per TID Info and SSC field.

The Per TID Info field is shown in Figure 11.12 and contains the TID value. Immediately following this is the SSC (Figure 11.10) corresponding to that TID.

11.2.1.6   BA

There are three variants of the BA frame: the uncompressed or basic form with the 128 octet Block Ack Bitmap and compatible with the form originally defined in the 802.11e amendment, the compressed form with the 8 octet Block Ack Bitmap, and the Multi-TID BA. The uncompressed and compressed forms are described here while the Multi-TID BA is described in Section 11.2.1.7.

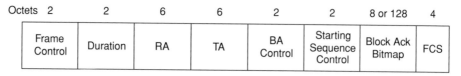

Figure 11.13 Basic BA frame. Reproduced with permission from IEEE (2007a) © IEEE.

B0	B1	B2	B3 B11	B12 B15
BA Ack Policy	Multi-TID	Compressed Bitmap	Reserved	TID/ NumTIDs

Figure 11.14 BA Control field. Reproduced with permission from IEEE (2007b) © IEEE.

The format of the basic Block Ack or BA frame is shown in Figure 11.13. The RA field is set to the address of the originator taken from the TA address of the BAR or QoS Data frame that solicited the BA. The TA field is the address of the recipient.

The BA Control field is shown in Figure 11.14 and the individual subfields have the following meaning:

- **BA Ack Policy**. This field is used under HT-delayed block ack. When set to 0 (Normal Acknowledgement) on an HT-delayed block ack session, the BA frame will solicit an ACK frame if correctly received by the originator. This is the same behavior as under delayed block ack. If set to 1 (No Acknowledgment) on an HT-delayed block ack session the BA will not solicit an ACK response.
- **Multi-TID**. This field is always set to 0 in the basic BA frame. If set to 1 then this is a Multi-TID BA and the frame format differs from the basic BA frame as described in Section 11.2.1.7 below.
- **Compressed Bitmap**. If set to 1, the BA frame contains the compressed or 8 octet Block Ack Bitmap. If set to 0, the BA frame contains the uncompressed or 128 octet Block Ack Bitmap.
- **TID/NumTIDs**. For the basic BA frame this field contains the TID for which this BA frame applies.

The Starting Sequence Control (SSC) field is shown in Figure 11.10. The Starting Sequence Number subfield is the sequence number of the first MSDU for which this BA is sent. If the BA was solicited by a BAR frame then the SSC is set to the same value as the SSC in that BAR frame. If the BA is solicited by an aggregate transmission containing QoS Data frames with Normal Ack ack policy then the SSC is set to the current value of WinStart, the starting sequence number of the recipient's scoreboard.

The Block Ack Bitmap is either 128 octets in the case of an uncompressed BA or 8 octets in the case of a compressed BA. The bitmap indicates the receive status of up to 64 MSDUs. In the case of the uncompressed bitmap, each MSDU is represented by a 16-bit word where the individual bits of the word represent the MSDU fragments, if any. In the case of the compressed bitmap each MSDU is represented by a single bit and signaling the receive status of individual fragments is not supported.

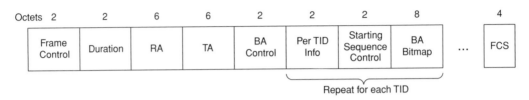

Figure 11.15 Multi-TID BA frame.

Figure 11.16 PS-Poll frame. Reproduced with permission from IEEE (2007a) © IEEE.

For the uncompressed bitmap, bit position n, if set to 1, acknowledges receipt of an MPDU with sequence control value equal to SSC + n. A value of 0 in bit position n indicates that an MPDU with sequence control value equal to SSC + n has not been received.

For the compressed bitmap, bit position n, if set to 1, acknowledges receipt of an MSDU with the sequence number that matches the SSC Sequence Number field value + n. A value of 0 in bit position n indicates that an MSDU with the sequence number that matches the SSC Sequence Number field value + n has not been received.

11.2.1.7   Multi-TID BA

The Multi-TID BA is a variant of the BA used under PSMP. The format of the Multi-TID BA frame is shown in Figure 11.15. The RA field is set to the address of the originator taken from the TA address of the Multi-TID BAR or QoS Data frames that solicited the Multi-TID BA. The TA field contains the address of the recipient station sending the Multi-TID BA.

The BA Control field is shown in Figure 11.14 and the individual subfields have the following meaning:

- **BA Ack Policy**. This field is set to 0 and is not applicable to PSMP where the Multi-TID BA is used since HT-immediate block ack is used.
- **Multi-TID**. This field is always set to 1 in a Multi-TID BA.
- **Compressed Bitmap**. This field is always set to 1 in a Multi-TID BA frame indicating that the BA Bitmaps are compressed.
- **TID/NumTIDs**. In a Multi-TID BA frame this field carries the number of TIDs for which this BA frame applies. For each TID, the body of the BA contains a Per TID Info field, SSC field, and BA Bitmap field.

11.2.1.8   PS-Poll

The Power Save Poll (PS-Poll) frame is used in the power save protocol defined in the original 802.11 standard. The frame format is shown in Figure 11.16. The BSSID is the

Octets 2   2   6   6   4

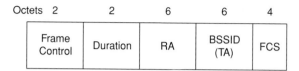

| Frame Control | Duration | RA | BSSID (TA) | FCS |

Figure 11.17 CF-End and CF-End+CF-Ack frame. Reproduced with permission from IEEE (2007a) © IEEE.

Octets 2   2   6   2   2   variable   4

| Frame Control | Duration | RA | Carried Frame Control | HT Control | Carried Frame | FCS |

Figure 11.18 Control Wrapper frame. Reproduced with permission from IEEE (2007b) © IEEE.

address of the AP and the TA field contains the address of the station sending the frame. The AID is the value assigned to the station transmitting the frame by the AP in the association response frame that established the station's current association.

11.2.1.9 CF-End and CF-End+CF-Ack

The format of the CF-End and CF-End+CF-Ack frames is shown in Figure 11.17. The BSSID field is the address of the AP. The RA field contains the broadcast group address. The Duration field is set to 0.

11.2.1.10 Control Wrapper

The Control Wrapper frame is a new control frame subtype that allows any control frame (other than the Control Wrapper frame itself) to be carried together with an HT Control field. The Control Wrapper frame was defined rather than extending the existing control frame types to include the HT Control field because there was concern that some legacy implementations do not consider the actual frame length when performing the CRC check. Instead, these implementations infer the frame length from the recognized frame subtype and would thus see an invalid CRC when the HT Control field was present. As a result they would not set their NAV and would defer for EIFS following the frame. A new frame subtype avoids this problem since these stations would not recognize the frame subtype, and would validate the CRC based on the actual length and set their NAV appropriately.

The Control Wrapper frame is shown in Figure 11.18. The RA field carries the same address as the RA of the carried frame. The Carried Frame Control field contains the value of the Frame Control field from the carried frame. The Carried Frame field contains the remaining fields of the control frame, i.e. between the RA and FCS field, and excluding the FCS field.

Table 11.11 Address field contents (IEEE, 2007b)

				Address 3		Address 4	
To DS	From DS	Address 1	Address 2	MSDU	A-MSDU	MSDU	A-MSDU
0	0	RA = DA	TA = SA	BSSID	BSSID	not present	
0	1	RA = DA	TA = BSSID	SA	BSSID	not present	
1	0	RA = BSSID	TA = SA	DA	BSSID	not present	
1	1	RA	TA	DA	BSSID	SA	BSSID

Octets 2	2	6	6	6	2	6	2	4	0-7955	4
Frame Control	Duration/ID	Address 1	Address 2	Address 3	Sequence Control	Address 4	QoS Control	HT Control	Frame Body	FCS

Figure 11.19 Data frame. Reproduced with permission from IEEE (2007b) © IEEE.

11.2.2   Data frames

The format of the data frame is shown in Figure 11.19. Not all fields are always present in the frame. The Address 4 field is only present if both the To DS and From DS fields are set to 1. The QoS Control field is only present in data frames of the QoS Data subtype. The HT Control field is only present in QoS Data frames and only when the Order bit in the Frame Control field is set and the frame is sent in a high throughput PHY PDU, i.e. HT Greenfield format or HT Mixed format.

The address fields in the data frame are set according to Table 11.11. Address 1 always carries the receiver address (RA) and Address 2 always carries the transmitter address (TA). The source address (SA) is the address within the DS where the MSDU originated. The destination address (DA) is the address within the DS where the MSDU is ultimately destined.

In the case of an A-MSDU, the SA and DA for each MSDU are carried in the subframe header within the body of the frame and thus the Address 3 and Address 4 fields in the MAC header carry the BSSID.

The Sequence Control field is defined in Figure 11.3 and the QoS Control field is defined in Section 11.1.5.

11.2.3   Management frames

The management frame format is shown in Figure 11.20. The Address 1 field contains the destination address on the wireless network for the management frame. This may be the address of an individual station or a group address. In the latter case the station receiving the management frame would validate the BSSID to ensure that the broadcast or multicast originated within the BSS to which the receiving station is a member. The exception to this is the Beacon frame, which any station may receive. The SA field contains the address of the station sending the frame.

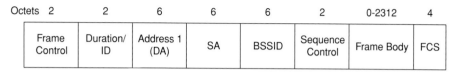

Octets 2	2	6	6	6	2	0-2312	4
Frame Control	Duration/ ID	Address 1 (DA)	SA	BSSID	Sequence Control	Frame Body	FCS

Figure 11.20 Management frame. Reproduced with permission from IEEE (2007a) © IEEE.

11.2.3.1 Beacon frame

The Beacon frame is broadcast periodically by the AP in an infrastructure BSS and stations in an IBSS. The Beacon frame is extended in 802.11n to include the following elements:

- HT Capabilities (always present)
- HT Information (always present)
- Secondary Channel Offset (present if spectrum management in a 20/40 MHz BSS is supported)
- Extended Capabilities element (present if 20/40 MHz BSS coexistence management is supported)
- 20/40 BSS Coexistence element (may be present)
- Overlapping BSS Scan Parameters element (may be present).

11.2.3.2 Association and Reassociation Request frame

The Association Request is sent by a station to the AP managing a BSS to request membership of the BSS. The Reassociation Request is sent by a station to the target AP when migrating from the source BSS to the target BSS in the same ESS or to the AP with which the station is associated to update the attributes with which the station initially associated with the BSS. The Association and Reassociation Request frames are extended to include the following elements:

- HT Capabilities element (always present)
- 20/40 BSS Coexistence element (may be present).

11.2.3.3 Association and Reassociation Response frame

The Association Response and Reassociation Response frames are sent by the AP in response to Association Request or Reassociation Request frames, respectively. The Association and Reassociation Response frames are extended to include the following elements:

- HT Capabilities element (always present)
- HT Information element (always present)
- 20/40 BSS Coexistence element (may be present)
- Overlapping BSS Scan Parameters element (may be present).

11.2.3.4   Disassociation frame

The Disassocation frame is sent by a station or by the AP to disassociate the station from the BSS. The Disassocation frame is not a request and contains no additional information.

11.2.3.5   Probe Request frame

The Probe Request frame is sent by a station to solicit a Probe Response frame from a specific AP or from multiple APs operating in a particular channel. The Probe Request frame is extended to include the HT Capabilities element.

11.2.3.6   Probe Response frame

The Probe Response frame is sent to a specific station in response to a Probe Request. The Probe Response frame is extended to include the following elements:

- HT Capabilities element (always present)
- HT Information element (always present)
- Secondary Channel Offset element (present if spectrum management is supported on a 20/40 MHz BSS)
- 20/40 BSS Coexistence element (may be present)
- Overlapping BSS Scan Parameters element (may be present).

11.2.3.7   Authentication frame

The Authentication frame is used during open system authentication by a station prior to association with an AP. The 802.11n amendment makes no changes to this frame.

11.2.3.8   Deauthentication frame

The Deauthentication frame is used by a station to signal that it wishes to be deauthenticated. The 802.11n amendment makes no changes to this frame.

11.2.3.9   Action and Action No Ack frames

The Action frame was introduced in the 802.11h amendment. This frame subtype essentially extends the number of management frame types available as the subtype for management frames in the Frame Control field is nearly exhausted. The Action No Ack frame was introduced in the 802.11n amendment for a similar purpose and is essentially the same as the Action frame except that it does not result in an ACK response from the receiving station.

The general format for the Frame Body of the Action and Action No Ack frames is shown in Figure 11.21. The Category field gives the general category to which the Action/Action No Ack frame belongs and the Action field gives the specific action taken by the Action/Action No Ack frame. A list of Category field values is given in Table 11.12.

Table 11.12 Action frame category values (IEEE, 2007a, 2007b)

Code	Meaning
0	Spectrum management
1	QoS
2	DLS
3	Block Ack
4	Reserved
5	Radio Measurement[1]
6	Fast BSS Transition[2]
7	HT
8	Public
9–126	Reserved
127	Vendor-specific
128–255	Error

[1] Assigned for 802.11k (Radio Resource Measurement), but not discussed here.
[2] Assigned for 802.11r (Fast BSS Transition), but not discussed here.

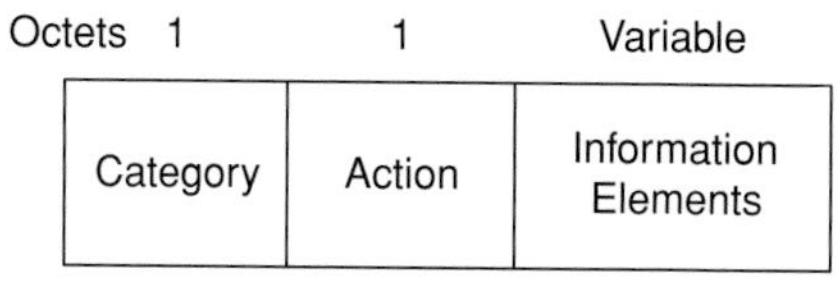

Octets   1   1   Variable

Category	Action	Information Elements

Figure 11.21 Frame body for Action and Action No Ack frame.

Spectrum management action frames

There are five Action frame formats defined for spectrum management as listed in Table 11.13.

QoS action frames

The Action frames defined to support QoS are listed in Table 11.14. These frames are used to set up, maintain, and delete traffic streams (TS) and to announce the schedule for the delivery of data and polls.

DLS Action frames

The DLS Action frames listed in Table 11.15 are used to manage the establishment and discontinuation of a direct link session (DLS) between peer MAC entities in a BSS.

Block Ack action frames

The Block Ack Action frames are used to manage the establishment and discontinuation of a block ack session. The Block Ack Action frames are listed in Table 11.16.

HT action frames

The HT Action frame category was introduced in the 802.11n amendment. The HT Action frames perform a variety of functions and are listed in Table 11.17.

Table 11.13 Spectrum Management Action field values (IEEE, 2007a)

Action	Name	Description
0	Measurement Request	Transmitted by a station requesting that another station perform channel measurements and return the information gathered
1	Measurement Report	Transmitted in response to a Measurement Request frame or autonomously by a station. This frame provides channel measurement information
2	TPC Request	Transmitted by a station requesting transmit power and link margin information from another station
3	TPC Report	Transmitted in response to a TPC Request frame, providing power and link margin information
4	Channel Switch Announcement	Transmitted by an AP in a BSS or station in an IBSS to advertise a channel switch
5	Extended Channel Switch Announcement	Transmitted by an AP in a BSS or station in an IBSS to advertise a channel switch
6–255	Reserved	

Table 11.14 QoS Action field values (IEEE, 2007a)

Action	Name	Description
0	ADDTS Request	The ADDTS Request frame is used to carry the TSPEC and optionally TCLAS elements to set up and maintain TSs
1	ADDTS Response	The ADDTS Response frame is transmitted in response to an ADDTS Request frame
2	DELTS	The DELTS frame is used to delete a TS
3	Schedule	The Schedule frame is transmitted by the HC to the non-AP station to announce the delivery schedule of data and polls
4–255	Reserved	

Table 11.15 DLS Action field values (IEEE, 2007a)

Action	Name	Description
0	DLS Request	The DLS Request frame is used to set up a direct link with a peer MAC in the same BSS
1	DLS Response	The DLS Response frame is sent in response to a DLS Request frame
2	DLS Teardown	The DLS Teardown frame is used to tear down a DLS session
3–255	Reserved	

Table 11.16 Block Ack Action field values (IEEE, 2007a)

Action	Name	Description
0	ADDBA Request	The ADDBA Request frame is sent from the originator of a block ack session to another station
1	ADDBA Response	The ADDBA Response frame is sent in response to an ADDBA Request frame
2	DELBA	The DELBA frame is sent by either the originator or responder in a block ack session to tear that session down
3–255	Reserved	

Table 11.17 HT Action field values (IEEE, 2007b)

Action	Name	Description
0	Notify Channel Width	This frame may be sent by both AP and non-AP stations to notify other stations of its current channel width, i.e. the channel width with which it can receive frames. A station may change its channel width to conserve power
1	SM Power Save	The SM Power Save Action frame is used to manage spatial multiplexing power saving state transitions. A station may conserve power by reducing the number of active receive chains and as a consequence reducing its ability to receive spatially multiplexed transmissions
2	PSMP	The Power Save Multi-Poll (PSMP) Action frame is used to manage PSMP scheduling. See Section 9.4
3	Set PCO Phase	The Set PCO Phase Action frame announces the transition between 20 MHz and 40 MHz phases of operation in a BSS. The operation of PCO is described in Section 10.4
4	CSI	The CSI (Channel State Information) Action frame is used to transfer channel state information from the measuring station to the transmit beamforming station. This frame may be sent as either an Action frame or Action No Ack frame
5	Non-compressed Beamforming	The Non-compressed Beamforming Action frame transfers uncompressed transmit beamforming vectors from the measuring station to the transmit beamforming station. This frame may be sent as either an Action frame or Action No Ack frame
6	Compressed Beamforming	The Compressed Beamforming Action frame transfers compressed transmit beamforming vectors from the measuring station to the transmit beamforming station. This frame may be sent as either an Action frame or Action No Ack frame
7	Antenna Selection Indices Feedback	The Antenna Selection Indices Feedback frame is used for antenna selection
8–255	Reserved	

Table 11.18 Public Action field values (IEEE, 2007b)

Action	Name	Description
0	20/40 BSS Coexistence Management	This frame is used to manage 20/40 MHz BSS coexistence. This frame contains the 20/40 BSS Coexistence element and zero or more 20/40 BSS Intolerant Channel Report elements
1–255	Reserved	

Public Action frames
The Public Action frame is defined to allow inter-BSS communication. One frame is currently defined in the 802.11n amendment for managing 20/40 MHz BSS coexistence as shown in Table 11.18.

11.3   Management Frame fields

This section describes some of the fields that may be present in various management frames. This is not an exhaustive list of all fields, but rather a list of fields relevant to the areas covered in this book.

Some fields in the management frame are identified by their position in the management frame. Other fields, particularly those added in later amendments, are information elements and use an identifier–length–value format. The identifier occupies one octet and identifies the information element. The length value also occupies one octet and allows implementations to parse the element without understanding its contents.

11.3.1   Fields that are not information elements

11.3.1.1   Capability Information field
The Capability Information field is 16 bits in length with fields as defined in Table 11.19. The use of the QoS, CF-Pollable, and CF-Poll Request subfields for both station and AP is defined in Tables 11.20 and 11.21 respectively.

11.3.2   Information elements

11.3.2.1   Extended Channel Switch Announcement element
The Extended Channel Switch Announcement element is shown in Figure 11.22. The Channel Switch Mode field indicates any restrictions on transmission until a channel switch. A value of 1 indicates that stations should stop transmitting until the channel switch occurs. A value of 0 does not restrict stations from transmitting.

The New Regulatory Class field is set to the number of the regulatory class after the channel switch. The regulatory classes are listed in Appendix 5.1. Note that the regulatory class also defines the channel width (20 MHz or 40 MHz).

Table 11.19 Capability Information field

Bits	Subfield	Definition	Encoding
B0	ESS	Indicates whether or not this is an infrastructure BSS	Set to 1 by an AP in a BSS
B1	IBSS	Used by stations in an IBSS	Set to 1 by a station in an IBSS. Always set to 0 by an AP
B2	CF-Pollable	Indicates CF-Poll capability	See Tables 11.20 and 11.21
B3	CF-Poll Request	Indicates CF-Poll capability	See Tables 11.20 and 11.21
B4	Privacy	Used by AP to indicate that encryption is required for all data frames within the BSS. Used by station establishing a direct link (DLS) to indicate that encryption is required on that link	1 = Encryption required 0 = Encryption not required
B5	Short Preamble	802.11b PHY related	
B6	PBCC	802.11g PHY related	
B7	Channel Agility	802.11b PHY related	
B8	Spectrum Management	Indicates whether or not spectrum management is supported	0 = not supported 1 = supported
B9	QoS	Indicates whether or not this is a QoS capable station (802.11e)	0 = non-QoS station 1 = QoS station
B10	Short Slot Time	Indicates support for short time slot with DSSS/CCK (802.11b) PHY	0 = not supported 1 = supported
B11	APSD	Used by the AP to indicate support for Automatic Power Save Delivery (APSD), an enhanced power save protocol introduced in 802.11e	0 = not supported 1 = supported
B12	Reserved		
B13	DSSS-OFDM	802.11g PHY related	
B14	Delayed Block Ack	Indicates support for the Delayed Block Ack protocol	0 = not supported 1 = supported
B15	Immediate Block Ack	Indicates support for the Immediate Block Ack Protocol	0 = not supported 1 = supported

Table 11.20 Station usage of QoS, CF-Pollable, and CF-Poll Request (IEEE, 2007a)

QoS	CF-Pollable	CF-Poll request	Meaning
0	0	0	Station is not CF-Pollable
0	0	1	Station is CF-Pollable, not requesting polling
0	1	0	Station is CF-Pollable, requesting to be polled
0	1	1	Station is CF-Pollable, requesting not to be polled
1	0	0	QoS station
1	0	1	Reserved
1	1	0	Reserved
1	1	1	Reserved

Table 11.21 AP usage of QoS, CF-Pollable, and CF-Poll Request (IEEE, 2007a)

QoS	CF-Pollable	CF-Poll request	Meaning
0	0	0	AP does not support PCF
0	0	1	AP includes PCF for delivery only (no polling)
0	1	0	AP includes PCF for delivery and polling
0	1	1	Reserved
1	0	0	QoS AP does not use CFP for delivery of unicast data frames
1	0	1	QoS AP uses CFP for delivery and does not poll non-QoS stations
1	1	0	QoS AP uses CFP for delivery and polls non-QoS stations
1	1	1	Reserved

Octets	1	1	1	1	1	1
	Element ID	Length	Channel Switch Mode	New Regulatory Class	New Channel Number	Channel Switch Count

Figure 11.22 Extended Channel Switch Announcement element IEEE (2007c).

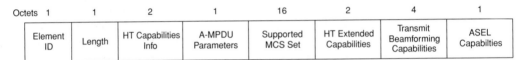

Octets	1	1	2	1	16	2	4	1
	Element ID	Length	HT Capabilities Info	A-MPDU Parameters	Supported MCS Set	HT Extended Capabilities	Transmit Beamforming Capabilities	ASEL Capabilties

Figure 11.23 HT Capabilities element. Reproduced with permission from IEEE (2007b) © IEEE.

The New Channel Number field is set to the number of the channel after the channel switch. The channel number is a channel from the station's new Regulatory Class as defined in Appendix 5.1.

The Channel Switch Count field indicates the number of Beacon transmissions until the station sending the Channel Switch Announcement element switches to the new channel. A value of 1 indicates that the switch occurs immediately before the next Beacon transmission time. A value of 0 indicates that the switch occurs any time after the frame containing the element is transmitted.

11.3.2.2   HT Capabilities element

The HT Capabilities information element (Figure 11.23) contains fields that are used to advertise the optional capabilities of the HT station or HT AP. The HT Capabilities element is present in the Beacon, Association Request, Association Response, Reassociation Request, Reassociation Response, Probe Request, and Probe Response frames.

HT Capabilities Info and HT Extended Capabilities fields
The HT Capabilities Info field and the HT Extended Capabilities field contain capability information bits. The separation into two fields, an HT Capabilities Info field and an HT

Extended Capabilities field, is an artifact of the standardization process. When the first 16-bit field was outgrown an additional field was added. However, since some companies were already developing products there was resistance to simply increasing the size of the HT Capabilities Info field. The subfields comprising the HT Capabilities Info and HT Extended Capabilities fields are given in Tables 11.22 and 11.23.

It is telling that while the original 802.11 Capabilities Information field is only 16 bits long and had sufficient reserved bits available for many of the follow-on amendments, the HT Capabilities Info field and HT Extended Capabilities field together are 32 bits in length. This speaks to the many optional capabilities added in the 802.11n amendment.

A-MPDU Parameters

The subfields for the A-MPDU Parameters field are defined in Table 11.24. These parameters were defined to prevent a transmitting station from overwhelming a receiver with limited buffering and processing capabilities for handling A-MPDUs. A receiver with limited buffering may specify a maximum A-MPDU length that the transmitter must honor. A receiver may also specify a minimum expected duration between MPDUs in an A-MPDU. This is to accommodate the fixed per packet processing in a receiver and in particular the decryption engines in some implementations that may have a fixed startup latency irrespective of the length of the MPDU.

Supported MCS Set field

The Supported MCS Set field indicates the MCSs supported by the station for transmit and receive. The subfields making up the Supported MCS Set field are defined in Table 11.25.

Transmit Beamforming Capabilities field

The subfields in the Transmit Beamforming Capabilities field are defined in Table 11.26.

Antenna Selection Capability field

The subfields making up the Antenna Selection Capability field are defined in Table 11.27.

11.3.2.3 HT Information element

The HT Information element (Figure 11.24) controls the behavior of HT STAs in the BSS. Some fields in the HT Information element change dynamically with changes in the BSS. For example, the association of a legacy station would require the AP prohibit the use of RIFS.

11.3.2.4 20/40 BSS Coexistence element

The format of the 20/40 BSS Coexistence element is shown in Figure 11.25. The Information Request field is used to indicate that a transmitting station is requesting that the recipient respond with a 20/40 BSS Coexistence Management frame. The Forty MHz Intolerant field, when set to 1, prohibits the receiving BSS from operating a 20/40 MHz

Table 11.22 HT Capabilities Info fields (IEEE, 2007b)

Bits	Subfield	Definition	Encoding
B0	LDPC Coding Capability	Support for receiving LDPC coded packets	0 = not supported 1 = supported
B1	Supported Channel Width Set	Channel widths the STA supports	0 = 20 MHz only operation 1 = 20 MHz and 40 MHz operation
B2–B3	SM Power Save	Spatial Multiplexing Power Save mode	0 = Static SM Power Save mode 1 = Dynamic SM Power Save mode 2 = reserved 3 = SM Power Save disabled
B4	Greenfield	Support for receiving HT Greenfield PPDU format	0 = not supported 1 = supported
B5	Short GI for 20 MHz	Support for receiving short GI in 20 MHz PPDU formats	0 = not supported 1 = supported
B6	Short GI for 40 MHz	Support for receiving short GI in 40 MHz PPDU formats	0 = not supported 1 = supported
B7	Tx STBC	Support for transmitting STBC PPDUs	0 = not supported 1 = supported
B8–B9	Rx STBC	Support for receiving STBC PPDUs	0 = no support 1 = support for 1 spatial stream 2 = support for 1 or 2 spatial streams 3 = support for 1, 2, or 3 spatial streams
B10	HT-delayed Block Ack	Support for HT-delayed block ack operation	0 = not supported 1 = supported
B11	Maximum A-MSDU length	Maximum supported A-MSDU length	0 = 3839 octets 1 = 7935 octets
B12	DSSS/CCK in 40 MHz	Use of DSSS/CCK in a 40 MHz capable BSS operating in 20/40 MHz mode	In Beacon and Probe Response frames: 0 = BSS does not allow use of DSSS/CCK in 40 MHz 1 = BSS does allow use of DSSS/CCK in 40 MHz Otherwise: 0 = STA does not use DSSS/CCK in 40 MHz 1 = STA uses DSSS/CCK in 40 MHz
B13	PSMP Support	Support for PSMP operation	In Beacon and Probe Response frames transmitted by an AP: 0 = AP does not support PSMP operation 1 = AP supports PSMP operation In Beacon frames transmitted by a non-AP station in an IBSS: always 0 Otherwise: 0 = STA does not support PSMP operation 1 = STA supports PSMP operation

Table 11.22 (*cont.*)

Bits	Subfield	Definition	Encoding
B14	40 MHz Intolerant	When sent by an AP, indicates whether other BSSs receiving this information are required to prohibit 40 MHz transmissions. When sent by a non-AP STA, indicates whether the AP associated with this STA is required to prohibit 40 MHz transmissions by all members of the BSS	When set by AP: 0 = AP allows use of 40 MHz transmissions in neighboring BSSs 1 = AP does not allow use of 40 MHz transmissions in neighboring BSSs When set by non-AP STA: 0 = indicates to associated AP that the AP is not required to restrict the use of 40 MHz transmissions within its BSS 1 = indicates to associated AP that the AP is required to restrict the use of 40 MHz transmissions within its BSS
B15	L-SIG TXOP protection support	Indicates support for L-SIG TXOP protection mechanism	0 = not supported 1 = supported

Table 11.23 HT Extended Capabilities fields (IEEE, 2007b)

Bits	Subfield	Definition	Encoding
B0	PCO	Support for PCO	0 = not supported 1 = supported A PCO capable AP sets this field to 1 to indicate that it can operate its BSS as PCO BSS. A PCO capable non-AP STA sets this field to 1 to indicate that it can operate as a PCO STA when the Transition Time field in its HT Extended Capabilities field meets the intended transition time of the PCO capable AP
B1–B2	PCO Transition Time	Indicates that the STA can switch between 20 MHz channel width and 40 MHz channel width within the specified time The value contained in this field is dynamic – the value of these bits may change at any time during the lifetime of the association of any STA	0 = no transition. The PCO STA does not change its operating channel width and is able to receive 40 MHz PPDUs during the 20 MHz phase 1 = 400 μs 2 = 1.5 ms 3 = 5 ms

(*cont.*)

Table 11.23 (*cont.*)

Bits	Subfield	Definition	Encoding
B3–B7	Reserved		
B8–B9	MCS feedback	Capable of providing MCS feedback	0 = STA does not provide MCS feedback 1 = reserved 2 = STA provides only unsolicited MCS feedback 3 = STA provides MCS feedback in response to MRQ as well as unsolicited
B10	+HTC Support	Indicates support for HT Control field	0 = not supported 1 = supported
B11	RD responder	Indicates support for acting as a reverse direction responder	0 = not supported 1 = supported
B12–B15	Reserved		

Table 11.24 A-MPDU Parameters field (IEEE, 2007b)

Bits	Subfield	Definition	Encoding
B0–B1	Maximum A-MPDU Length	Indicates the maximum length of an A-MPDU that the station can receive	0 = 8191 octets 1 = 16 383 octets 2 = 32 767 octets 3 = 65 535 octets
B2–B4	Minimum MPDU start spacing	Determines the minimum time between the start of adjacent MPDUs within an A-MPDU measured at the PHY SAP	0 = no restriction 1 = 0.25 μs 2 = 0.5 μs 3 = 1 μs 4 = 2 μs 5 = 4 μs 6 = 8 μs 7 = 16 μs
B5–B7	Reserved		

Table 11.25 Supported MCS Set field (IEEE, 2007b)

Bits	Subfield	Definition	Encoding
B0–B76	Rx MCS Bitmask	Defines a bitmap of supported MCS values	Array indexed by MCS such that B0 corresponds to MCS 0 and B76 corresponds to MCS 76
B77–B79	Reserved		
B80–B89	Highest Supported Data Rate	The highest data rate that the station is able to receive in units of 1 Mbps	1 = 1 Mbps. . .1023 = 1023 Mbps
B90–B95	Reserved		

(*cont.*)

Table 11.25 (*cont.*)

Bits	Subfield	Definition	Encoding
B96	Tx MCS Set Defined	Indicates whether or not a Tx MCS Set is defined	0 = no TX MCS set defined 1 = Tx MCS set is defined
B97	Tx Rx MCS Set Not Equal	Indicates whether or not the Tx MCS set (if defined) is equal to the Rx MCS set	0 = Tx MCS set is the same as the Rx MCS set 1 = the Tx MCS set differs from the Rx MCS Set by the number of spatial streams supported and/or whether or not unequal modulation is supported
B98–B99	Tx Maximum Number Spatial Streams Supported	If Tx Rx MCS Set Not Equal is set to 1 then this field gives the maximum number of spatial streams supported and by implication the portion of the Rx MCS set that is not supported for transmit	0 = 1 spatial stream 1 = 2 spatial streams 2 = 3 spatial streams 3 = 4 spatial streams
B100	Tx Unequal Modulation Supported	If Tx Rx MCS Set Not Equal is set to 1 then this field indicates whether or not unequal modulation is supported for transmission and by implication the portion of the Rx MCS set that is not supported for transmit	0 = unequal modulation not supported 1 = unequal modulation supported
B101–B127	Reserved		

Table 11.26 Transmit Beamforming Capabilities field (IEEE, 2007b)

Bits	Subfield	Definition	Encoding
B0	Implicit TxBF Receiving Capable	Indicates whether this station can receive TxBF steered frames using implicit feedback	0 = not supported 1 = supported
B1	Receive Staggered Sounding Capable	Indicates whether this station can receive staggered sounding frames	0 = not supported 1 = supported
B2	Transmit Staggered Sounding Capable	Indicates whether this station can transmit staggered sounding frames	0 = not supported 1 = supported
B3	Receive NDP Capable	Indicates whether this receiver can interpret Null Data Packets as sounding frames	0 = not supported 1 = supported
B4	Transmit NDP Capable	Indicates whether this receiver can transmit Null Data Packets as sounding frames	0 = not supported 1 = supported
B5	Implicit TxBF Capable	Indicates whether this station can apply implicit transmit beamforming	0 = not supported 1 = supported

(*cont.*)

Table 11.26 (*cont.*)

Bits	Subfield	Definition	Encoding
B6–B7	Calibration	Indicates that this station can participate in a calibration procedure initiated by another station that is capable of generating an immediate sounding PPDU and can provide a CSI Report in response to the receipt of a Sounding PPDU	0 = not supported 1 = station can respond to calibration request using the CSI Report but cannot initiate calibration 2 = reserved 3 = station can both initiate and respond to a calibration request
B8	Explicit CSI TxBF Capable	Indicates whether this station can apply transmit beamforming using CSI explicit feedback	0 = not supported 1 = supported
B9	Explicit Non-compressed Steering Capable	Indicates whether this station can apply transmit beamforming using a non-compressed beamforming feedback matrix	0 = not supported 1 = supported
B10	Explicit Compressed Steering Capable	Indicates whether this station can apply transmit beamforming using a compressed beamforming feedback matrix	0 = not supported 1 = supported
B11–B12	Explicit TxBF CSI Feedback	Indicates whether this receiver can return CSI explicit feedback	0 = not supported 1 = supported with delayed feedback 2 = supported with immediate feedback 3 = supported with delayed and immediate feedback
B13–B14	Explicit Non-compressed Beamforming	Indicates whether this receiver can return non-compressed beamforming feedback matrix explicit feedback	0 = not supported 1 = supported with delayed feedback 2 = supported with immediate feedback 3 = supported with delayed and immediate feedback
B15–B16	Explicit Compressed Beamforming	Indicates whether this receiver can return compressed beamforming feedback matrix explicit feedback	0 = not supported 1 = supported with delayed feedback 2 = supported with immediate feedback 3 = supported with delayed and immediate feedback
B17–B18	Minimal Grouping	Indicates the minimal grouping used for explicit feedback reports	0 = no grouping 1 = groups of 1, 2 2 = groups of 1, 4 3 = groups of 1, 2, 4
B19–B20	CSI Number of Beamformer Antennas Supported	Indicates the maximum number of beamformer antennas the beamformee can support when CSI feedback is required	0 = 1 Tx antenna sounding 1 = 2 Tx antenna sounding 2 = 3 Tx antenna sounding 3 = 4 Tx antenna sounding

(*cont.*)

Table 11.26 (cont.)

Bits	Subfield	Definition	Encoding
B12–B22	Non-compressed Steering Number of Beamformer Antennas Supported	Indicates the maximum number of beamformer antennas the beamformee can support when non-compressed beamforming feedback matrix is required	0 = 1 Tx antenna sounding 1 = 2 Tx antenna sounding 2 = 3 Tx antenna sounding 3 = 4 Tx antenna sounding
B23–B24	Compressed Steering Number of Beamformer Antennas Supported	Indicates the maximum number of beamformer antennas the beamformee can support when compressed beamforming feedback matrix is required	0 = 1 Tx antenna sounding 1 = 2 Tx antenna sounding 2 = 3 Tx antenna sounding 3 = 4 Tx antenna sounding
B25–B26	CSI Max Number of Rows Beamformer Supported	Indicates the maximum number of rows of the CSI explicit feedback from the beamformee or calibration responder or Tx ASEL responder that a beamformer or calibration initiator or Tx ASEL initiator can support when CSI feedback is required	0 = single row of CSI 1 = 2 rows of CSI 2 = 3 rows of CSI 3 = 4 rows of CSI
B27-B28	Channel Estimation Capability	Indicates the maximum number of space-time streams (columns of the MIMO channel matrix) for which channel dimensions can be simultaneously estimated. When staggered sounding is supported this limit applies independently to both the data portion and to the extension portion of the Long Training fields	0 = 1 space-time stream 1 = 2 space-time streams 2 = 3 space-time streams 3 = 4 space-time streams If the reception of staggered sounding is not supported, the value indicated by this field is equal to the maximum number of supported space-time streams If the reception of staggered sounding is supported:[1] 0 = (1,0), (1,1) 1 = (1,0), (2,0), (1,1), (1,2), (2,1), (2,2) 2 = (1,0), (2,0), (3,0), (1,1), (1,2), (1,3), (2,1), (2,2), (3,1) 3 = (1,0), (2,0), (3,0), (4,0), (1,1), (1,2), (1,3), (2,1), (2,2), (3,1)
B29–B31	Reserved		

[1] The notation (a, b) indicates: a: supported channel estimation dimensions using long training symbol(s) that will be used for demodulating data symbols. b: supported channel estimation dimensions using long training symbol(s) that will not be used for demodulating data symbols.

Table 11.27 Antenna Selection Capabilities field (IEEE, 2007b)

Bits	Subfield	Definition	Encoding
B0	Antenna Selection Capability	Indicates whether this station supports Antenna Selection	0 = not supported 1 = supported
B1	Explicit CSI Feedback Based Tx ASEL Capable	Indicates whether this station has Tx ASEL capability based on explicit CSI feedback	0 = not supported 1 = supported
B2	Antenna Indices Feedback Based Tx ASEL	Indicates whether this station has Tx ASEL capability based on antenna indices feedback	0 = not supported 1 = supported
B3	Explicit CSI Feedback Capable	Indicates whether this station can compute CSI and feedback in support of Antenna Selection	0 = not supported 1 = supported
B4	Antenna Indices Feedback Capable	Indicates whether this station can conduct antenna indices selection computation and feedback the results in support of Antenna Selection	0 = not supported 1 = supported
B5	Rx ASEL Capable	Indicates whether this station has Rx Antenna Selection capability	0 = not supported 1 = supported
B6	Transmit Sounding PPDUs Capable	Indicates whether this station can transmit sounding PPDUs for Antenna Selection training per request	0 = not supported 1 = supported
B7	Reserved		

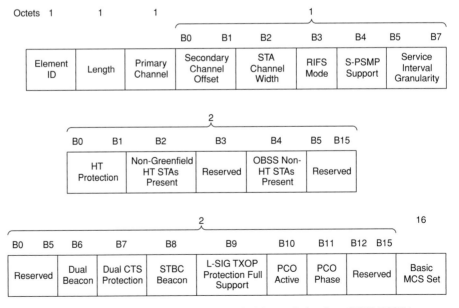

Figure 11.24 HT Information element. Reproduced with permission from IEEE (2007b) © IEEE.

Table 11.28 HT Information fields (IEEE, 2007b)

Field	Definition	Encoding
Primary Channel	Indicates the channel number of the primary channel	Channel number of primary channel
Secondary Channel Offset	Indicates the offset of the secondary channel relative to the primary channel	0 = no secondary channel 1 = secondary channel is above the primary channel 2 = reserved 3 = secondary channel is below the primary channel
STA Channel Width	Defines the channel widths that may be used to transmit to the STA (subject also to its Supported Channel Width Set)	0 = 20 MHz channel width 1 = allow use of any channel width in Supported Channel Width Set
RIFS Mode	Indicates whether or not RIFS is permitted within the BSS	0 = RIFS is prohibited 1 = RIFS is permitted
S-PSMP Support	Indicates support for scheduled PSMP	0 = not supported 1 = supported
Service Interval Granularity	Duration of the shortest Service Interval for scheduled PSMP	0 = 5 ms 1 = 10 ms 2 = 15 ms 3 = 20 ms 4 = 25 ms 5 = 30 ms 6 = 35 ms 7 = 40 ms
HT Protection	Protection requirements for HT transmissions	Set to 0 (HT members with same channel width, HT non-members) if: • this is a 20 MHz BSS and all stations in the BSS or detected in the primary channel are 20 MHz HT stations, or • this is a 20/40 MHz BSS and all stations in the primary and secondary channel are HT stations and all stations that are members of the BSS are 20/40 MHz HT stations Set to 1 (HT members, non-HT non-members) if: • there is at least one non-HT station detected in either the primary or the secondary channel or both, that is not a member of this BSS, and • all stations that are members of the BSS are HT stations *(cont.)*

Table 11.28 (*cont.*)

Field	Definition	Encoding
		Set to 2 (HT members with different channel widths, HT non-members) if: • this is a 20/40 MHz BSS, and • all stations detected in the primary or secondary channel are HT stations, and • all members of the BSS are HT stations, but at least one member is a 20 MHz HT station
		Set to 3 (Non-HT members) otherwise, i.e. one or more non-HT STAs are members of this BSS
Non-Greenfield HT STAs Present	Indicates if any HT STAs that are not Greenfield capable have associated Determines when a non-AP STA should use Greenfield protection Present in Beacon and Probe response frames transmitted by an AP. Otherwise reserved	Set to 0 if all HT STAs that are associated are Greenfield capable Set to 1 if one or more HT STAs that are not Greenfield capable are associated
Reserved		
OBSS Non-HT STAs present	Indicates whether use of protection for non-HT STAs in overlapping BSS is determined desirable. Present in Beacon and Probe response frames transmitted by an AP. Otherwise reserved	Set to 1 when use of protection for non-HT STAs by overlapping BSS is determined to be desirable. Set, for example, when • one or more non-HT STAs are associated • a non-HT BSS is overlapping • a management frame (excluding a Probe Request) is received indicating only non-HT rates in the supported rate set Set to 0 otherwise
Dual Beacon	Indicates whether the AP transmits an STBC beacon	0 = STBC beacon not transmitted 1 = STBC beacon is transmitted by the AP
Dual CTS Protection	Indicates whether Dual CTS Protection is used by the AP to set NAV at STAs that do not support STBC and at STAs that can associate only through the secondary (STBC frame) beacon	0 = DualS CTS protection is not required 1 = Dual CTS protection is required

(*cont.*)

Table 11.28 (*cont.*)

Field	Definition	Encoding
STBC Beacon	Indicates whether the beacon containing this element is the primary or secondary STBC beacon. The STBC beacon has half a beacon period shift relative to the primary beacon. Defined only in beacon transmission. Otherwise reserved	0 = primary beacon 1 = STBC beacon
L-SIG TXOP Protection Full Support	Indicates whether all HT STAs in the BSS support L-SIG TXOP Protection	0 = one or more HT STAs in the BSS do not support L-SIG TXOP Protection 1 = all HT STAs in the BSS support L-SIG TXOP Protection
PCO Active	Indicates whether PCO is active in the BSS. Present in Beacon and Probe Response frames. Otherwise reserved	0 = PCO not active in the BSS 1 = PCO active in the BSS
PCO Phase	Indicates the PCO phase of operation. Defined only in Beacon and Probe Response frames when PCO Active is 1. Otherwise reserved	0 = switch to or continue 20 MHz phase 1 = switch to or continue 40 MHz phase
Basic MCS Set	Indicates the MCS values that are supported by all HT STAs in the BSS. Present in Beacon and Probe Response frames. Otherwise reserved	The Basic MCS Set is a bitmap of size 128 bits. Bit 0 corresponds to MCS 0. A bit is set to 1 to indicate support for that MCS and 0 otherwise

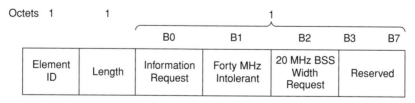

Figure 11.25 20/40 BSS Coexistence element. Reproduced with permission from IEEE (2007b) © IEEE.

BSS. The 20 MHz BSS Width Request field, when set to 1, prohibits the receiving AP from operating its BSS as a 20/40 MHz BSS.

This element is present in the 20/40 BSS Coexistence Management frame and may also be present in Beacon, Association Request, Reassociation Request, Association Response, Reassociation Response, Probe Request and Probe Response frames.

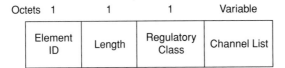

Octets	1	1	1	Variable
	Element ID	Length	Regulatory Class	Channel List

Figure 11.26 20/40 BSS Intolerant Channel Report element. Reproduced with permission from IEEE (2007b) © IEEE.

Octets	1	1	2	2	2	2	2	2	2
	Element ID	Length	OBSS Scan Passive Dwell	OBSS Scan Active Dwell	BSS Width Trigger Scan Interval	OBSS Scan Passive Total Per Channel	OBSS Scan Active Total Per Channel	BSS Channel Transition Delay Factor	OBSS Scan Activity Threshold

Figure 11.27 Overlapping BSS Scan Parameters element. Reproduced with permission from IEEE (2007b) © IEEE.

11.3.2.5   20/40 BSS Intolerant Channel Report element

The format of the 20/40 BSS Intolerant Channel Report element is shown in Figure 11.26. This element contains a list of channels on which the station has found conditions that disallow the operation of a 20/40 MHz BSS. The Regulatory Class field contains the regulatory class in which the channel list is valid. The Channel List field contains one or more octets where each octet provides a channel number. The regulatory classes and associated channel numbers are listed in Appendix 5.1.

This element may appear zero or more times in the 20/40 BSS Coexistence Management frame.

11.3.2.6   Overlapping BSS Scan Parameters element

The format of the Overlapping BSS Scan Parameters element is shown in Figure 11.27. This element is used by the AP in a 20/40 MHz BSS to set the parameters controlling OBSS scanning in member stations.

References

IEEE (2007a). *IEEE Std 802.11™-2007, IEEE Standard for Information Technology – Telecommunications and Information Exchange Between Systems – Local and Metropolitan Networks – Specific Requirements. Part 11: Wireless LAN Medium Access Control (MAC) and Physical Layer (PHY) Specifications.*

IEEE (2007b). *IEEE P802.11n™/D3.00, Draft Amendment to STANDARD for Information Technology – Telecommunications and Information Exchange Between Systems – Local and Metropolitan Networks – Specific Requirements. Part 11: Wireless LAN Medium Access Control (MAC) and Physical Layer (PHY) Specifications: Amendment 4: Enhancements for Higher Throughput.*

IEEE (2007c). *IEEE P802.11y™/D6.0, Draft Amendment to STANDARD for Information Technology – Telecommunications and Information Exchange Between Systems – Local and Metropolitan Networks – Specific Requirements. Part 11: Wireless LAN Medium Access Control (MAC) and Physical Layer (PHY) Specifications: Amendment 3: 3650–3700 MHz Operation in USA.*

Part III

Transmit beamforming

12 Transmit beamforming

The capability to perform adaptive transmit beamforming is provided in the 802.11n standard. With transmit beamforming (TxBF), we apply weights to the transmitted signal to improve reception. The weights are adapted from knowledge of the propagation environment or channel state information (CSI). Since by definition transmit beamforming weights are derived from channel information, spatial expansion as defined in Section 6.2 is not considered transmit beamforming.

The key advantage with transmit beamforming is the ability to significantly improve link performance to a low cost, low complexity device. This advantage is illustrated in Figure 12.1, which depicts a beamforming device with four antennas. Such a device could be an AP or a home media gateway. The device at the other end of the link has only two antennas, typical of a small client device. Such a system would benefit from 4×2 transmit beamforming gain from device A to device B. However, when transmitting from device B to device A, the system gain would be matched with 2×4 SDM with MRC as described in Section 6.1. Therefore link performance would be balanced in both directions.

In Figure 12.2, the generic MIMO system is modified to illustrate the application of beamformer weights to the transmitted signal. To simplify notation, the system description is given in terms of the frequency domain for a single subcarrier. Transmit beamforming as described is applied to each subcarrier in the frequency band. It is assumed that, on transmit, the frequency domain data would be transformed into a time domain waveform as described in Section 4.2. Furthermore, the receive procedure to generate frequency domain samples is described in Section 4.2.4.

The general system model is written as follows:

$$Y_{N_{RX}} = \sqrt{\rho/N_{TX}} \cdot H_{N_{RX} \times N_{TX}} V_{N_{TX} \times N_{SS}} X_{N_{SS}} + Z_{N_{RX}} \qquad (12.1)$$

where X is the transmitted data with dimension N_{SS} (number of spatial streams) $\times 1$; V is the transmit weighting matrix with dimension N_{TX} (number of transmit antennas) $\times$ N_{SS}; Y is the received signal with dimension N_{RX} (number of receive antennas) $\times 1$; H is the channel fading matrix with dimension $N_{RX} \times N_{TX}$; Z is the additive white Gaussian noise (AWGN) defined as Normal(0,1) with dimension $N_{RX} \times 1$; and ρ is the SNR.

The 802.11n standard does not dictate a specific approach for determining the transmitter weighting matrix. However, the most common approach is using singular value decomposition to calculate the transmitter weights.

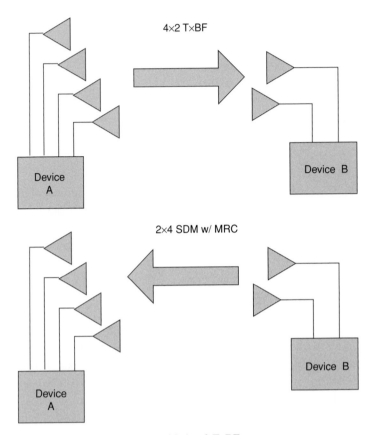

4×2 TxBF

Device A

Device B

2×4 SDM w/ MRC

Device A

Device B

Figure 12.1 System advantage with 4×2 TxBF.

12.1   Singular value decomposition

The singular value decomposition (SVD) of the channel matrix H is as follows:

$$H_{N \times M} = U_{N \times N} S_{N \times M} V^*_{M \times M} \qquad (12.2)$$

where V and U are unitary matrices, S is a diagonal matrix of singular values, and V^* is the Hermitian (complex conjugate transpose) of V. The definition of a unitary matrix is

$$VV^* = V^*V = I \qquad (12.3)$$

The diagonal values in S are non-negative and ordered in decreasing order.

The following properties are useful in determining the singular values and vectors of H:

$$\det (H^*H - \lambda I) = 0$$
$$(H^*H - \lambda I)v = 0 \qquad (12.4)$$
$$(HH^* - \lambda I)u = 0$$

The eigenvalues λ are the square of the singular values of H.

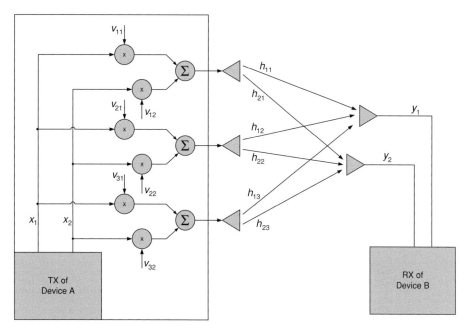

Figure 12.2 MIMO system with transmit beamforming.

To illustrate a singular value decomposition, we calculate the U, S, and V matrices for the example where

$$H = \begin{bmatrix} 2 & 1 \\ 1 & 2 \end{bmatrix}.$$

Therefore, H^*H is equal to

$$\begin{bmatrix} 5 & 4 \\ 4 & 5 \end{bmatrix}.$$

To find the eigenvalues of H^*H, we solve the equation

$$\det\left(\begin{bmatrix} 5 & 4 \\ 4 & 5 \end{bmatrix} - \lambda I\right) = 0$$

by the following steps:

$$\det\left(\begin{bmatrix} 5-\lambda & 4 \\ 4 & 5-\lambda \end{bmatrix}\right) = 0$$

$$(5-\lambda)^2 - 16 = 0$$

$$(\lambda - 9)(\lambda - 1) = 0$$

$$\lambda = 9, 1$$

Therefore the matrix S is equal to

$$\begin{bmatrix} 3 & 0 \\ 0 & 1 \end{bmatrix}.$$

Next we solve for the unitary V matrix,

$$\begin{bmatrix} v_1 & v_3 \\ v_2 & v_4 \end{bmatrix},$$

using

$$\left(\begin{bmatrix} 5 & 4 \\ 4 & 5 \end{bmatrix} - \lambda I \right) v = 0.$$

With λ equal to 9, the result is

$$\begin{bmatrix} -4 & 4 \\ 4 & -4 \end{bmatrix} \begin{bmatrix} v_1 \\ v_2 \end{bmatrix} = 0.$$

From this we determine that v_1 is equal to v_2. Next with λ equal to 1, then

$$\begin{bmatrix} 4 & 4 \\ 4 & 4 \end{bmatrix} \begin{bmatrix} v_3 \\ v_4 \end{bmatrix} = 0$$

and v_3 is equal to $-v_4$. We similarly solve for the unitary U matrix,

$$\begin{bmatrix} u_1 & u_3 \\ u_2 & u_4 \end{bmatrix}$$

using

$$\left(\begin{bmatrix} 5 & 4 \\ 4 & 5 \end{bmatrix} - \lambda I \right) u = 0.$$

The result is the same, u_1 is equal to u_2, and u_3 is equal to $-u_4$.

Using the property of unitary matrices, we solve for the elements of V as follows:

$$\begin{bmatrix} v_1 & v_3 \\ v_2 & v_4 \end{bmatrix} \begin{bmatrix} v_1 & v_2 \\ v_3 & v_4 \end{bmatrix} = \begin{bmatrix} 1 & 0 \\ 0 & 1 \end{bmatrix}$$

$$\begin{bmatrix} v_1 & v_3 \\ v_1 & -v_3 \end{bmatrix} \begin{bmatrix} v_1 & v_1 \\ v_3 & -v_3 \end{bmatrix} = \begin{bmatrix} 1 & 0 \\ 0 & 1 \end{bmatrix}$$

$$\begin{bmatrix} v_1^2 + v_3^2 & v_1^2 - v_3^2 \\ v_1^2 - v_3^2 & v_1^2 + v_3^2 \end{bmatrix} = \begin{bmatrix} 1 & 0 \\ 0 & 1 \end{bmatrix}$$

From $v_1^2 - v_3^2 = 0$, we determine that v_1^2 is equal to v_3^2. And from $v_1^2 + v_3^2 = 1$, we arrive at the result of $v_1, v_3 = -1/\sqrt{2}$. The same steps are taken to solve for the elements of U, which also has the result that $u_1, u_3 = -1/\sqrt{2}$. Therefore,

$$V = \begin{bmatrix} -1/\sqrt{2} & -1/\sqrt{2} \\ -1/\sqrt{2} & 1/\sqrt{2} \end{bmatrix}$$

$$U = \begin{bmatrix} -1/\sqrt{2} & -1/\sqrt{2} \\ -1/\sqrt{2} & 1/\sqrt{2} \end{bmatrix}$$

$$S = \begin{bmatrix} 3 & 0 \\ 0 & 1 \end{bmatrix}$$

using Eq. (12.2).

Beyond this simple 2×2 example, SVD is computed numerically. LAPACK provides routines to solve SVD (Anderson *et al.*, 1999). The SVD function in Matlab® uses LAPACK subroutines.

12.2   Transmit beamforming with SVD

For this section, we assume the transmitter and receiver have full knowledge of the channel state information. Subsequent sections discuss feedback mechanisms to acquire the channel state information. Therefore, given knowledge of H, the matrix V is calculated by SVD according to Eq. (12.2). Subsequently, the first N_{SS} columns of V are used as transmit weights in Eq. (12.1).

The motivation behind using the matrix V calculated by SVD is that it results in maximum likelihood performance with a linear receiver, greatly simplifying receiver design. We prove this as follows.

The maximum likelihood estimate of X from the received signal Y described by Eq. (12.1) is given by the following equation, as discussed in Section 3.7:

$$\hat{X} = \arg \min_{X} \|Y - H \cdot V \cdot X\| \tag{12.5}$$

With $H = USV^*$, this is expanded as given below:

$$\hat{X} = \arg \min_{X} \|Y - U \cdot S \cdot X\| \tag{12.6}$$

Since U is a unitary matrix we factor this equation as follows:

$$\hat{X} = \arg \min_{X} \|U^*Y - U^*U \cdot S \cdot X\|$$
$$= \arg \min_{X} \|U^*Y - S \cdot X\| \tag{12.7}$$

Ideally, the process of multiplying Y by U^* diagonalizes the result such that the spatial streams can be separated:

$$\hat{X} = \arg \min_{X} \sum_{i=1}^{N_{SS}} |(U^*Y)_i - S_i \cdot X_i|^2$$
$$= \arg \min_{X_i} |(U^*Y)_i - S_i \cdot X_i|^2 \qquad i = 1, \ldots, N_{SS} \tag{12.8}$$

With one last factoring step, the final result is as follows:

$$\hat{X} = \arg \min_{X_i} S_i^2 |S_i^{-1} (U^*Y)_i - X_i|^2 \qquad i = 1, \ldots, N_{SS} \tag{12.9}$$

This final result describes a simple receiver for each spatial stream, as discussed in Section 4.1.6. The $S_i^{-1} (U^*Y)_i$ term in Eq. (12.9) is equivalent to the ZF receiver. The $|S_i^{-1} (U^*Y)_i - X_i|$ term is the demapping operation. The subcarrier weighting factor is given by the term S_i^2. However, if a low-complexity BICM demapping operation is used,

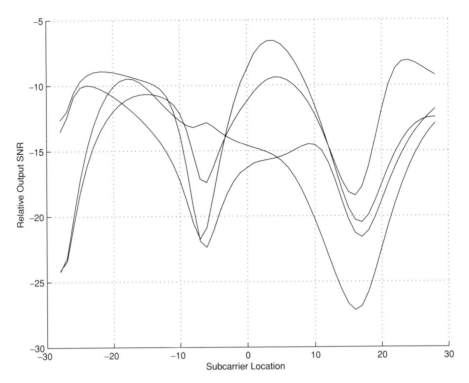

Figure 12.3 Output SNR for basic MIMO/OFDM system.

ML performance is only achieved for BPSK and QPSK. Higher order modulation with coupling between bits is degraded as compared to ML.

12.3   Eigenvalue analysis

In typical indoor WLAN channels, the wide bandwidth has large frequency selective variations due to multipath fading. A basic MIMO/OFDM system has an output SNR with such variations across the band. Figure 12.3 illustrates such a system having a ZF receiver with four spatial streams for one instantiation of channel model D. Each curve represents the relative output SNR for each spatial stream as a function of subcarrier location for a 20 MHz transmission. The output SNR is calculated based on Eq. (3.33). Each spatial stream has a variation in the output SNR by more than 15 dB. In addition, on average the output SNR for each spatial stream is roughly the same. In terms of selecting a proper MCS for this transmission, an MCS with lower order modulation and/or lower code rate would have to be selected to contend with the fading characteristics across all spatial streams.

Now we consider the signal transmitted with beamformer weights as described by Eq. (12.1). Furthermore, the V matrix from an SVD of the channel H is selected as the transmit weighting matrix. The output SNR for a ZF receiver based on Eq. (3.33) is rewritten as follows:

$$\text{SNR}_i = \frac{1}{\text{diag}_i\left(\left(\frac{\rho}{N_{\text{TX}}} \cdot (HV)^* HV\right)^{-1}\right)} \qquad i = 1, \ldots, N_{\text{SS}} \qquad (12.10)$$

If we replace the channel H with its SVD equivalent expansion in Eq. (12.10), the result is the following expression for output SNR that is only dependent on the singular values:

$$\text{SNR}_i = \text{diag}_i\left(\frac{\rho}{N_{\text{TX}}} \cdot S^{\text{T}} S\right)$$

$$= \frac{\rho}{N_{\text{TX}}} \cdot S_i^2 \qquad i = 1, \ldots, N_{\text{SS}} \qquad (12.11)$$

It can be shown that Eq. (12.11) also applies to MMSE and that ideally with SVD-based TxBF a ZF receiver and MMSE receiver are equivalent. The output SNR for MMSE is given by Eq. (3.32), and is a function of the diagonal terms of the MSE. With SVD-based TxBF, the diagonal terms of the MSE are described as follows:

$$J_i = \left(\frac{\rho}{N_{\text{TX}}} W (HV)(HV)^* W^*\right)_i + (WW^*)_i - \left(2\sqrt{\frac{\rho}{N_{\text{TX}}}} \text{Re}\{W(HV)\}\right)_i + 1 \qquad (12.12)$$

with the noise covariance matrix equal to the identity matrix since we defined the noise term as Normal(0,1). With SVD-based TxBF, the MMSE receiver weights are modified as follows:

$$W = \sqrt{\frac{\rho}{N_{\text{TX}}}} (HV)^* \left(\frac{\rho}{N_{\text{TX}}} (HV)(HV)^* + I\right)^{-1} \qquad (12.13)$$

We replace the W into Eq. (12.12). Afterwards, the three components that were a function of W can be rewritten as a function of the singular values S:

$$\left(\sqrt{\frac{\rho}{N_{\text{TX}}}} W(HV)\right)_i = \frac{\frac{\rho}{N_{\text{TX}}} S_i^2}{\frac{\rho}{N_{\text{TX}}} S_i^2 + 1} \qquad (12.14)$$

$$\left(\frac{\rho}{N_{\text{TX}}} W(HV)(HV)^* W^*\right)_i = \left(\frac{\frac{\rho}{N_{\text{TX}}} S_i^2}{\frac{\rho}{N_{\text{TX}}} S_i^2 + 1}\right)^2 \qquad (12.15)$$

$$(WW^*)_i = \frac{\frac{\rho}{N_{\text{TX}}} S_i^2}{\left(\frac{\rho}{N_{\text{TX}}} S_i^2 + 1\right)^2} \qquad (12.16)$$

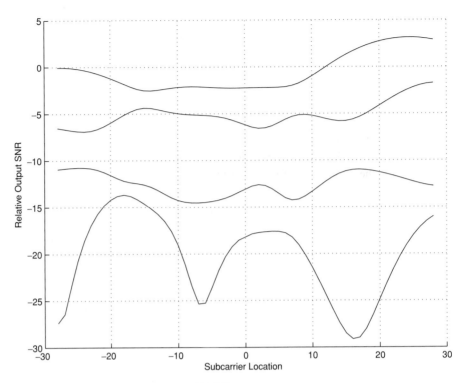

Figure 12.4 Output SNR for SVD based TxBF system.

If we replace the terms in Eq. (12.12) with those given by Eqs. (12.14)–(12.16), the expression for the diagonal terms of the MSE is simplified to

$$J_i = \frac{1}{\left(\dfrac{\rho}{N_{\mathrm{TX}}}\right) S_i^2 + 1} \tag{12.17}$$

When this expression for MSE is substituted into the output SNR equation for MMSE given by Eq. (3.32), the result is equivalent to the output SNR for ZF in Eq. (12.11).

With SVD, the output SNR is a function of the singular values of channel H. Furthermore, since the singular values in the matrix S are by definition in decreasing order, the largest singular value for each subcarrier is associated with the first spatial stream, the next largest singular value is associated with the second spatial stream, and so forth. Figure 12.4 illustrates the output SNR for an SVD-based TxBF system using the same channel and receiver as the example given in Figure 12.3.

Two effects are evident. First, the output SNR of the individual spatial streams is clearly separated across the band. The first spatial stream, the curve with the highest output SNR, corresponds to the largest eigenvalue. For this particular channel instantiation, with beamforming the average output SNR of the first spatial stream is more than 10 dB larger than without. Even the second spatial stream has a larger average output SNR than without beamforming.

Without beamforming, a lower MCS is selected for all spatial streams that meet the average channel condition over all the spatial streams. With beamforming, and the subsequent spatial separation of output SNR, we may apply a different MCS to each spatial stream to improve the overall throughput. The first spatial stream supports an MCS with high order modulation and code rate, the second spatial stream supports an MCS with a lower order modulation and code rate, and so forth. By matching each spatial stream to the appropriate MCS, the overall data rate with beamforming is higher than without.

A simplification to per-spatial stream MCS selection is that the same MCS is selected for the strong spatial streams and no MCS is applied to the weaker spatial streams, resulting in fewer transmitted spatial streams. This approach works well if the strongest spatial streams have similar output SNR, or exceed the SNR requirements for the highest modulation order and code rate. In our example, perhaps only two spatial streams (with four transmit antennas) would be transmitted. Such an approach also applies when the transmitter has more antennas than the receiver. If in this example the receiver only had two antennas, two spatial streams could be supported with the SNR performance of the two strongest spatial streams illustrated in Figure 12.4.

The second effect is that the first few spatial streams are relatively flat across the band. They exhibit improved coded performance more closely related to an AWGN channel rather than a frequency selective channel (Nanda *et al.*, 2005). Another benefit of spectral flatness is that selecting the same modulation order and code rate for each subcarrier location of the spatial stream approaches the throughput performance of adaptive bit loading (selecting a unique modulation order and code rate for each subcarrier location) (Nanda *et al.*, 2005).

The difference in output SNR, based on MMSE, of basic MIMO and SVD-based TxBF is more generally illustrated in Figure 12.5. The probability distribution of the output SNR for each spatial stream is given for channel model D, with NLOS conditions, using an input SNR of 25 dB. The distribution is attained over all subcarriers and over a large number of channel instantiations.

For the basic MIMO system, illustrated by the dashed line, there is no purposeful separation in output SNR between different spatial streams. Therefore the results for all the spatial streams are combined into a single curve.

With SVD-based TxBF, the separation in output SNR between the different spatial streams is quite evident as demonstrated by the four solid lines. At a probability of 90%, with TxBF all but one spatial stream has an output SNR that exceeds that of a basic MIMO system. The three largest spatial streams have an output SNR that is 18.5 dB, 13 dB, and 6.5 dB greater than a basic MIMO system.

The benefit of TxBF may also be demonstrated using capacity analysis. The formula for capacity based on output SNR is given by Eq. (3.34). For basic MIMO, the output SNR given for a ZF receiver by Eq. (3.33) or for an MMSE receiver by Eq. (3.32) is substituted into Eq. (3.34).

For TxBF, we substitute the output SNR given by Eq. (12.11) into Eq. (3.34). This is equivalent to the capacity formula given by Eq. (3.11), further proving that SVD-based TxBF provides Shannon capacity in ideal conditions.

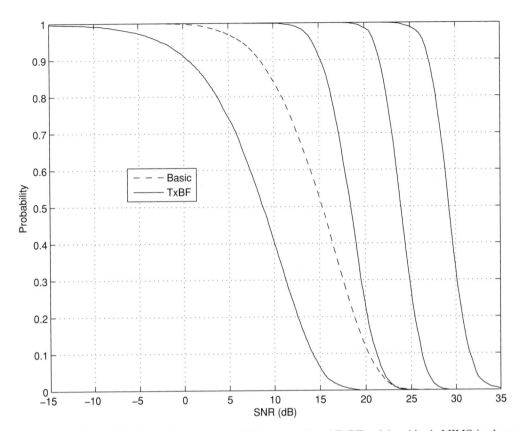

Figure 12.5 Comparison of output SNR between 4 × 4 TxBF and 4 × 4 basic MIMO in channel model D, NLOS with 25 dB input SNR.

Capacity is computed with the output SNR data of Figure 12.5 and illustrated in Figure 12.6. At a probability of 90%, the capacity in terms of bits/symbol/subcarrier is 24 with an input SNR of 25 dB in channel model D. This reduces to 14 bits/symbol/subcarrier for basic MIMO with an MMSE receiver.

12.4   Unequal MCS

As described in the previous section, with SVD based TxBF, applying an MCS per spatial stream improves overall throughput. MCS 33–76 (which apply to both 20 and 40 MHz and both 800 ns and 400 ns GI) provide unequal MCS combinations for two, three, and four spatial streams. All the unequal MCSs have the same code rate for each spatial stream; only the modulation order is different. This was done to minimize receiver complexity. As illustrated in the transmitter block diagram in Figure 4.21, the FEC encoder is prior to the stream parser, whereas the constellation mapper is after the stream parser. This basic structure is unchanged with the same code rate for each spatial

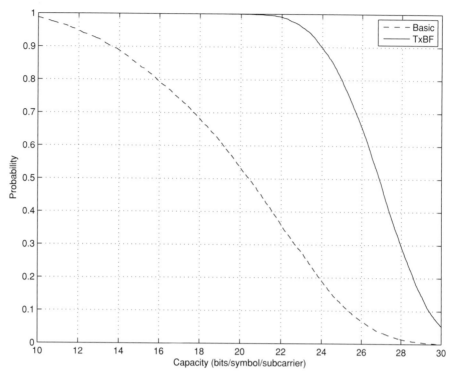

Figure 12.6 Comparison of capacity between 4 × 4 TxBF and 4 × 4 basic MIMO in channel model D with 25 dB input SNR.

stream. With the constellation mapper after the stream parser, only the stream parser must change to enable per-stream modulation mapping.

In order to best match each particular instantiation of the channel, it is desirable to have maximum flexibility in assignment of the modulation order to spatial stream. On the other hand, link adaptation becomes more and more difficult as the granularity of the MCSs becomes finer. Typically the quality of the channel state information is not good enough to support the ability to select between MCS with slight differences in modulation order. Link adaptation is discussed further in Section 12.12. Furthermore, the larger the MCS set the more complexity is incurred in designing and testing the added MCSs.

The MCS set for 20 MHz and 40 MHz is given in Appendix 12.1. We observe that in order to reduce the size of the MCS set, BPSK is not used for unequal MCSs. BPSK on a particular spatial stream gives marginal refinement between not using a spatial stream and QPSK. Furthermore, only code rates of 1/2 and 3/4 are used for unequal MCSs. Similarly, code rates of 2/3 and 5/6 give marginal refinement as well since they are only used in conjunction with 64-QAM.

As an example, without TxBF the channel may only support an average output SNR for QPSK on two streams (MCS 9). However, as illustrated in Figure 12.4, with TxBF

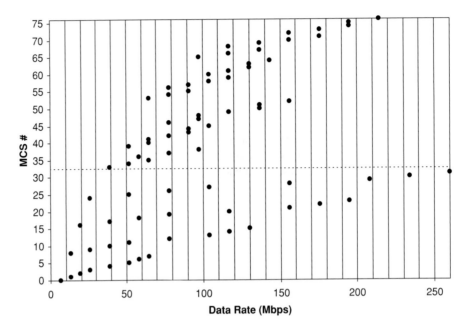

Figure 12.7 20 MHz MCSs grouped by data rate.

the spatial stream associated with the largest eigenvalue has a 5 dB larger output SNR and supports a higher modulation order of at least 16-QAM (MCS 33). This results in a 50% increase in data rate.

An alternative approach to viewing the added flexibility in MCS selection provided by the unequal MCS set is to group the MCSs by data rate. This allows us to see the number of MCSs with the same data rate. This grouping is illustrated in Figure 12.7 for 20 MHz and Figure 12.8 for 40 MHz. The data rates are based on 800 ns GI only. The MCSs above 32 are the unequal MCSs, signified by being above the dashed line.

As an example, for 78 Mbps with 20 MHz mode, there are eight possible MCSs. Three of the possible choices are equal MCSs. The other five MCSs are unequal. These are listed in Table 12.1. With two streams, the channel would need to support the average output SNR for 16-QAM on both streams. However, with TxBF, an alternative of selecting unequal MCS 37 with 64-QAM and QPSK is possible.

As mentioned above, the stream parser must change to enable per-stream modulation mapping. The basic operation of stream parsing was described in Section 4.2.3.3, and then expanded to include two encoders in Section 5.1.4.3. Building upon Eq. (4.23) in Section 4.2.3.3, the block of bits, s, assigned to each spatial stream is modified as a function of each spatial stream as given in Eq. (12.18) (IEEE, 2007):

$$s(i_{SS}) = \max\left\{1, \frac{N_{BPSCS}(i_{SS})}{2}\right\} \qquad (12.18)$$

Table 12.1 MCSs for 78 Mbps with 20 MHz mode

| | Modulation | | | | |
MCS Index	Stream 1	Stream 2	Stream 3	Stream 4	Code rate
12	16-QAM	16-QAM	×	×	$\frac{3}{4}$
19	16-QAM	16-QAM	16-QAM	×	$\frac{1}{2}$
26	QPSK	QPSK	QPSK	×	$\frac{3}{4}$
37	64-QAM	QPSK	×	×	$\frac{3}{4}$
42	64-QAM	16-QAM	QPSK	×	$\frac{1}{2}$
46	16-QAM	QPSK	QPSK	×	$\frac{3}{4}$
54	16-QAM	16-QAM	QPSK	QPSK	$\frac{1}{2}$
56	64-QAM	QPSK	QPSK	QPSK	$\frac{1}{2}$

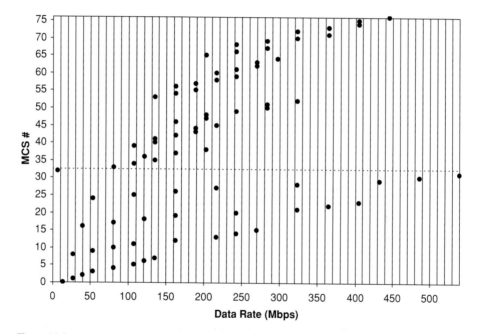

Figure 12.8 40 MHz MCSs grouped by data rate.

where $N_{\text{BPSCS}}(i_{\text{SS}})$ is the number of coded bits per single subcarrier for each spatial stream, or equivalently the modulation order, N_{SS} is the number of spatial streams, and $i_{\text{SS}} = 1, \ldots N_{\text{SS}}$.

With two encoders, redefining Eq. (5.9) in Section 5.1.4.3, a block of S bits from the output of each encoder is alternately used:

$$S = \sum_{i_{\text{SS}}=1}^{N_{\text{SS}}} s(i_{\text{SS}}) \tag{12.19}$$

The new equation for the output of the stream parser with two encoders is as follows:

$$j = \left\lfloor \frac{k}{s(i_{SS})} \right\rfloor \mod N_{ES}$$

$$i = \sum_{l=1}^{i_{SS}-1} s(l) + S \cdot \left\lfloor \frac{k}{N_{ES} \cdot s(i_{SS})} \right\rfloor + k \mod s(i_{SS}) \qquad (12.20)$$

for $1 \leq i_{SS} \leq N_{SS}$ and $k = 0, 1, \ldots, N_{CBPSS}(i_{SS}) - 1$; and where N_{ES} is the number of encoders; $\lfloor v \rfloor$ is the largest integer less than or equal to v, corresponding to the *floor* function; and $v \mod w$ is the remainder from the division of the integer v by the integer w; $N_{CBPSS}(i_{SS})$ is the number of coded bits per spatial stream.

Since with unequal MCS the parameters s, N_{BPSCS}, and N_{CBPSS} are a function of the spatial stream index, the equations describing interleaving in Section 4.2.3.4 are modified by replacing those parameters with $s(i_{SS})$, $N_{BPSCS}(i_{SS})$, and $N_{CBPSS}(i_{SS})$, respectively. This includes Eqs. (4.25)–(4.27).

12.5   Receiver design

The received signal using an SVD based beamforming transmitter is described by Eqs. (12.1) and (12.2). Ideally, the equation for the received signal can be rewritten by replacing $H \cdot V$ with $U \cdot S$, as follows:

$$Y = H \cdot V \cdot X + Z$$
$$= U \cdot S \cdot X + Z \qquad (12.21)$$

A common approach to receiver design was to filter the received signal by U^*, as follows:

$$R = U^* \cdot Y$$
$$= U^* \cdot U \cdot S \cdot X + U^* \cdot Z$$
$$= S \cdot X + \tilde{Z} \qquad (12.22)$$

Since U is unitary, the properties of noise matrix Z remain unchanged when filtered by U. After filtering by U, a standard ZF or MMSE receiver may be used.

This approach requires knowledge of the SVD at both the transmit and receive sides of the system. The transmitter requires knowledge of V and the receiver requires knowledge of U. In addition, the receiver should know when filtering with the matrix U, that the transmitter is actually beamforming with the corresponding matrix V.

It is a common misconception that it is necessary to have knowledge of U at the receiver and to filter by U prior to a ZF or MMSE receiver. It was demonstrated in (Lebrun *et al.*, 2002) that it is not required to filter the receive signal by U prior to a ZF receiver. By not filtering with U, the U matrix is incorporated in the matrix inversion of the ZF receiver. However, U is unitary, and the matrix inversion of U by a ZF function is equivalent to a Hermitian operation, resulting in no noise enhancement (Lebrun *et al.*, 2002). In (Lebrun *et al.*, 2002) the performance of a system with and without filtering by U prior to a ZF receiver was demonstrated by simulation to be equivalent even in a Doppler channel where a mismatch in U, V, and H occurred due to delayed CSI used in the simulation.

Based on this finding, our basic linear receiver with ZF or MMSE described in Section 3.6 ideally achieves the full benefit of TxBF. Note the use of the word ideally in the previous sentence. Implementation issues in a TxBF system should not be underestimated, involving sounding of the channel and feedback of the CSI. These are described in the subsequent sections and require the receiver to mitigate degradations.

As briefly mentioned in Section 4.2.2.1, subcarrier smoothing of the channel estimate should not be employed in conjunction with SVD-based TxBF. Random phase differences between adjacent subcarriers may occur. This is due to the fact that the columns of U and V are unique up to a per-column phase factor (Sadowsky *et al.*, 2005). Therefore, when applying SVD-based TxBF, the transmitter should set the Smoothing bit in the HT-SIG to 0. If the receiver has implemented subcarrier smoothing, that bit should always be checked prior to application of subcarrier smoothing.

12.6   Channel sounding

In order to determine the weights for TxBF, knowledge of the CSI is required. The channel needs to be sounded between the two devices (devices A and B in Figure 12.1) participating in TxBF. The basic concept of sounding is for device A to transmit a packet to device B. As part of the standard receiver operations, device B estimates the channel from the HT-LTF in the preamble portion of the packet. Device B calculates the channel estimate for each spatial stream corresponding to the long training symbols in the HT-LTF.

Consider a poor channel which only supports a single stream with basic transmission. Applying TxBF may enable transmission of two spatial streams. Prior to utilizing TxBF, a packet needs to be sent to sound the channel. The full dimensionality of the channel is equivalent to the number of antennas at the transmitter and the number of antennas at the corresponding receiver. However, since the channel only supports a single stream with basic transmission, only one long training symbol is sent and the receiver can only estimate the channel for each receiver antenna for one spatial stream. TxBF weights for two streams can not be computed since only CSI for one stream is available.

Two solutions to this problem are provided for in the 802.11n standard. Two types of what is termed a sounding packet are described. The first type is called a null data packet (NDP). This type of packet contains no Data field, as illustrated in Figure 12.9. In an NDP, an MCS is selected corresponding to the number of spatial streams equivalent to the channel dimensionality to be sounded. For example if we wish to transmit with TxBF weights for two spatial streams, any MCS 8–15 would be selected in the HT-SIG of the NDP. Since no data is being transmitted, any number of spatial streams can be sounded with an NDP as long as the HT-SIG is properly decoded.

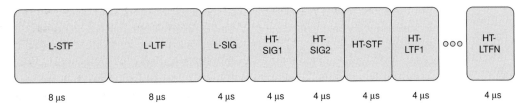

| L-STF | L-LTF | L-SIG | HT-SIG1 | HT-SIG2 | HT-STF | HT-LTF1 | ∘∘∘ | HT-LTFN |

| 8 µs | 8 µs | 4 µs | 4 µs | 4 µs | 4 µs | 4 µs | | 4 µs |

Figure 12.9 NDP with MF preamble.

When transmitting an NDP, the Not Sounding field in the HT-SIG is set to 0. In addition, the Length field is also always set to 0. An NDP may be composed of an MF preamble or a GF preamble.

With an NDP, a packet may be received with an MCS that the receiver does not support. For example, if the transmitter requires sounding over four antennas, it sets the MCS field to a value between 24 and 31. A receiver that does not support that many spatial streams should not terminate processing of the packet without first checking if the Length field is set to zero, indicating sounding with an NDP.

The second approach to channel sounding expands the number of long training symbols in the HT-LTF of a packet and is referred to as the staggered preamble. If the dimensionality of the channel is larger than the number of long training symbols used for channel estimation of the data, additional long training symbols may be included in the HT-LTF. For example, if we wish to transmit the data with single stream MCS 0, yet sound over two spatial streams, two long training symbols would be transmitted in the HT-LTF. The additional streams to be sounded are termed extension spatial streams in IEEE (2007).

The number of long training symbols required for data detection (N_{DLTF}) is given in Eq. (4.20). The number of additional long training symbols (N_{ELTF}) required for the number of extension spatial streams (N_{ESS}) is given as follows:

$$N_{\text{ELTF}} = \begin{cases} N_{\text{ESS}} & \text{if } N_{\text{ESS}} = 0, 1, 2 \\ 4 & \text{if } N_{\text{ESS}} = 3 \end{cases} \qquad (12.23)$$

The maximum number of extension spatial streams is three, since there is at least one data spatial stream in a data packet.

The total number of long training symbols, $N_{\text{LTF}} = N_{\text{DLTF}} + N_{\text{ELTF}}$, for all combinations of the number of data spatial streams (N_{SS}) and N_{ESS} is given in Table 12.2. Note that in some cases where either the number of data spatial streams or the number of extension spatial streams is three, the total number of long training symbols may be five.

To signify that a data packet is to be used for sounding, the Not Sounding field in the HT-SIG is set to 0. This would be set even if N_{ESS} and N_{ELTF} are equal to zero. Furthermore, the Number of Extension Spatial Streams field in HT-SIG is set as given by N_{ESS} in Table 12.2.

Table 12.2 Total number of long training symbols for data and extension spatial streams

N_{SS}	N_{DLTF}	N_{ESS}	N_{ELTF}	N_{LTF}
1	1	0	0	1
1	1	1	1	2
1	1	2	2	3
1	1	3	4	5
2	2	0	0	2
2	2	1	1	3
2	2	2	2	4
3	4	0	0	4
3	4	1	1	5
4	4	0	0	4

As presented in Section 4.2.4.3 for MF packet type and Section 5.4.4 for GF packet type, the transmit time computation given in Eqs. (4.32) and (5.19) excludes sounding packets. Actually, Eqs. (4.32) and (5.19) apply to sounding packets by modifying the definition of a few of the parameters. If sounding with extension HT-LTFs is used, N_{LTF} in Eqs. (4.32) and (5.19) is defined by Table 12.2. With NDP, N_{SYM} in Eqs. (4.32) and (5.19) is zero.

Limitations on applying beamforming matrices to sounding packets are addressed in Section 12.7.1.

As dictated by the standard, sounding packets are an optional feature. It is not required for a receiver to be able to process an NDP or a data packet with extension HT-LTFs.

12.7   Channel state information feedback

We have described the general TxBF system in Figure 12.2, and then detailed the use of SVD to compute the beamforming weights in Section 12.2. This discussion assumed full knowledge of the channel state information. Two methods to sound the channel in order to measure the CSI were described in Section 12.6. We now present two approaches for CSI feedback: implicit and explicit.

12.7.1   Implicit feedback

Implicit feedback is based on the reciprocity relationships for electromagnetism developed by H. A. Lorentz in 1896 (Smith, 2004). Specifically to the field of communications, antenna analysis based on these relationships results in the fact that the far field beam patterns for an antenna are equivalent on transmission and reception (Smith, 2004). Therefore, given that in 802.11n the same frequency carrier is used for both link directions (a TDD system), the propagation environment from one device to another will

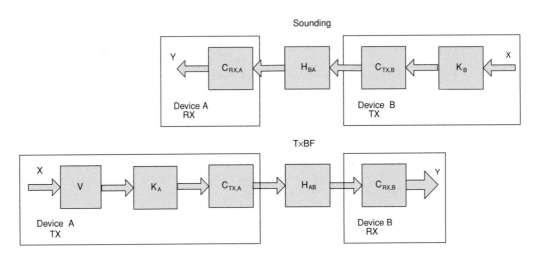

Figure 12.10 TxBF system model with implicit feedback.

be reciprocal. Ideally, the channel state information measured at either end of the link would be equivalent.

It is important to note that interference is not reciprocal. Furthermore, the channel between the digital baseband of one device to the digital baseband of another device includes transmitter and receiver RF distortions/impairments which are not reciprocal. The RF distortions are represented by matrices $C_{TX,A}$ and $C_{RX,A}$ for device A for the transmitter and receiver, respectively, in Figure 12.10. Similarly, the RF distortions for device B are represented by matrices $C_{TX,B}$ and $C_{RX,B}$. Though not noted in the figure, such a frequency domain system model exists for each subcarrier in the frequency band and each matrix may be different from subcarrier to subcarrier. The coupling between transmitter chains and the coupling between receiver chains is assumed to be low. Therefore the distortion is modeled by diagonal matrices containing complex values on the diagonal entries. All non-diagonal terms that are associated with antenna cross coupling are zero. This represents only gain and phase differences between the antenna chains. The propagation environment in the link direction from device A to device B is denoted as H_{AB} and from device B to device A as H_{BA}. Channel H_{AB} is equivalent to the transpose of H_{BA} to within a complex scaling factor.

The composite channel when transmitting from device A to device B is given by Eq. (12.24):

$$\tilde{H}_{AB} = C_{RX,B}H_{AB}C_{TX,A} \tag{12.24}$$

The composite channel when transmitting from device B to device A is given by Eq. (12.25):

$$\tilde{H}_{BA} = C_{RX,A}H_{BA}C_{TX,B} \tag{12.25}$$

As illustrated by Figure 12.10, two steps are performed when transmit beamforming with implicit feedback. The first step is that device B transmits a sounding packet to device A. Device A estimates the CSI from the HT-LTFs. With RF distortions, the CSI is based on $\tilde{H}_{BA}$. In the second step, device A computes the beamforming matrix V from the CSI and applies it when transmitting a packet to device B. The beamforming matrix V is derived from $\tilde{H}_{BA}$ and may be mismatched to the effective channel $\tilde{H}_{AB}$ between device A and device B.

To eliminate mismatch between the beamforming matrix V and the channel $\tilde{H}_{AB}$, calibration may be performed. The calibration coefficients are represented in Figure 12.10 by matrices K_A for device A and K_B for device B. Both matrices are diagonal with complex valued elements. The goal of calibration is to restore reciprocity by making the channel from A to B equivalent to the channel from B to A, to within a complex scaling factor. This requirement is expressed in Eq. (12.26) (IEEE, 2007):

$$\tilde{H}_{AB} K_A = \delta \left(\tilde{H}_{BA} K_B \right)^{T} \tag{12.26}$$

Reciprocity is restored by values for K_A and K_B as given in Eq. (12.27) (IEEE, 2007):

$$\begin{aligned} K_A &= \alpha_A \left(C_{TX,A} \right)^{-1} C_{RX,A} \\ K_B &= \alpha_B \left(C_{TX,B} \right)^{-1} C_{RX,B} \end{aligned} \tag{12.27}$$

The values α_A and α_B are complex scaling factors.

If calibration is applied at both devices, the channel from device A to device B is:

$$\tilde{H}_{AB} K_A = \alpha_A C_{RX,B} H_{AB} C_{RX,A} \tag{12.28}$$

and the transpose of the channel from device B to device A is:

$$\begin{aligned} \left(\tilde{H}_{BA} K_B \right)^{T} &= \left(\alpha_B C_{RX,A} H_{BA} C_{RX,B} \right)^{T} \\ &= \alpha_B C_{RX,B} H_{BA}^{T} C_{RX,A} \end{aligned} \tag{12.29}$$

Since channel H_{AB} is equivalent to the transpose of H_{BA} to within a complex scaling factor, we see that with calibration the two channels are equivalent and reciprocity is restored.

The calibration coefficients between the two devices may be computed as follows. To compute K_A, device A transmits a sounding packet to device B. This allows device B to compute the channel estimate $\tilde{H}_{AB}$. Next, device B transmits a sounding packet to device A. This allows device A to compute the channel estimate $\tilde{H}_{BA}$. We are assuming that channel H_{AB} is equivalent to the transpose of H_{BA} to within a complex scaling factor. Therefore these two transmissions must occur within as short a time interval as possible to minimize the change in the channel between the two transmissions. The last step in the calibration packet exchange sequence is for device B to send the channel estimate $\tilde{H}_{AB}$ back to device A, which is not time critical. At this point device A has both channel estimates $\tilde{H}_{AB}$ and $\tilde{H}_{BA}$ and can compute the calibration coefficient K_A.

To more clearly visualize one possible approach to computing K_A, we expand the matrices $\tilde{H}_{AB}$ and $\tilde{H}_{BA}$ as given below:

$$
\tilde{H}_{AB} = \begin{bmatrix}
c_{RX,B,1}h_{AB,11}c_{TX,A,1} & c_{RX,B,1}h_{AB,12}c_{TX,A,2} & \cdots & c_{RX,B,1}h_{AB,1M}c_{TX,A,M} \\
c_{RX,B,2}h_{AB,21}c_{TX,A,1} & c_{RX,B,2}h_{AB,22}c_{TX,A,2} & \cdots & c_{RX,B,2}h_{AB,2M}c_{TX,A,M} \\
\vdots & \vdots & \ddots & \vdots \\
c_{RX,B,N}h_{AB,N1}c_{TX,A,1} & c_{RX,B,N}h_{AB,N2}c_{TX,A,2} & \cdots & c_{RX,B,N}h_{AB,NM}c_{TX,A,M}
\end{bmatrix}
$$

$$(12.30)$$

$$
\tilde{H}_{BA} = \begin{bmatrix}
c_{RX,A,1}h_{BA,11}c_{TX,B,1} & c_{RX,A,1}h_{BA,12}c_{TX,B,2} & \cdots & c_{RX,A,1}h_{BA,1N}c_{TX,B,N} \\
c_{RX,A,2}h_{BA,21}c_{TX,B,1} & c_{RX,A,2}h_{BA,22}c_{TX,B,2} & \cdots & c_{RX,A,2}h_{BA,2N}c_{TX,B,N} \\
\vdots & \vdots & \ddots & \vdots \\
c_{RX,A,M}h_{BA,M1}c_{TX,B,1} & c_{RX,A,M}h_{BA,M2}c_{TX,B,2} & \cdots & c_{RX,A,M}h_{BA,MN}c_{TX,B,N}
\end{bmatrix}
$$

$$(12.31)$$

If we divide the elements of the first column of $\tilde{H}_{BA}$ by the elements of the first row of $\tilde{H}_{AB}$ we get the following:

$$
K_{A,i,i} = \frac{c_{RX,A,i}h_{BA,i1}c_{TX,B,1}}{c_{RX,B,1}h_{AB,1i}c_{TX,A,i}} \qquad i = 1, 2, \ldots, M \qquad (12.32)
$$

Since $h_{BA,i1}$ is equivalent to $h_{AB,1i}$ and will cancel, Eq. (12.32) results in the diagonal elements of K_A where α_A is equal to $c_{TX,B,1}/c_{RX,B,1}$ as given in Eq. (12.27). Even though the method by which K_A was derived results in α_A being a function of the distortion of device B, these calibration coefficients may be used when beamforming to any other devices. Since α_A is a scalar value, this is compensated by the equalizer at any other device.

With implicit beamforming, an AP would typically beamform to a low complexity client device to enhance link performance. To minimize complexity, the client device would not be calibrated. Referring to Figure 12.10, when device B is not calibrated, the calibration matrix K_B is neither computed nor applied. Therefore, full reciprocity is not achieved, except in the case where device B has only one antenna. In the single antenna case, $C_{TX,B}$ and $C_{RX,B}$ are both scalars that do not require calibration. However, we will show in Section 12.9 that implicit feedback beamforming is very insensitive to distortions at device B receiving the beamformed transmission.

In the case where the system is operating with bi-directional beamforming between devices A and B, then similar steps as above are taken in order to calculate the calibration coefficients for device B, K_B. In this case device B transmits a sounding packet to device A, enabling device A to estimate $\tilde{H}_{BA}$. Device A transmits a sounding packet to device B, enabling device B to estimate $\tilde{H}_{AB}$. After device A sends the estimates of $\tilde{H}_{BA}$ to device B, device B can compute K_B.

It is not necessary to use the calibration method provided in the standard. Another approach is that at each device, the Tx chains are calibrated to themselves and the Rx chains are calibrated to themselves. By doing so, C matrix is an identity matrix multiplied by a complex scalar. Therefore it is not necessary to calibrate the receiver to the transmitter. With the technology trending towards the entire transceiver on a chip, this almost occurs by design. With symmetric design for each chain, the gain and phase for the traces to each antenna are very similar. It is feasible that devices may be calibrated by design.

During the development of the implicit feedback beamforming section of the 802.11n standard, questions were raised regarding how often calibration was necessary. Claims were made that in a typical indoor environment with reasonable controls on environmental conditions, calibration would be required very infrequently. On the other hand, others were concerned that internal temperature changes in the electronics, carrier frequency changes, and gain changes would require frequent calibration. Questions were also raised whether a diagonal matrix was a reasonable model for the RF distortions. Or, would coupling between antennas cause non-diagonal elements in the RF distortion matrix, in which case calibration with a diagonal matrix would not be sufficient to restore channel reciprocity. Little evidence was produced to substantiate either claims or concerns.

Another issue with implicit feedback is beamforming to devices with more receive RF chains than transmit RF chains. For example, consider in Figure 12.10 device B has M_B transmit antennas and N_B receive antennas with M_B less than N_B. When device B transmits a sounding packet to device A, the maximum number of spatial streams that can be sounded is M_B, regardless of the receive capability of device B. It is not possible to measure CSI between devices A and B for receive antennas $M_B + 1$ to N_B at device B.

The effect on TxBF is as follows. The TxBF performance is no worse than if device B had M_B receive antennas. In essence, device A is beamforming to the first M_B receive antennas at device B. The additional $M_B + 1$ to N_B receive antennas at device B behave as receive diversity antennas, as described in Section 6.1, which should improve receive performance.

When beamforming with implicit feedback, there are restrictions on the beamforming matrices when transmitting a sounding packet. With bi-directional TxBF, it is advantageous to use beamformed data packets as sounding packets. However, applying beamforming matrices on a sounding packet may impair the ability to estimate the CSI. If we restrict the beamforming matrix to be unitary when sounding, there is no impact on the ability to estimate the CSI and compute proper beamforming matrices. Since the V matrix of an SVD computation is unitary, this is a mild restriction.

Sounding packets for calibration is further restricted, by not permitting beamforming. Typically with calibration, the number of spatial streams is fewer than the number of transmit antennas. Therefore a unique spatial expansion matrix is defined in the 802.11n standard. The Q matrix for each subcarrier k in Eq. (4.31) is specified as given in Eq. (12.33) (IEEE, 2007):

$$Q_k = C_{CSD}(k)P_{CAL} \qquad (12.33)$$

where $C_{CSD}(k)$ is the diagonal cyclic shift matrix in which the diagonal elements contain frequency-domain representation of the cyclic shifts given in Table 4.3; and P_{CAL} is the unitary matrix, which is specified for the number of transmit antennas ranging from one to four as follows:

$$P_{CAL} = 1 \qquad \text{(one transmit antenna)}$$

$$P_{CAL} = \frac{\sqrt{2}}{2} \begin{bmatrix} 1 & -1 \\ 1 & 1 \end{bmatrix} \qquad \text{(two transmit antennas)}$$

$$P_{CAL} = \frac{\sqrt{3}}{3} \begin{bmatrix} 1 & 1 & 1 \\ 1 & e^{-j2\pi/3} & e^{-j4\pi/3} \\ 1 & e^{-j4\pi/3} & e^{-j2\pi/3} \end{bmatrix} \qquad \text{(three transmit antennas)}$$

$$P_{CAL} = \frac{1}{2} \begin{bmatrix} 1 & -1 & 1 & 1 \\ 1 & 1 & -1 & 1 \\ 1 & 1 & 1 & -1 \\ -1 & 1 & 1 & 1 \end{bmatrix} \qquad \text{(four transmit antennas)}$$

12.7.2   Explicit feedback

With an explicit feedback mechanism, the device performing TxBF is the same device that transmits the sounding packet. Device A sends a sounding packet to device B. Device B transmits the CSI or beamforming weights to device A. Device A uses the feedback to transmit a packet with TxBF to device B. Ideally, the channel remains the same between the time the sounding packet is sent and the time the beamformed packet is sent. Unlike implicit feedback, no calibration is required. Furthermore, the channel dimensions between the beamformer's transmit antennas and beamformee's receive antennas is always matched between sounding and beamforming.

On the other hand, with explicit feedback we must define the format of the feedback. In 802.11n, three types explicit feedback are specified: CSI, non-compressed beamforming weights, and compressed beamforming weights. Since the explicit feedback will reduce the system efficiency, quantization and subcarrier grouping techniques are provided to minimize the overhead.

12.7.2.1   CSI feedback

Explicit feedback of CSI has many uses beyond TxBF. For network management, a central controller may have APs collect CSI information by sending sounding packets to client devices and client devices transmitting the CSI back to the AP. The CSI may be used to generate a profile of the propagation environment.

CSI may also be used to assist with link adaptation. Link adaptation algorithms may incorporate information about the channel regarding the multipath fading, delay spread, channel correlation, or the number of streams the channel can support. These properties of the channel can be extracted from the CSI.

A device may wish to beamform to another device which does not have the ability to compute beamforming weights. Simple, low cost devices may not wish to incur the complexity of beamforming weight computation. In such a situation, the device serving as the beamformee could send CSI to the device serving as the beamformer. In which case, the beamformer computes the beamforming weights from the CSI. Collection of CSI by the beamformee is relatively simple, since channel estimation is required by all devices in order to process a packet.

CSI feedback is also required for calibration for TxBF with implicit feedback. As was described in Section 12.7.1, the last step in the calibration exchange sequence is for device B to send the channel estimate to device A. This step uses the explicit CSI feedback mechanism.

There is a CSI matrix with elements comprised of complex values for all data and pilot subcarriers. The number of rows and columns of each CSI matrix corresponds to the number of receive antennas at the recipient of the sounding packet and the number of HT-LTFs in the sounding packet, respectively.

The amount of CSI for feedback grows with the dimensionality of the system resulting in an increasing amount of overhead. For example, a 4×4 system with 40 MHz requires CSI for 16 complex elements for each of the 114 subcarriers. For the least quantization distortion, the 802.11n standard defines eights bits for each real and eight bits for each imaginary component of the complex element for CSI feedback. This results in 3648 bytes.

To reduce the overhead associated with CSI feedback, the number of bits used to represent the real and imaginary parts of each CSI value may also be set to four, five, and six. This allows the amount of feedback to be reduced by as much as half. However, using fewer bits degrades the quality of the CSI with more quantization distortion. Depending on the sensitivity to CSI, the performance of certain mechanisms may be impacted.

Another approach to reducing CSI feedback overhead is to not send a CSI matrix for every data or pilot subcarrier. Instead, subcarriers are grouped together and a CSI matrix is only transmitted for each group. Groupings of two subcarriers and four subcarriers are permitted, reducing the overhead to a half and a quarter, respectively. Depending on the channel, the performance may be impacted using subcarrier grouping.

12.7.2.2   Non-compressed beamforming weights feedback

As described in the previous section, CSI feedback requires the beamformer to compute beamforming matrices. With non-compressed beamforming weights, after receiving the sounding packet, the beamformee computes the beamforming weights. Any method may be used to compute the beamforming weights, since upon receiving the feedback, the beamformer applies the weights without any further modification. This approach gives the beamformee control over the technique used for beamforming, be it SVD or another algorithm.

The overhead of the feedback with non-compressed beamforming weights is identical to that of CSI feedback. Similarly, non-compressed beamforming weights may be quantized. In 802.11n draft 2.0 the quantization choices are eight, six, five, or four bits for each real and imaginary component of the complex element of the beamforming matrix. This was changed to eight, six, four, or two bits in 802.11n draft 3.0.

12.7.2.3   Compressed beamforming weights feedback

Non-compressed beamforming matrices may require a large number of bits to represent the complex values with limited quantization loss, due to the necessary dynamic range. If the technique used to compute the beamforming weights results in unitary matrices, polar coordinates may be used to reduce the number of bits required for beamforming weights feedback. For example, the matrix V in SVD is unitary. The 802.11n standard uses Givens rotations to perform a planar rotation operation on a unitary matrix V.

The Givens rotation matrix is represented by Eq. (12.34) (IEEE, 2007b):

$$G_{li}(\psi) = \begin{bmatrix} I_{i-1} & 0 & 0 & 0 & 0 \\ 0 & \cos(\psi_{l,i}) & 0 & \sin(\psi_{l,i}) & 0 \\ 0 & 0 & I_{l-i-1} & 0 & 0 \\ 0 & -\sin(\psi_{l,i}) & 0 & \cos(\psi_{l,i}) & 0 \\ 0 & 0 & 0 & 0 & I_{M-l} \end{bmatrix} \tag{12.34}$$

where I_m is an $m \times m$ identity matrix, and the terms $\cos(\psi_{li})$ and $\sin(\psi_{li})$ are located at row l and column i. A useful property of the Givens rotation matrix is that it is orthogonal. Furthermore, when we multiply a matrix by a Givens rotation matrix, only rows i and l are affected. Therefore we can decompose the problem to just the two elements involved.

The Givens rotation problem may be expressed as follows:

$$\begin{bmatrix} \cos(\psi) & \sin(\psi) \\ -\sin(\psi) & \cos(\psi) \end{bmatrix} \begin{bmatrix} x_1 \\ x_2 \end{bmatrix} = \begin{bmatrix} y \\ 0 \end{bmatrix} \tag{12.35}$$

where x_1 and x_2 are real values from the matrix to which we are applying the planar rotation. A solution to this problem is as follows:

$$\psi = \mathrm{acos}\left(\frac{x_1}{\sqrt{x_1^2 + x_2^2}} \right) = \mathrm{asin}\left(\frac{x_2}{\sqrt{x_1^2 + x_2^2}} \right)$$

$$y = \sqrt{x_1^2 + x_2^2} \tag{12.36}$$

The function *planerot* in Matlab® performs the Givens plane rotation.

To decompose the beamforming matrix V into polar values, we apply to it a sequence of Givens rotations. However, since the beamforming matrix V may comprise complex values, preprocessing steps are required before applying Givens rotations to the matrix V. A diagonal matrix D_i is derived for a matrix V such that the elements of column i of $D_i^* V$ are all non-negative real numbers, given by Eq. (12.37) (IEEE, 2007):

$$D_i = \begin{bmatrix} I_{i-1} & 0 & \cdots & \cdots & 0 \\ 0 & e^{j\phi_{i,i}} & 0 & \cdots & 0 \\ \vdots & 0 & \ddots & 0 & \vdots \\ \vdots & \vdots & 0 & e^{j\phi_{M-1,i}} & 0 \\ 0 & 0 & \cdots & 0 & 1 \end{bmatrix} \tag{12.37}$$

The angles $\phi_{l,i}$ may be computed as follows:

$$\phi_{l,i} = \text{angle}(V_{l,i}) \tag{12.38}$$

Since the last element of D_i is always 1, the elements in the last row of V are not altered. Therefore a prior step of multiplying by $\tilde{D}$ is required to make the last row of $V\tilde{D}^*$ consist of non-negative real values:

$$\tilde{D} = \begin{bmatrix} e^{j\theta_1} & 0 & \cdots & & 0 \\ 0 & e^{j\theta_2} & 0 & \cdots & 0 \\ \vdots & 0 & \ddots & 0 & \vdots \\ \vdots & \vdots & 0 & e^{j\theta_{N-1}} & 0 \\ 0 & 0 & \cdots & 0 & e^{j\theta_N} \end{bmatrix}$$

The angles θ_i may be computed as follows:

$$\theta_i = \text{angle}(V_{M,i}) \tag{12.39}$$

Therefore, an $M \times N$ beamforming matrix V is decomposed into a sequence of $\tilde{D}$, D_i, and $G_{li}(\psi_{l,i})$ matrices given by Eq. (12.40) (IEEE, 2007):

$$V = \tilde{D} \prod_{i=1}^{\min(N,M-1)} \left[D_i \prod_{l=i+1}^{M} G_{li}^T (\psi_{l,i}) \right] \times \tilde{I}_{M \times N} \tag{12.40}$$

where $\tilde{I}_{M \times N}$ is an $M \times N$ identity matrix with extra rows or columns filled with zeros when M is not equal to N. The feedback consists of the angles $\phi_{l,i}$ and $\psi_{l,i}$. It can be shown that it is not necessary for the beamformer to have the angles θ_i when reconstructing the matrix V.

Without the angles θ_i, the beamformer is only able to reconstruct a matrix $\tilde{V}$, which is equal to $V\tilde{D}^*$. However, receiver performance at the beamformee is equivalent to $U^*HV\tilde{D}^*$, which is equal to $S\tilde{D}^*$. Since output SNR is proportional to $S\tilde{D}^* (S\tilde{D}^*)^*$, which is equal to SS^T, the output SNR is unchanged with $\tilde{V}$ as compared to V.

The following steps are performed to derive the $\tilde{D}$, D_i, and $G_{li}(\psi_{l,i})$ matrices. As an example, consider a matrix V with the dimensions 4×2. In the first step we derive $\tilde{D}$ with angles θ_i equal to angle($V_{4,i}$):

$$\tilde{D} = \begin{bmatrix} e^{j\theta_1} & 0 \\ 0 & e^{j\theta_2} \end{bmatrix} \tag{12.41}$$

In the next step we derive D_1 from the $V\tilde{D}^*$, where the angles $\phi_{l,1}$ are computed as angle($[V\tilde{D}^*]_{l,1}$):

$$D_1 = \begin{bmatrix} e^{j\phi_{1,1}} & 0 & 0 & 0 \\ 0 & e^{j\phi_{2,1}} & 0 & 0 \\ 0 & 0 & e^{j\phi_{3,1}} & 0 \\ 0 & 0 & 0 & 1 \end{bmatrix} \tag{12.42}$$

Next we compute Givens rotation matrices $G_{21}(\psi_{2,1})$, $G_{31}(\psi_{3,1})$, and $G_{41}(\psi_{4,1})$. The matrix $G_{21}(\psi_{2,1})$ is computed by setting x_1 equal to element $[D_1^* V \tilde{D}^*]_{1,1}$ and x_2 equal to element $[D_1^* V \tilde{D}^*]_{2,1}$ in Eq. (12.35) and solving for $\psi_{2,1}$. After which, the matrix $G_{21}(\psi_{2,1})$ is populated as follows:

$$G_{21}(\psi_{2,1}) = \begin{bmatrix} \cos(\psi_{2,1}) & \sin(\psi_{2,1}) & 0 & 0 \\ -\sin(\psi_{2,1}) & \cos(\psi_{2,1}) & 0 & 0 \\ 0 & 0 & 1 & 0 \\ 0 & 0 & 0 & 1 \end{bmatrix} \qquad (12.43)$$

Similarly, the matrix $G_{31}(\psi_{3,1})$ is computed by setting x_1 equal to element $[G_{21}(\psi_{2,1})D_1^* V \tilde{D}^*]_{1,1}$ and x_2 equal to element $[G_{21}(\psi_{2,1})D_1^* V \tilde{D}^*]_{3,1}$ in Eq. (12.35) and solving for $\psi_{3,1}$. After which, the matrix $G_{31}(\psi_{3,1})$ is populated as follows:

$$G_{31}(\psi_{3,1}) = \begin{bmatrix} \cos(\psi_{3,1}) & 0 & \sin(\psi_{3,1}) & 0 \\ 0 & 1 & 0 & 0 \\ -\sin(\psi_{3,1}) & 0 & \cos(\psi_{3,1}) & 0 \\ 0 & 0 & 0 & 1 \end{bmatrix} \qquad (12.44)$$

And finally for this step, the matrix $G_{41}(\psi_{4,1})$ is computed by setting x_1 equal to element $[G_{31}(\psi_{3,1})G_{21}(\psi_{2,1})D_1^* V \tilde{D}^*]_{1,1}$ and x_2 equal to element $[G_{31}(\psi_{3,1})G_{21}(\psi_{2,1})D_1^* V \tilde{D}^*]_{4,1}$ in Eq. (12.35) and solving for $\psi_{4,1}$. After which, the matrix $G_{41}(\psi_{4,1})$ is populated as follows:

$$G_{41}(\psi_{4,1}) = \begin{bmatrix} \cos(\psi_{4,1}) & 0 & 0 & \sin(\psi_{4,1}) \\ 0 & 1 & 0 & 0 \\ 0 & 0 & 1 & 0 \\ -\sin(\psi_{4,1}) & 0 & 0 & \cos(\psi_{4,1}) \end{bmatrix} \qquad (12.45)$$

At the end of this step, the non-diagonal elements of the first row and column of the composite matrix are now zero. The diagonal element is one. The composite matrix is represented as follows:

$$\begin{bmatrix} \cos(\psi_{4,1}) & 0 & 0 & \sin(\psi_{4,1}) \\ 0 & 1 & 0 & 0 \\ 0 & 0 & 1 & 0 \\ -\sin(\psi_{4,1}) & 0 & 0 & \cos(\psi_{4,1}) \end{bmatrix} \begin{bmatrix} \cos(\psi_{3,1}) & 0 & \sin(\psi_{3,1}) & 0 \\ 0 & 1 & 0 & 0 \\ -\sin(\psi_{3,1}) & 0 & \cos(\psi_{3,1}) & 0 \\ 0 & 0 & 0 & 1 \end{bmatrix} \cdot$$
$$\begin{bmatrix} \cos(\psi_{2,1}) & \sin(\psi_{2,1}) & 0 & 0 \\ -\sin(\psi_{2,1}) & \cos(\psi_{2,1}) & 0 & 0 \\ 0 & 0 & 1 & 0 \\ 0 & 0 & 0 & 1 \end{bmatrix} \begin{bmatrix} e^{j\phi_{1,1}} & 0 & 0 & 0 \\ 0 & e^{j\phi_{2,1}} & 0 & 0 \\ 0 & 0 & e^{j\phi_{3,1}} & 0 \\ 0 & 0 & 0 & 1 \end{bmatrix}^* \cdot$$
$$V \begin{bmatrix} e^{j\theta_1} & 0 \\ 0 & e^{j\theta_2} \end{bmatrix}^* = \begin{bmatrix} 1 & 0 \\ 0 & \\ 0 & \hat{V}_2 \\ 0 & \end{bmatrix} = V_2 \qquad (12.46)$$

The process starts again by deriving D_2 from the matrix V_2, where the angles $\phi_{l,2}$ are computed as angle($[V_2]_{l,2}$):

$$D_2 = \begin{bmatrix} 1 & 0 & 0 & 0 \\ 0 & e^{j\phi_{2,2}} & 0 & 0 \\ 0 & 0 & e^{j\phi_{3,2}} & 0 \\ 0 & 0 & 0 & 1 \end{bmatrix} \qquad (12.47)$$

Next we compute Givens rotation matrices $G_{32}(\psi_{3,2})$ and $G_{42}(\psi_{4,2})$. The matrix $G_{32}(\psi_{3,2})$ is computed by setting x_1 equal to element $[D_2^* V_2]_{2,2}$ and x_2 equal to element $[D_2^* V_2]_{3,2}$ in Eq. (12.35) and solving for $\psi_{3,2}$. After which, the matrix $G_{32}(\psi_{3,2})$ is populated as follows:

$$G_{32}(\psi_{3,2}) = \begin{bmatrix} 1 & 0 & 0 & 0 \\ 0 & \cos\left(\psi_{3,2}\right) & \sin\left(\psi_{3,2}\right) & 0 \\ 0 & -\sin\left(\psi_{3,2}\right) & \cos\left(\psi_{3,2}\right) & 0 \\ 0 & 0 & 0 & 1 \end{bmatrix} \qquad (12.48)$$

Similarly, the matrix $G_{42}(\psi_{4,2})$ is computed by setting x_1 equal to element $[G_{32}(\psi_{3,2})D_2^*]_{2,2}$ and x_2 equal to element $[G_{32}(\psi_{3,2})D_2^*]_{4,2}$ in Eq. (12.35) and solving for $\psi_{4,2}$. After which, the matrix $G_{42}(\psi_{4,2})$ is populated as follows:

$$G_{42}(\psi_{4,2}) = \begin{bmatrix} 1 & 0 & 0 & 0 \\ 0 & \cos\left(\psi_{3,2}\right) & 0 & \sin\left(\psi_{4,2}\right) \\ 0 & 0 & 1 & 0 \\ 0 & -\sin\left(\psi_{4,2}\right) & 0 & \cos\left(\psi_{3,2}\right) \end{bmatrix} \qquad (12.49)$$

At the end of this final step, the composite matrix is equal to $\tilde{I}_{4\times2}$:

$$\begin{bmatrix} 1 & 0 & 0 & 0 \\ 0 & \cos\left(\psi_{4,2}\right) & 0 & \sin\left(\psi_{4,2}\right) \\ 0 & 0 & 1 & 0 \\ 0 & -\sin\left(\psi_{4,2}\right) & 0 & \cos\left(\psi_{3,2}\right) \end{bmatrix} \begin{bmatrix} 1 & 0 & 0 & 0 \\ 0 & \cos\left(\psi_{3,2}\right) & \sin\left(\psi_{3,2}\right) & 0 \\ 0 & -\sin\left(\psi_{3,2}\right) & \cos\left(\psi_{3,2}\right) & 0 \\ 0 & 0 & 0 & 1 \end{bmatrix} \cdot$$

$$\begin{bmatrix} 1 & 0 & 0 & 0 \\ 0 & e^{j\phi_{2,2}} & 0 & 0 \\ 0 & 0 & e^{j\phi_{3,2}} & 0 \\ 0 & 0 & 0 & 1 \end{bmatrix}^* \begin{bmatrix} \cos\left(\psi_{4,1}\right) & 0 & 0 & \sin\left(\psi_{4,1}\right) \\ 0 & 1 & 0 & 0 \\ 0 & 0 & 1 & 0 \\ -\sin\left(\psi_{4,1}\right) & 0 & 0 & \cos\left(\psi_{4,1}\right) \end{bmatrix} \cdot$$

$$\begin{bmatrix} \cos\left(\psi_{3,1}\right) & 0 & \sin\left(\psi_{3,1}\right) & 0 \\ 0 & 1 & 0 & 0 \\ -\sin\left(\psi_{3,1}\right) & 0 & \cos\left(\psi_{3,1}\right) & 0 \\ 0 & 0 & 0 & 1 \end{bmatrix} \begin{bmatrix} \cos\left(\psi_{2,1}\right) & \sin\left(\psi_{2,1}\right) & 0 & 0 \\ -\sin\left(\psi_{2,1}\right) & \cos\left(\psi_{2,1}\right) & 0 & 0 \\ 0 & 0 & 1 & 0 \\ 0 & 0 & 0 & 1 \end{bmatrix} \cdot$$

$$\begin{bmatrix} e^{j\phi_{1,1}} & 0 & 0 & 0 \\ 0 & e^{j\phi_{2,1}} & 0 & 0 \\ 0 & 0 & e^{j\phi_{3,1}} & 0 \\ 0 & 0 & 0 & 1 \end{bmatrix}^* V \begin{bmatrix} e^{j\theta_1} & 0 \\ 0 & e^{j\theta_2} \end{bmatrix}^* = \begin{bmatrix} 1 & 0 \\ 0 & 1 \\ 0 & 0 \\ 0 & 0 \end{bmatrix} \qquad (12.50)$$

Table 12.3 Angles for compressed beamforming matrices (IEEE, 2007)

$M \times N$	Number of angles	Angles
2×1	2	$\phi_{1,1}, \psi_{2,1}$
2×2	2	$\phi_{1,1}, \psi_{2,1}$
3×1	4	$\phi_{1,1}, \phi_{2,1}, \psi_{2,1}, \psi_{3,1}$
3×2	6	$\phi_{1,1}, \phi_{2,1}, \psi_{2,1}, \psi_{3,1}, \phi_{2,2}, \psi_{3,2}$
3×3	6	$\phi_{1,1}, \phi_{2,1}, \psi_{2,1}, \psi_{3,1}, \phi_{2,2}, \psi_{3,2}$
4×1	6	$\phi_{1,1}, \phi_{2,1}, \phi_{3,1}, \psi_{2,1}, \psi_{3,1}, \psi_{4,1}$
4×2	10	$\phi_{1,1}, \phi_{2,1}, \phi_{3,1}, \psi_{2,1}, \psi_{3,1}, \psi_{4,1}, \phi_{2,2}, \phi_{3,2}, \psi_{4,2}$
4×3	12	$\phi_{1,1}, \phi_{2,1}, \phi_{3,1}, \psi_{2,1}, \psi_{3,1}, \psi_{4,1}, \phi_{2,2}, \phi_{3,2}, \psi_{3,2}, \psi_{4,2}, \phi_{3,3}, \psi_{4,3}$
4×4	12	$\phi_{1,1}, \phi_{2,1}, \phi_{3,1}, \psi_{2,1}, \psi_{3,1}, \psi_{4,1}, \phi_{2,2}, \phi_{3,2}, \psi_{3,2}, \psi_{4,2}, \phi_{3,3}, \psi_{4,3}$

At this point the explicit feedback angles $\phi_{l,i}$ and $\psi_{l,i}$ are transmitted from the beamformee to the beamformer. The beamformer computes a TxBF weighting matrix $\tilde{V}$, which is given in Eq. (12.51). As a reminder, there may be a unique matrix $\tilde{V}$ for each subcarrier. A matrix $\tilde{V}$ is computed as follows, utilizing the property that the Givens rotation matrices are orthogonal matrices:

$$
\begin{aligned}
\tilde{V} &= V \tilde{D}^* \\
&= D_1 G_{21}^{\mathrm{T}}\left(\psi_{2,1}\right) G_{31}^{\mathrm{T}}\left(\psi_{3,1}\right) G_{41}^{\mathrm{T}}\left(\psi_{3,1}\right) D_2 G_{32}^{\mathrm{T}}(\psi_{3,2}) G_{42}^{\mathrm{T}}(\psi_{4,2}) \tilde{I}_{4\times 2}
\end{aligned} \quad (12.51)
$$

Table 12.3 summarizes the angles used for compressed beamforming feedback for all possible dimensions of the matrix V.

After converting to polar coordinates, the angles are quantized. The angles ϕ are quantized between 0 and 2π and the angles ψ are quantized between 0 and $\pi/2$, as given by Eq. (12.52) (IEEE, 2007):

$$
\begin{aligned}
\phi &= \pi \left(\frac{1}{2^{b+2}} + \frac{k}{2^{b+1}} \right) & k &= 0, 1, \ldots, 2^{b+2} - 1 \\
\psi &= \pi \left(\frac{1}{2^{b+2}} + \frac{k}{2^{b+1}} \right) & k &= 0, 1, \ldots, 2^{b} - 1
\end{aligned} \quad (12.52)
$$

where $(b + 2)$ is the number of bits used to quantize ϕ and b is the number of bits used to quantize ψ. The value b may be set to 1, 2, 3, or 4.

The total number of bits transmitted per subcarrier for the matrix V is given by the number of angles multiplied by the number of bits used for quantization. For example, with 4×4 there are twelve angles. Therefore the total number of bits is 12, 24, 36, or 48, respectively. For the same dimension matrix V, non-compressed beamforming requires between 128 and 256 bits per subcarrier.

Grouping of subcarriers may be used to further reduce the feedback overhead with compressed beamforming matrices, similar to the other explicit feedback techniques. Groups of two and four subcarriers are permitted.

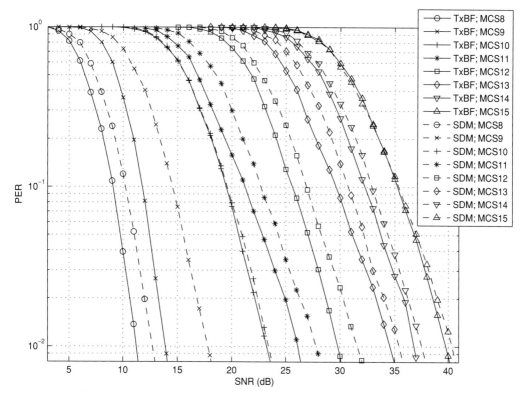

Figure 12.11 Comparison between TxBF and basic SDM for 2 × 2, two streams, channel model D, NLOS.

12.8 Improved performance with transmit beamforming

We present results based on transmit beamforming PHY simulations to compare performance between TxBF, STBC, SE, and basic SDM. The simulations are performed in channel model D, with NLOS conditions, as described in Section 3.5. Physical layer impairments were included in the simulation, as described in Section 3.5.4. The equalizer is based on MMSE. Synchronization, channel estimation, and phase tracking are included in the simulation. The TxBF simulations are performed with beamforming weights calculated by SVD, with the CSI determined from a noisy channel estimate. The beamformer weights have no compression, no subcarrier grouping, and no feedback delay.

We first compare TxBF to basic SDM for square system dimensions, 2 × 2 and 4 × 4. The waterfall curves comparing MCS 8 through MCS 15 are given in Figure 12.11 for a 2 × 2 system. The waterfall curves comparing MCS 24 through MCS 31 are given in Figure 12.12 for a 4 × 4 system. The solid lines represent TxBF and the dashed lines represent basic SDM. In both cases we have a system where the number of transmit antennas equals the number of receive antennas and is also equal to the number of spatial streams.

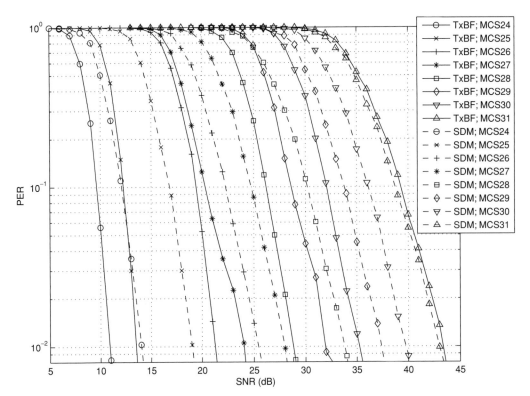

Figure 12.12 Comparison between TxBF and basic SDM for 4 × 4, four streams, channel model D, NLOS.

For such a system configuration the gain varies between significant and negligible. For 2 × 2, we see up to 4 dB gain at a PER of 1% at the lower MCSs. But for the higher MCSs the gain reduces to zero. For 4 × 4, the gain is as high as 5 dB at a PER of 1% for some of the lower MCSs. But again, the gain vanishes for the MCS 31.

The true benefit of transmit beamforming is achieved when the number of transmit antennas exceeds the number of receive antennas and the number of spatial streams. As described in Section 12.3, with SVD the stronger spatial streams are utilized whereas the weaker spatial streams are disregarded. For example in Figure 12.4, with a 4 × 2 TxBF system, the strongest two spatial streams would be used and the weaker two would not, maximizing output SNR.

Simulation results comparing 4 × 2 TxBF with 4 × 2 SE are given in Figure 12.13. The solid lines represent TxBF and the dashed lines represent SE. With MCS 8, the gain at a PER of 1% is 5 dB. The gain increases to 11 dB with MCS 15. With TxBF, the required SNR of MCS 15 decreases to 28 dB, well within the capability of typical receivers.

Next we compare 4 × 2 TxBF to 4 × 2 STBC. The results are given in Figure 12.14. The solid lines represent TxBF and the dashed lines represent STBC. In Chapter 6, Figure 6.15 illustrated that STBC provides 6 dB gain over SE for MCS 15. TxBF

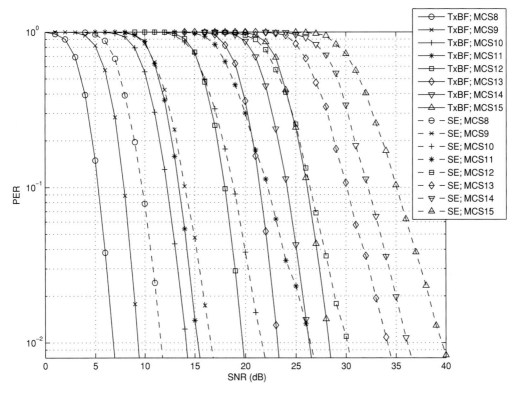

Figure 12.13 Comparison between TxBF and SE for 4 × 2, two streams, channel model D, NLOS.

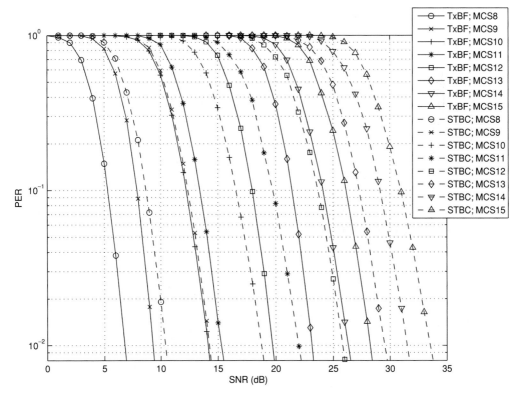

Figure 12.14 Comparison between TxBF and STBC for 4 × 2, two streams, channel model D, NLOS.

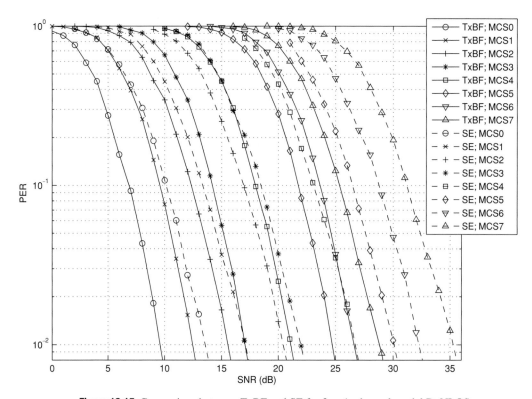

Figure 12.15 Comparison between TxBF and SE for 2 × 1, channel model D, NLOS.

provides an additional 5 dB of gain beyond the gain afforded by STBC for MCS 15. For other MCSs, the gain of TxBF over STBC ranges from 4.5 to 6.5 dB at a PER of 1% depending on the MCS.

The following two figures illustrate the performance of 2 × 1 TxBF. As discussed in Chapter 6, such a device configuration enables robust performance to low power, low cost devices such as handheld. In such a system, the AP has two antennas and the handheld device has a single antenna. Figure 12.15 compares 2 × 1 TxBF to 2 × 1 SE. The solid lines represent TxBF and the dashed lines represent SE. At a PER equal to 1%, the gain of TxBF over SE ranges from 4 dB at MCS 0 to 6.5 dB at MCS 7.

In Chapter 6, we discussed the use of STBC for use with 2 × 1 configurations. In Figure 12.16, we compare the performance of TxBF to STBC with a 2×1 system. The solid lines represent TxBF and the dashed lines represent STBC. At a PER equal to 1%, the gain of TxBF over STBC ranges from 2 dB at the higher MCSs to 3 dB at the lower MCSs.

A summary of the performance of TxBF as compared to SDM or SE is given in Table 12.4. The gain of TxBF is as large as 11 dB over SE with 4 × 2 and MCS 15. However there is no gain with 4 × 4 and MCS 31. The gain of TxBF over STBC ranges from approximately 2 dB to 5 dB.

Waterfall curves demonstrate the PER versus SNR performance for individual MCSs. In an actual system, devices incorporate link adaptation schemes to select the proper

Table 12.4 Summary of TxBF performance compared to SDM/SE and STBC

		Required SNR (dB) @PER = 1%			TxBF Gain (dB)	
		SDM or SE	STBC	TxBF	over SE	over STBC
2 × 1	MCS 0	13.6	12.8	9.6	4.0	3.2
	MCS 7	35.3	31.1	28.8	6.5	2.3
2 × 2	MCS 8	12.7		11.2	1.5	
	MCS 15	40.4		39.8	0.6	
4 × 2	MCS 8	11.6	10.4	6.8	4.8	3.6
	MCS 15	39.6	33.5	28.3	11.3	5.2
4 × 4	MCS 24	12.5		10.9	1.6	
	MCS 31	43.0		43.3	−0.3	

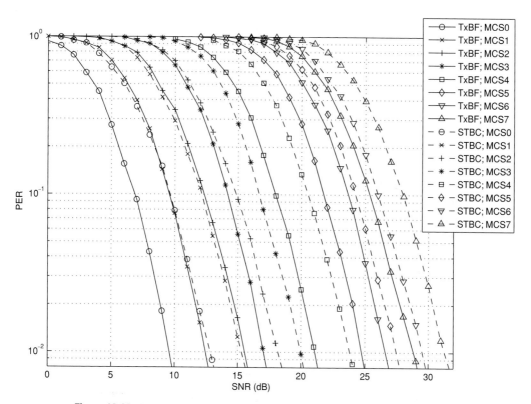

Figure 12.16 Comparison between TxBF and STBC for 2 × 1, channel model D, NLOS.

MCS. Ideally, the highest rate MCS is selected which is supported by the current channel and SNR. In a similar manner, it is possible to simulate PHY over-the-air throughput incorporating ideal link adaptation. Such simulation results provide an upper bound on PHY throughputperformance.

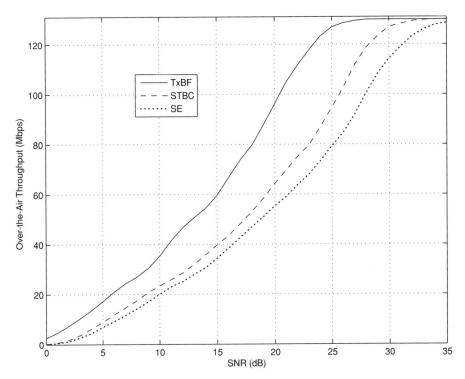

Figure 12.17 Throughput comparison between TxBF, STBC, and SE for 4 × 2, channel model D, NLOS.

In the following throughput results, a Monte Carlo simulation is performed whereby for each channel instantiation, an MCS is selected that results in a successful packet transmission and maximizes the data rate. The noise and channel vary from packet to packet. Unlike a waterfall curve, no packet errors occur in such a simulation except when packet failures occur with all MCSs. The throughput at each SNR is reported as the PHY rate averaged over all the packets and their corresponding MCS. The simulation parameters are the same as in previous results in this section.

The results in Figure 12.17 compare the PHY over-the-air throughput for 4 × 2 TxBF, 4 × 2 STBC, and 4 × 2 SE. In general the gain of TxBF or STBC over SE increases with throughput. Furthermore, the gain of TxBF over STBC also increases with throughput. To extract the performance metric of gain in dB, we select a particular throughput and compare the required SNR for each feature. As the throughput approaches 130 Mbps, the gain of TxBF over SE is 9 dB and the gain of STBC over SE is 4 dB. At a throughput of 20 Mbps, the gain of TxBF over SE reduces to 4 dB and the gain of STBC over SE is only 1.25 dB.

An alternate approach to measure gain is by examining the increase in throughput at a particular SNR. At a fairly high SNR of 25 dB, TxBF provides a 60% increase in throughput over SE and STBC provides a 20% increase in throughput over SE. At a low

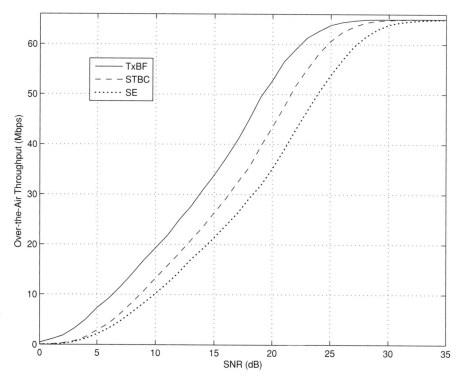

Figure 12.18 Throughput comparison between TxBF, STBC, and SE for 2 × 1, channel model D, NLOS.

SNR of 10 dB, the throughput with TxBF increase by 70% over SE. The throughput improvement of STBC over SE reduces slightly to 17%.

In all conditions, TxBF provides substantial performance benefit with respect to STBC and SE with a 4 × 2 configuration.

The results in Figure 12.18 compare the PHY over-the-air throughput for 2 × 1 TxBF, 2 × 1 STBC, and 2 × 1 SE. As the throughput approaches 65 Mbps, the gain of TxBF over SE is 5 dB and the gain of STBC over SE is 2.5 dB. At a throughput of 10 Mbps, the gain of TxBF over SE reduces to 3.5 dB and the gain of STBC over SE is again only 1.25 dB. At a higher SNR of 25 dB, TxBF provides a 19% increase in throughput over SE and STBC provides a 13% increase in throughput over SE. At a lower SNR of 10 dB, the throughput with TxBF is almost double that of SE. The throughput improvement of STBC over SE is only 30%.

In all conditions, TxBF provides reasonable performance benefit with respect to STBC and SE with a 2 × 1 configuration. Even at lower SNRs, TxBF improves throughput and reduces the required SNR, whereas STBC performance converges to that of SE.

The throughput benefit of TxBF over SDM/SE and STBC is summarized in Table 12.5. An SNR equal to 5 dB represents a link with a longer range between devices. An SNR equal to 20 dB represents a link with a shorter range between devices. At a lower SNR, throughputs are lower resulting in dramatic percentage improvements for

Table 12.5 Summary of TxBF over-the-air throughput compared to SDM/SE and STBC

		Over-the-Air Throughput (Mbps)			Improvement (%)	
		SDM/SE	STBC	TxBF	over SE	over STBC
2×1	SNR = 5 dB	2.1	2.9	7.3	248	152
	SNR = 20 dB	35.2	43.5	52.8	50	21
4×2	SNR = 5 dB	6.9	8.9	17.2	149	93
	SNR = 20 dB	55.4	64.3	96.2	74	50

TxBF. The absolute improvement is roughly 5 Mbps at an SNR equal to 5 dB. At an SNR equal to 20 dB, the throughput increase afforded by TxBF ranges from 21% to 74%.

12.9 Degradations

For implicit feedback beamforming, calibration errors degrade performance. For explicit feedback beamforming, quantization, subcarrier grouping, and compression used to reduce the overhead from feedback also potentially degrade performance. Furthermore, in either feedback method, delay in the feedback or delay in applying the weights relative to when the CSI was measured degrades performance if the channel varies in time.

A semi-analytic capacity formulation is used to quantify the impairments. First, the estimate of the transmitted signal given for a basic MIMO system in Eq. (3.29) is rewritten for SVD-based TxBF with non-ideal beamforming matrices $\hat{V}$, as follows:

$$\hat{X} = W \cdot Y$$
$$= \sqrt{\rho/N_{TX}} \cdot W \cdot H \cdot \hat{V} \cdot X + W \cdot Z \qquad (12.53)$$

Subsequently, the expression for the MSE is given by

$$J_{N_{SS} \times N_{SS}} = \frac{\rho}{N_{TX}} W \left(H\hat{V} \right) \left(H\hat{V} \right)^* W^* + W \Phi_Z W^* - 2 \sqrt{\frac{\rho}{N_{TX}}} \mathrm{Re} \left\{ W \left(H\hat{V} \right) \right\} + I \qquad (12.54)$$

where Φ_Z is the noise covariance matrix and I is the identity matrix. The receiver weights for MMSE, W, are given as follows:

$$W = \sqrt{\frac{\rho}{N_{TX}}} \left(H\hat{V} \right)^* \left(\frac{\rho}{N_{TX}} \left(H\hat{V} \right) \left(H\hat{V} \right)^* + \Phi_Z \right)^{-1} \qquad (12.55)$$

The output SNR for MMSE is computed by replacing Eq. (12.54) into Eq. (3.32). Subsequently the capacity for TxBF with MMSE with distortion is determined by replacing the output SNR into Eq. (3.34).

To determine the effect of delay on transmit beamforming, Monte-Carlo simulations were performed in which two channel matrices were derived separated by a specified delay based on channel model D, with NLOS conditions, and with the Doppler model defined in Section 3.5.3. The Doppler spread in the 5 GHz band is approximately 6 Hz

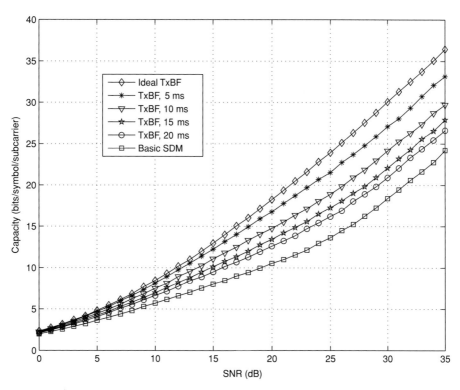

Figure 12.19 Impact of delay on capacity of 4 × 4 TxBF.

and in the 2.4 GHz band is approximately 3 Hz. One channel matrix is used as H in Eq. (12.53), and the other channel matrix is used to compute the matrix $\hat{V}$ in Eq. (12.53). The computation for capacity at a given input SNR is repeated for a large number of channel instantiations in order to generate the probability distribution shown in Figure 12.6. The value for capacity is selected at 90% probability. This procedure is repeated for a range of input SNRs.

The results for feedback delays of 5, 10, 15, and 20 ms are given in Figure 12.19. Curves for ideal TxBF, which we have proven is equivalent to Shannon capacity, and basic SDM with MMSE are also included in the figure for comparison. A 4 × 4 system is modeled. With every additional 5 ms of delay, the capacity decreases by a few bits/symbol/subcarrier. With 20 msec of delay, TxBF is only 3 bits/symbol/subcarrier better than basic SDM. Another way to analyze the system is in terms of required input SNR. For example, at 15 bits/symbol/subcarrier, ideal TxBF (which is equivalent to Shannon capacity) only requires 17 dB SNR. Basic SDM with MMSE requires 28 dB SNR. With 5 ms delay, TxBF gain degrades by 1 dB, with 10 ms another 2 dB, with 15 ms another 2 dB, and with 20 ms also another 2 dB. At this point the gain of TxBF with 20 ms delay is only 3 dB.

Subcarrier grouping and quantization are used to reduce the overhead for explicit feedback with non-compressed beamforming weights. This causes a loss of information

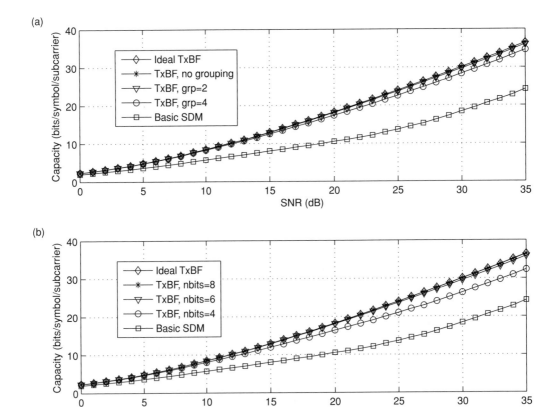

Figure 12.20 Impact of subcarrier grouping and quantization on capacity for 4 × 4 explicit feedback beamforming with uncompressed weights.

in the feedback. To model this degradation, similar simulations were performed as for delay. However, in this case, for each iteration of the simulation, a single new channel instantiation is generated and an SVD-based beamforming matrix V is calculated. Quantization or subcarrier grouping is applied to this matrix resulting in the matrix $\hat{V}$. The effect of subcarrier grouping is illustrated in Figure 12.20(a). In channel model D, with NLOS conditions, a subcarrier grouping of two has no degradation. Even a subcarrier grouping of four has a small loss of 1 dB. For this channel model, subcarrier grouping provides an effective method of reducing the feedback overhead.

Degradation due to quantization is illustrated in Figure 12.20(b). With six or eight bits there is no measurable loss. With four bits, the required SNR increases by three dB.

A reasonable compromise between feedback overhead and degradation is to use six bits for quantization and a subcarrier grouping of two. The size of the feedback is reduced to 3/8 of the feedback with eight bits and no subcarrier grouping.

Further reduction in the overhead of explicit feedback is achieved with compression. To analyze the degradation, similar simulations were performed as for non-compressed weights. However, in this case, the compression operation is performed on the matrix V

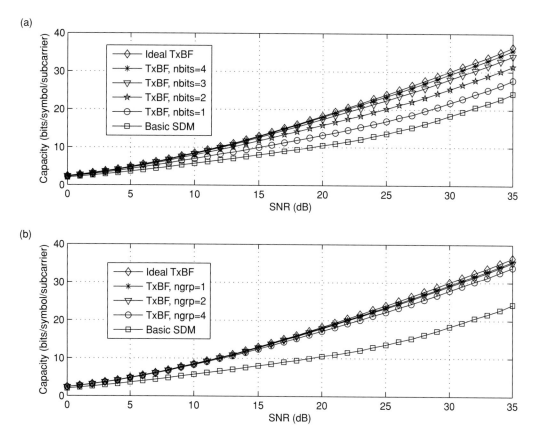

(a)

(b)

Figure 12.21 Impact of compression and subcarrier grouping for 4 × 4 explicit feedback beamforming with compressed weights.

followed by subcarrier grouping resulting in the matrix $\hat{V}$. The effect of compression as a function of the number of bits used to represent the phase values is illustrated in Figure 12.21(a). In channel model D with four bits, there is 1 dB degradation with respect to input SNR. With three bits, this increases to 2 dB. With two bits, this increases to 3 dB. And with one bit, the performance degrades by 5 dB.

The effect of subcarrier grouping is illustrated in Figure 12.21(b). The results are given with four bits of compression. Similar to non-compressed beamforming weights, in channel model D, a subcarrier grouping of two has no degradation and a subcarrier grouping of four has a small loss of roughly 1 dB.

A reasonable compromise between feedback overhead and degradation is to use four bits to represent the phase values and a subcarrier grouping of two. The size of the feedback is reduced to less than a tenth of the non-compressed feedback with eight bits and no subcarrier grouping. Moreover, the size of the feedback is reduced to a quarter of the non-compressed feedback with six bits and a subcarrier grouping of two.

The main source of degradation in implicit beamforming is imperfect calibration. We analyze the impact of the calibration error in two parts, the calibration error at the beamformer and calibration error at the beamformee. The beamformer is device A and the beamformee is device B in Figure 12.10.

The receive RF distortion at the beamformee, $C_{RX,B}$ in Eq. (12.24), affects the reception of a signal transmitted with basic SDM in the same way as TxBF. Receiver equalization with ZF or MMSE mitigates the distortions equally comparing the two systems. Therefore we do not include this term when analyzing calibration error.

We demonstrate that the transmit RF distortion at the beamformee, $C_{TX,B}$ in Eq. (12.25), has minimal effect on TxBF performance. We start with the effective channel from the beamformer to the beamformee with ideal calibration given in Eq. (12.28) (and not including the receive RF distortion at the beamformee):

$$\tilde{H}_{AB} K_A \approx H_{AB} C_{RX,A} \tag{12.56}$$

From Eq. (12.25) we derive the following equality between the channels:

$$\tilde{H}_{BA}^T = C_{TX,B} H_{AB} C_{RX,A} \tag{12.57}$$

Therefore the effective channel from the beamformer to the beamformee can be rewritten as follows:

$$\tilde{H}_{AB} K_A = C_{TX,B}^{-1} C_{TX,B} H_{AB} C_{RX,A}$$
$$= C_{TX,B}^{-1} \tilde{H}_{BA}^T \tag{12.58}$$

With beamforming, the resulting channel is given by

$$\tilde{H}_{AB} K_A V = C_{TX,B}^{-1} \tilde{H}_{BA}^T V_{BA}$$
$$= C_{TX,B}^{-1} U_{BA} S_{BA} \tag{12.59}$$

If transmit RF distortion at the beamformee is phase only with no gain variations, $C_{TX,B}$ is unitary. This proves that phase distortion at the transmitter of the beamformee has no effect on TxBF performance.

We model the impact of gain error distortion at the transmitter of the beamformee by simulation as previously described. For each instantiation, a new channel model D matrix is computed and a new distortion matrix is computed with each diagonal element uniformly distributed in decibels. Three curves with distortion are illustrated in Figure 12.22. The curve marked with "*" illustrates the degradation with the elements in the distortion matrix uniformly distributed between ±2 dB. For the curves with triangles and circles, the extent of the distortion is increased to 5 dB and 10 dB, respectively. The degradation with ±2 dB and ±5 dB variation is negligible. With ±10 dB gain variation, the TxBF performance is degraded by only 1 dB.

By specification the edge subcarriers may only deviate by +2/−4 dB, and the inner subcarriers by ±2 dB. In the worst case, gain variation between the antennas due to this spectral deviation is 6 dB. However, with small integrated RF devices, the spectral variation for each antenna should be similar by the nature of the design. In addition, the absolute gain on each transmit RF chain of the device should be comparable to maximize the total transmit power. Therefore, subcarrier gain distortions between transmit RF chains should be modest. Degradation to TxBF performance without calibration of the beamformee should be less than 1 dB.

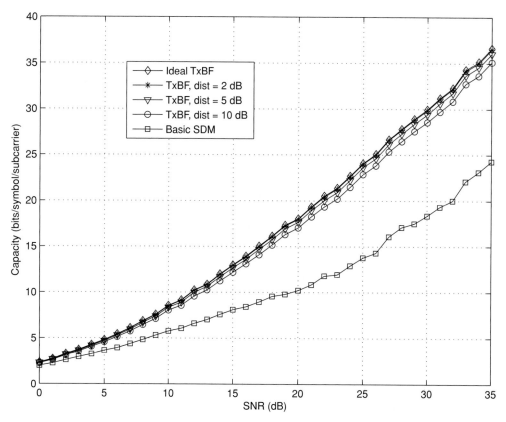

Figure 12.22 Impact of gain error distortion at the transmitter of the beamformee on 4×4 implicit feedback beamforming.

On the other hand, TxBF performance is sensitive to both gain and phase calibration errors at the beamformer. In the following analysis to isolate the impact of beamformer calibration error, we assume that the distortion at the beamformee is small and may be neglected. Therefore, we start with the effective channel from the beamformer to the beamformee given in Eq. (12.24) and apply non-ideal calibration, as follows:

$$\tilde{H}_{AB}K_A \approx H_{AB}C_{TX,A}K_A \tag{12.60}$$

From Eq (12.25) we determine the following equality between the channel in both directions without the distortion from the beamformee:

$$\tilde{H}_{BA}^{T} = H_{AB}C_{RX,A} \tag{12.61}$$

The effective channel from the beamformer to the beamformee with non-ideal beam-forming is rewritten as follows:

$$\tilde{H}_{AB}K_A = H_{AB}C_{RX,A}C_{RX,A}^{-1}C_{TX,A}K_A$$
$$= \tilde{H}_{BA}^{T} \cdot \left(C_{RX,A}^{-1}C_{TX,A}K_A \right) \tag{12.62}$$

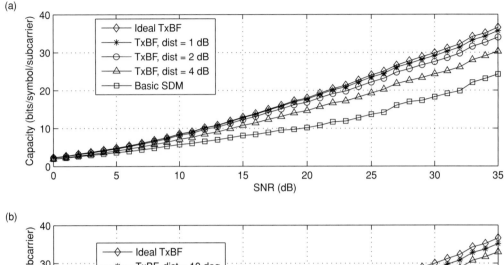

Figure 12.23 Impact of beamformer calibration error on 4 × 4 implicit feedback beamforming.

The term in the parentheses in Eq. (12.62) is the calibration error. With beamforming, the resulting channel is given by

$$\tilde{H}_{AB}K_A V = \tilde{H}_{BA}^{T} \cdot E \cdot V_{BA} \tag{12.63}$$

where E is the calibration error.

As before, a Monte-Carlo simulation is performed with a new channel and new calibration error for each instantiation. Figure 12.23 illustrates the results for channel model D with a 4 × 4 configuration. Figure 12.23(a) provides results for gain calibration error, and Figure 12.23(b) provides results for phase calibration error. As demonstrated, calibration gain error should be limited to less than 2 dB otherwise the SNR degrades by more than 1 dB. Similarly, with a calibration phase error of less than 10 degrees the required SNR increases by less than 1 dB.

With a system configuration of 4 × 2, the sensitivity to calibration is reduced as compared to a 4 × 4 system. As illustrated in Figure 12.24, a calibration gain error of less than 5 dB only increases the required SNR by less than 1 dB. And with a 4 × 2 configuration, the system is very tolerant of calibration phase error.

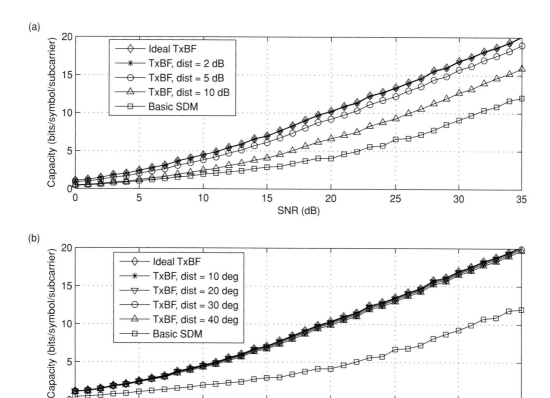

Figure 12.24 Impact of beamformer calibration error on 4 × 2 implicit feedback beamforming.

Finally, a 4 × 1 system configuration is very insensitive to calibration errors. The results for a 4 × 1 system are given in Figure 12.25. The performance does not degrade with a calibration gain error of less than 5 dB, nor with a calibration phase error of less than 60 degrees.

12.10   MAC considerations

The frame sequences for transmit beamforming are by necessity flexible to accommodate the many options and implementation capabilities. These options and capabilities include:

1. Support for implicit or explicit feedback
2. Support for receiving and transmitting staggered or NDP format sounding PPDUs
3. In the case of explicit feedback, the ability of the beamformee to return immediate feedback (within SIFS) or ability to only return feedback after a longer processing time

(a)

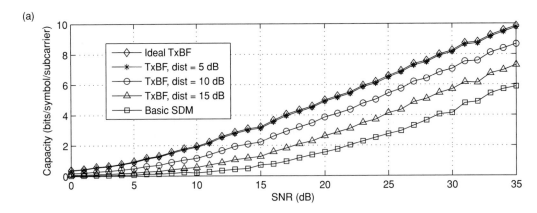

(b)

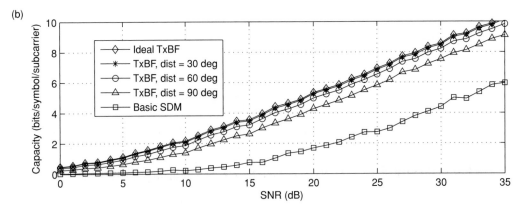

Figure 12.25 Impact of beamformer calibration error on 4 × 1 implicit feedback beamforming.

4. In the case of explicit feedback, the ability to return feedback as CSI, compressed beamforming or non-compressed beamforming
5. In the case of implicit feedback, the ability to participate in a calibration exchange

12.10.1   Sounding PPDUs

Channel sounding that exercises the full dimensionality of the channel is necessary for both implicit and explicit feedback beamforming. Two PPDU formats are defined for channel sounding: the regular or staggered PPDU, which carries a MAC frame, and the null data packet (NDP), which does not carry a MAC frame. Both formats are described in Section 12.6.

The regular or staggered sounding PPDU is simply a normal PPDU or a PPDU with additional HT-LTFs that is used to sound the channel. It serves the dual purpose of sounding the channel and carrying a MAC frame. The NDP is only used to sound the channel and because it does not carry a MAC frame it must be used in a sequence from which the addressing can be determined.

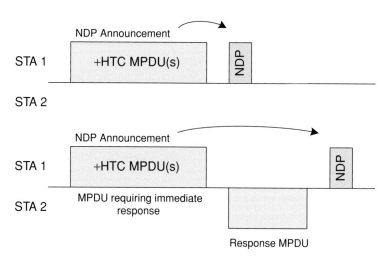

Figure 12.26 Basic NPD frame sequence.

12.10.1.1 NDP as sounding PPDU

The NDP sounding PPDU has no payload and thus contains no MAC frame. To use an NDP it must be part of a sequence such that the addressing and other MAC related information can be obtained from a MAC frame in a preceding PPDU. The two sequences shown in Figure 12.26 are possible. The NDP frame may follow another PPDU in a burst sequence where the preceding PPDU carries one or more MPDUs which contain the HT Control field with the NDP Announcement bit set to 1. If the NDP Announcement PPDU solicits an immediate response then the NDP itself follows the response PPDU.

The permitted sequences have an impact on the reverse direction protocol. An MPDU carrying an HTC field with the NDP Announcement bit set cannot also provide a reverse direction grant (see Section 9.3).

12.10.1.2 NDP use for calibration and antenna selection

The NDP can also be used in calibration for implicit feedback beamforming. With calibration, channel sounding is required in both link directions closely spaced in time. A special sequence to perform calibration is defined and described later.

The NDP may also be used for antenna selection, a technique for selecting the optimal receive and transmit antennas when there are more antennas than receive and transmit paths. When performing antenna selection multiple NDP transmissions may be needed since up to 8 antennas may be present.

12.10.2 Implicit feedback beamforming

A station that supports implicit feedback beamforming as a beamformer sets the following fields in the Transmit Beamforming Capability field of the HT Capabilities element:

1. It sets the Implicit TxBF Capable subfield to 1 to indicate that it is capable of generating beamformed transmissions based on implicit feedback.

2. It sets the Implicit TxBF Receive Capable subfield to 1 to indicate that it is capable of receiving beamformed transmissions based on implicit feedback. A beamformer must also be able to act as a beamformee.

3. It sets either or both of the Receive Staggered Sounding Capable and Receive NDP Capable subfields to 1 to indicate which forms of sounding it supports.

4. It sets the Calibration subfield to 3 to advertise full calibration support.

A station that supports implicit feedback beamforming as a beamformee only sets the following fields in the Transmit Beamforming Capability field of the HT Capabilities element:

1. It sets the Implicit TxBF Receiving Capable subfield to 1 to indicate that it is capable of receiving beamformed transmissions based on implicit feedback.

2. It sets either or both the Transmit Staggered Sounding Capable and Transmit NDP Capable subfields to 1 to indicate which form or forms of sounding it supports.

The requirements for a beamformee are relatively light and essentially require only that the beamformee be capable of transmitting a sounding PPDU in response to a TxBF sounding request. A TxBF sounding request is carried by a PPDU with one or more MPDUs in the PPDU having the TRQ bit in the HT Control field set to 1. Note that even if implicit beamforming is supported on both sides of a link it is still possible the station capabilities may preclude its operation. For instance the beamformer and beamformee may support different sounding PPDU formats.

A beamformee may optionally also support calibration to improve beamforming effectiveness by allowing the beamformer to account for the differences in the receive and transmit paths at both ends of the link. The requirements for calibration on the part of the beamformee are also relatively light. The beamformee should be able to participate in the calibration exchange and send channel state information to the beamformer following the exchange.

12.10.2.1 Calibration

A beamformer should calibrate for differences between its own receive and transmit paths. A beamformer may also calibrate for the differences between the receive and transmit paths of both the beamformer and beamformee through the calibration procedure described here. The calibration procedure involves computing correction matrices that effectively ensure that the observed channel matrices in the two directions of the link are transposes of each other and thus that the channel is reciprocal. This is done through a calibration exchange that involves sending sounding PPDUs closely spaced in time to sound the channel in both directions. The calibration responder then returns channel state information (CSI) for the initiator to responder direction to the initiator.

A station that supports beamforming using implicit feedback and is capable of initiating calibration sets the Calibration subfield of the Transmit Beamforming Capabilities field to 3 to indicate full support for calibration. A station that is capable of responding to a calibration request sets the Calibration subfield to 1 (can respond to calibration request but not initiate calibration) or 3 (can both initiate and respond to calibration request)

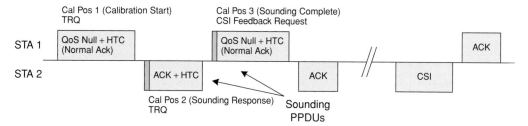

Figure 12.27 Calibration exchange.

depending on its capability. A station that is capable of responding to a calibration request will return CSI and sets the CSI Max Number of Rows Beamformer Supported subfield.

Calibration exchange using staggered sounding PPDUs

The calibration exchange is illustrated in Figure 12.27 and proceeds as follows. The calibration initiator transmits a QoS Null + HTC data frame (a data frame that includes the HT Control field) in which the TRQ field is set to 1 to solicit sounding and the Calibration Position field is set to 1 to indicate that it is the calibration start frame. The data frame is sent with Normal Ack policy so that the calibration responder responds with an ACK frame. The ACK frame is sent in a Control Wrapper frame to include an HT Control field and sent using a sounding PPDU in response to the TRQ from the calibration initiator. The HT Control field has TRQ set to 1 to solicit a sounding PPDU in turn and the Calibration Position field set to 2 to indicate a sounding response.

The calibration initiator uses the sounding PPDU to measure the channel state in the responder to initiator direction and then sends its own sounding PPDU containing a QoS Null +HTC data frame. This frame includes a CSI Feedback Request and has the Calibration Position field set to 3 to indicate sounding complete. The data frame has Normal Ack policy and the calibration responder returns an ACK if the frame is correctly received.

The responder uses the sounding PPDU to measure the channel state in the initiator to responder direction and, in a separate TXOP, the calibration responder returns the CSI in a CSI Action management frame.

Calibration exchange using NDP

The calibration exchange can also be performed using the NDP as shown in Figure 12.28. There are special rules for the use of NDP in calibration. The calibration initiator begins the sequence by sending a QoS Null +HTC data frame with the Calibration Position field set to 1 indicating calibration start. The CSI/Steering field is set to 1 to indicate that CSI feedback is expected. The NDP Announcement field is set to 1 to indicate that an NDP will follow.

The calibration responder returns an ACK for the QoS Null data frame. The ACK is wrapped in a Control Wrapper frame to include the HT Control field and the Calibration Position field is set to 2 to indicate a calibration response.

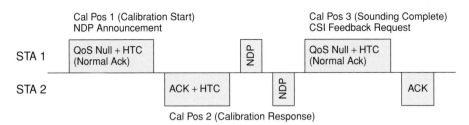

Figure 12.28 Calibration exchange using NDP.

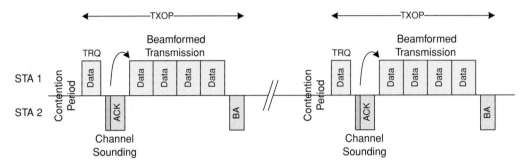

Figure 12.29 Implicit feedback beamforming sequence.

On correctly receiving the ACK, the calibration initiator sends its announced NDP. The calibration responder sends an NDP in turn. Note that the NDP returned by the calibration responder is not announced. Instead, the NDP is expected by virtue of the fact that the calibration responder previously sent a calibration response. This is a difference from the NDP rules for sounding for beamforming to allow for the bidirectional NDP exchange.

Following the NDP exchange, the calibration initiator sends a QoS Null +HTC data frame which includes a CSI Feedback Request and has the Calibration Position field set to 3 to indicate sounding complete. The calibration responder returns an ACK on correctly receiving the frame.

In a separate TXOP not shown in the diagram, the calibration responder returns the CSI for the initiator to responder direction in a CSI Action frame.

12.10.2.2 Sequences using implicit feedback

To sound the channel, the beamformer sends a PPDU containing one or more +HTC MPDUs which have the TRQ bit set to 1. If the PPDU requires a response then either the PPDU containing the response must be a sounding PPDU or the response PPDU contains an NDP announcement and is followed by an NDP. If the PPDU carrying the sounding request does not require an immediate response then the beamformee must sound the channel in a TXOP the beamformee obtains either using a sounding PPDU or NDP.

An example implicit beamforming sequence is shown in Figure 12.29. In this sequence the beamformer obtains a TXOP and performs a short frame exchange for collision

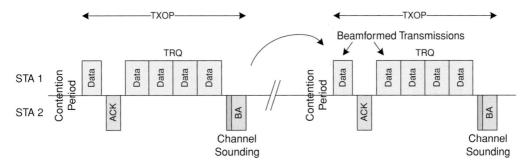

Figure 12.30 Implicit feedback beamforming with relaxed timing on applying beamforming weights.

detect. The frame sent to the beamformee in this case is a QoS Data +HTC frame with the TRQ bit set to 1. If the beamformee successfully receives the frame it responds with an ACK. The ACK is carried in a sounding PPDU in response to the TRQ. The beamformer measures the channel state with the received sounding PPDU and applies the appropriate beamforming weights to the aggregate transmission that follows.

The example sequence shows a channel sounding exchange followed immediately with the application of the beamforming weights to the aggregate transmission all occurring in the same TXOP. This has the advantage that the channel does not significantly change from the time the channel measurement was made to the time the beamforming weights are applied. However, in practice it may be difficult to perform the beamforming calculations in the SIFS turnaround time.

The sounding exchange in this example could also be performed using RTS/CTS. However, because the CTS may be carried in a staggered sounding PPDU it may not be broadly received, negating the key benefit from using RTS/CTS.

To relax the timing associating with receiving the sounding PPDU and applying the beamforming weights, the beamformer may use a sequence similar to that shown in Figure 12.30. Here the beamformer sends a training request (TRQ) in the data aggregate and the beamformee sends the BA response carried in a sounding PPDU. Alternatively, the beamformee could send the BA with an NDP Announcement followed by an NDP. The sounding PPDU is used by the beamformer to calculate beamforming weights for frames sent in a subsequent TXOP.

Sounding the channel at the end of the TXOP also has the advantage that the short frame exchange at the start of the TXOP is unburdened with the need to perform sounding and thus may be a broadly receivable RTS/CTS exchange.

12.10.3 Explicit feedback beamforming

With explicit feedback beamforming the beamformer receives feedback from the beamformee in one of three forms:

- **Channel state information (CSI).** The beamformee sends the MIMO channel coefficients.

- **Non-compressed beamforming**. The beamformee sends calculated beamforming matrices.
- **Compressed beamforming**. The beamformee sends compressed beamforming matrices.

The details of these forms of feedback have been discussed in earlier sections of this chapter. The beamformee advertises the particular forms it supports in the Transmit Beamforming Capability field in the HT Capabilities element.

The feedback data itself is returned in an Action or Action No Ack management frame of a type corresponding to the form of feedback provided, i.e. CSI Action frame, Non-compressed Beamforming Action frame, or Compressed Beamforming Action frame. The feedback may be aggregated with a control frame response or other data frames to reduce overhead. In this case the CSI is returned in an Action No Ack management frame subtype so that an ACK response does not need to be returned. In the general discussion that follows, the management frame used to return the feedback is simply referred to as a CSI/BF frame with the understanding that it is one of these specific frame types.

The beamformee may be limited in the timeliness with which it is able to send feedback in response to sounding. The beamformee advertises one of the following capabilities in the Transmit Beamforming Capability field in the HT Capabilities element:

- **Immediate**. The beamformee is capable of sending a feedback response SIFS after receiving a sounding PPDU either as a separate response or as part of an aggregate response.
- **Delayed**. The beamformee is not capable of sending a feedback response SIFS after receiving a sounding PPDU. The response will be sent in a TXOP that the beamformee obtains.
- **Immediate and delayed**. The beamformee is capable of both immediate and delayed behaviors.

The beamformee sends CSI, compressed, or non-compressed beamforming feedback in response to a request from the beamformer. The request is made using the CSI/Steering field in the HT Control field of a MAC frame, where the HT Control field also carries an NDP Announcement or the MAC frame itself is carried in a sounding PPDU. The feedback request in the CSI/Steering field also indicates the type of feedback requested: CSI, non-compressed beamforming, or compressed beamforming. See Section 11.1.6 for details on the HT Control field.

Different types of MAC frames can carry the feedback request, however, the following rules apply. If the request is carried with a RTS frame, then the beamformee responds with CTS but delays transmission of the CSI/BF frame until the beamformee's next transmission in the current TXOP. The CSI/BF frame may then be aggregated with that response, which would typically be an ACK or BA response.

If the request is carried in a data frame or data aggregate that requires an ACK or BA response and the beamformee is able to provide immediate feedback, then both the response frame and the CSI/BF frame may be aggregated in an A-MPDU. If, however,

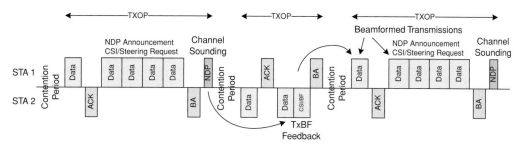

Figure 12.31 Example sequence using NDP, sounding with delayed explicit feedback.

the sounding itself is performed by an NDP, then the beamformee must obtain a TXOP to send the CSI/BF frame.

12.10.3.1 Sequences using explicit feedback

An example sequence using NDP sounding with delayed explicit feedback is shown in Figure 12.31. In this sequence sounding is done at the end of the first TXOP. The peer station gains a TXOP and returns CSI, uncompressed or compressed feedback.

The originating station on gaining a subsequent TXOP applies beamforming weights based on the feedback. This example assumes a traffic pattern where the bulk of the data transfer is in one direction (STA 1 to STA 2), perhaps a TCP transfer with the reverse direction comprising short TCP ack frames. In the sequence shown it is assumed that there is regular reverse direction data, which would be the case with TCP, and that the beamforming feedback is piggybacked with that data. The sequence is similar without reverse direction data, although in this case the reverse direction TXOP would be gained purely to transfer the CSI/BF frame which would add considerable overhead.

With this sequence there is a lapse in time between the sounding of the channel and the application of the beamforming weights. This delay is indeterminate and dependent on use of the channel by other stations. However, with a typical TXOP of 1.5 or 3 ms in duration the application of the beamforming weights should occur within 10 ms of channel sounding if the beamformer is the dominant user of the channel.

The additional overhead associated with this sequence amounts to roughly one NDP per beamformer TXOP and the time needed to transfer the CSI/BF frame in the beamformee TXOP.

An example sequence using the staggered sounding PPDU and immediate feedback is shown in Figure 12.32. In this sequence the beamformee returns feedback aggregated with the BA in the same TXOP as the sounding PPDU. The feedback is used in a subsequent TXOP. With this sequence there is no dependency on the beamformee obtaining a TXOP to return the CSI/BF frame.

12.10.3.2 Differences between NDP and staggered sounding

It is worth looking at some of the differences between the use of NDP and the staggered format for sounding the channel. In the NDP sequence the beamforming feedback is returned in a TXOP obtained by the beamformee. With the staggered sounding format

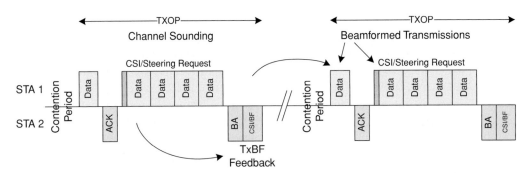

Figure 12.32 Example sequence using staggered sounding PPDU with immediate explicit feedback.

it is possible for the beamformee to return the beamforming feedback together with the BA response frame provided it is capable of generating the response in time, i.e. it supports immediate feedback. The beamformee that cannot meet the timing for the feedback response may still send the feedback in a subsequent TXOP as in the NDP case.

Sending immediate feedback following a staggered format PPDU has a downside beside the difficult implementation issues with meeting the turnaround time. The CSI/BF frame is aggregated with the BA which means that it must use the same MCS. Sending the PPDU carrying the BA and CSI/BF frame using a robust MCS means that the overhead may be high relative to the case where the CSI/BF frame is aggregated with data in a separate TXOP.

Sending the CSI/BF frame in a separate TXOP means that the beamformer is dependent on the beamformee obtaining a TXOP prior to the beamformer obtaining its TXOP in which it wishes to employ beamforming. However, this dependency is not strong, in the sense that the beamformer could simply either not use beamforming or use older beamforming data should it obtain a TXOP prior to the beamformee obtaining a TXOP.

Another difference between the use of staggered and NDP sounding is that with the staggered format channel sounding occurs early in the TXOP whereas with NDP it occurs at the end of the TXOP. An implementation receiving a staggered format PPDU may update the channel estimate through the data portion although perhaps not through all the dimensions sounded in the preamble. The use of NDP thus has the advantage that the channel estimate is obtained closer in time to its actual use in the next TXOP obtained by the beamformer.

12.11   Comparison between implicit and explicit

In comparing implicit feedback with explicit feedback, there are three major categories of differences: hardware complexity and limitations, overhead, and dependency of the beamformer on the beamformee. In the following we summarize each of these differences.

The most prominent hardware difference between implicit and explicit is that the implementation of transmit beamforming with explicit feedback is completely digital in nature. Beamforming weights are computed by the beamformee in digital baseband, and applied to the signal by the beamformer in digital baseband. On the other hand, implicit feedback includes both analog and digital aspects. Specifically, implicit feedback requires calibration. This raises the issue of stability of the analog RF over time and temperature. In addition, questions have been raised regarding calibration stability over frequency, transmit power, and receiver gain. The calibration procedure assumes a diagonal RF distortion matrix and diagonal calibration matrix, which has been questioned with regards to antenna coupling.

With regards to channel sounding, all beamformee receive antennas are sounded with explicit feedback. However, with implicit feedback only as many receive antennas as transmit antennas of the beamformee can be sounded. For example, a beamformee with one transmit antenna and two receive antennas is only able to sound one receive antenna. On the other hand, with explicit feedback, transmit beamforming is limited to four transmit antennas, whereas with implicit feedback, transmit beamforming can be performed over as many transmit antennas as contained in the beamformer.

An additional point about hardware complexity involves the computation of the weights. With implicit feedback, the weight computation is centralized at the beamformer. On the other hand, with explicit feedback with compressed and non-compressed weights, the weight computation is distributed among the beamformees. If the beamformer is beamforming to multiple beamformees, a distributed mechanism may be preferable.

Comparing the feedback overhead between implicit feedback and explicit feedback is straightforward. Explicit feedback requires the feedback of the CSI or weights, which is not required with implicit feedback. A further minor point: sounding with implicit feedback only requires as many HT-LTFs as spatial streams, whereas explicit feedback requires an HT-LTF for each transmit antenna of the beamformee. For example, with a 4×2 beamforming system, explicit feedback requires a sounding packet with four HT-LTFs and implicit feedback only requires a sounding packet with two HT-LTFs.

As a final comparison, with explicit feedback the beamformer is completely dependent on the beamformee to return the antenna weights. The beamformer is not able to perform transmit beamforming to a beamformee that does not perform this task. Furthermore, even if the beamformee does provide feedback, it may not be able to compute the weights fast enough for immediate feedback, which incurs longer delay between the sounding and the use of the weights. With implicit feedback, the beamformer can transmit beamform to any client devices which send it a compatible sounding packet.

12.12   Fast link adaptation

Link adaptation is the process by which the transmitter selects the optimal MCS with which to send data to a particular receiver. Link adaptation algorithms are implementation specific, however, they are generally based on the measured packet error rate (PER).

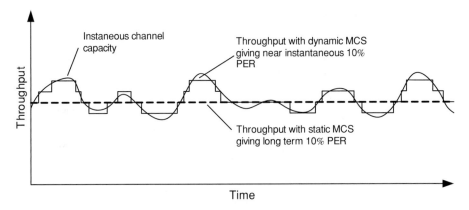

Figure 12.33 MCS selection with changing channel changes.

Most algorithms monitor the PER and adjust the MCS to track an optimal long term average that balances the reduced overhead from sending shorter packets with a higher MCS with the increased overhead from retransmissions due to the increased PER from the higher MCS.

Determining the PER by necessity means monitoring packet errors over a period that is long in comparison with the duration of a packet. For example, to very roughly measure a 10% PER requires that the transmitter send ten packets of which one is in error. Because of this, link adaptation based on PER adapts slowly to changing channel conditions. In many environments the channel is changing with time as the stations move or with changes in the environment itself, such as the 50 Hz or 60 Hz ionizing cycle in fluorescent bulbs, the movement of objects in the environment, or changes in external noise sources. These changing conditions may occur on time scales faster than PER can be measured. As a result the link adaptation algorithm is choosing an MCS that is the long term optimal MCS and not the instantaneously optimal MCS.

To see how more closely tracking the channel changes might improve performance consider Figure 12.33, which abstractly shows instantaneous capacity reflecting changing channel conditions, with throughput based on an MCS selected to realize a long term 10% PER and throughput based on an MCS selected to realize an instantaneous 10% PER. When the MCS is selected to satisfy a long term PER, packet errors occur primarily where the instantaneous capacity drops below the long term average. In the extreme case, where all packet errors occur during poor channel conditions, these conditions would account for 10% of the time. In a less extreme case, shorter periods of poor channel conditions would still account for most of the packet loss.

If the selected MCS were to track the channel changes then the same PER could be achieved with a higher average throughput by selecting a more robust MCS when the channel conditions are poor but taking advantage of periods where the channel conditions are good to increase the data rate. Since the periods during which channel conditions are poor are relatively short and since more data can be sent when channel conditions are good, performance overall is improved.

12.12.1   MCS feedback

One mechanism by which fast link adaptation can be achieved is to have the receiver participate in the MCS selection process by providing regular feedback. The 802.11n specification amendment adds the MCS feedback fields in the HT Control field as a mechanism for providing this feedback. The receiver continuously monitors the quality of the received transmissions or the characteristics of the channel itself and provides suggestions on the optimal MCS to take advantage of the channel conditions. The transmitter takes the suggested MCS and combines it with the knowledge it has (for example transmit power amplifier backoff) and derives an MCS that should optimally use the link.

The 802.11n specification does not specify the technique by which the receiver derives an MCS suggestion. A good assumption would be that the suggested MCS is the MCS that, in the view of the receiver, would optimize for throughput. However, throughput is dependent on the sequencing algorithms used and alternate optimizing points could be envisioned, such as optimizing for delay by targeting a low PER to avoid retransmission. It is likely that an actual link adaptation algorithm based on MCS feedback needs to adjust the suggested MCS adaptively and perhaps factor for receivers that consistently err on the high or low side with their MCS suggestion.

For all forms of beamforming, MCS selection could be made by the transmitter based on knowledge of the channel state. With implicit feedback beamforming and CSI based explicit feedback beamforming the transmitter has direct knowledge of the channel state. With compressed and non-compressed explicit beamforming, the transmitter receives indirect knowledge of the channel in the form of an SNR value for each spatial stream. The transmitter may thus use the channel state knowledge for MCS selection. In some case where there is interference at the receiver the transmitter may benefit from MCS feedback.

12.12.2   MCS feedback using the HT Control field

A station may receive MCS feedback in three ways:

- **Immediate**. A station sends a request for MCS feedback and receives an immediate response. This approach allows the requester to receive and apply the feedback within the same TXOP.
- **Delayed**. A station sends a request for MCS feedback and a delayed response occurs when the responder transmits the response in a subsequent TXOP obtained by the responder.
- **Unsolicited**. A station receives MCS feedback independent of any request for such feedback.

The MCS feedback mechanism is supported in the HT Control field, which may be present in QoS Data frames and may also be present in control frames such as ACK and BA when encapsulated in a Control Wrapper frame. Frames with the HT Control field present are referred to as +HTC frames.

To request feedback, a station sets the MRQ (MCS request) field to 1 in the HT Control field and chooses a value between 0 and 6 for the MSI (MCS request sequence identifier). The MSI is used to correlate the response with the request in the case of a delayed response and the value chosen is implementation dependent. In the MCS feedback response, the responder sets the MSFI (MCS feedback sequence identifier) to the value of the MSI in the corresponding MCS request. When the responder provides unsolicited MCS feedback, the MFSI value is set to 7.

If the HT Control field is included in more than one MPDU in an aggregate then the MRQ and MSI fields are set to the same values and act effectively as a single request. The HT Control field should be included in all frames making up an aggregate to improve robustness.

The MCS request should be sent in a staggered sounding PPDU or it should be sent with the NDP Announcement field set to 1 with an NDP transmission to follow. The number of HT-LTFs in the sounding PPDU or in the NDP is determined by the total number of spatial dimensions to be sounded, including any extra spatial dimensions beyond those used by the data portion of the PPDU.

On receipt of an MCS request, the responder should compute an MCS estimate. The responder may choose to send the response frame with any of the following combination of MFB (MCS feedback) and MFSI:

- MFB = 127, MFSI = 7. No information is provided for the immediately proceeding request or for any other pending request.
- MFB = 127, MFSI in range 0 to 6. The responder is unable to provide feedback.
- MFB in range 0 to 126, MFSI in range 0 to 6. The responder is providing feedback for the previously received request with MSI equal to MFSI.
- MFB in range 0 to 126, MFSI = 7. The responder is providing unsolicited feedback.

Hardware constraints may limit the number of outstanding MCS requests a responder can handle. When a new MCS request arrives, either from a different requester or from the same requester but with a different MSI value, the responder may choose to ignore the request or discard a current request and begin computation on the new request. If the responder discards a pending MCS estimate computation it should return a response with MFB set to 127 and MFSI set to the same value as the corresponding MSI in the MCS request.

The responder is constrained in the MCS suggestion it can make. It cannot make an MCS suggestion that includes more special streams than are supported by the requester. It should also not make an MCS suggestion for unequal modulation unless the requester indicates that it is capable of unequal modulation in the Tx Unequal Modulation Supported bit in the Supported MCS Set field (see the HT Capabilities element and 'Supported MCS Set field', p. 291).

References

Anderson, E., Bai, Z., Bischof, C., *et al.* (1999). *LAPACK User's Guide*, 3rd edn. Philadelphia: SIAM, available at: www.netlib.org/lapack/lug/lapack_lug.html.

IEEE (2007). *IEEE P802.11n™/D3.00, Draft Amendment to STANDARD for Information Technology – Telecommunications and Information Exchange Between Systems – Local and Metropolitan Networks – Specific Requirements. Part 11: Wireless LAN Medium Access Control (MAC) and Physical Layer (PHY). Amendment 4: Enhancements for Higher Throughput.*

Lebrun, G., Ying, T., and Faulkner, M. (2002). MIMO transmission over a time-varying channel using SVD. *IEEE GLOBECOM'02*, **1**, 414–18.

Nanda, S., Walton, R., Ketchum, J., *et al.* (2005). A high-performance MIMO OFDM wireless LAN. *IEEE Communications Magazine*, February, 101–9.

Sadowsky, J. S., Yamaura, T., and Ketchum, J. (2005). *WWiSE Preambles and MIMO Beamforming?*, IEEE 802.11–05/1635r1.

Smith, G. S. (2004). A direct derivation of a single-antenna reciprocity relation for the time domain. *IEEE Transactions on Antennas and Propagation*, **52**(6), 1568–77.

Appendix 12.1: Unequal MCS

This appendix gives unequal MCS tables for 20 MHz and 40 MHz. For 20 MHz, the number of data subcarriers is 52 and the number of pilot subcarriers is 4 in an OFDM symbol for an individual spatial stream. For 40 MHz, the number of data subcarriers is 108 and the number of pilot subcarriers is 6 in an OFDM symbol for an individual spatial stream.

Table 12.6 describes the symbols used in the subsequent tables.

Table 12.6 Symbols used for MCS parameters (IEEE, 2007)

Symbol	Explanation
R	Code rate
N_{TBPS}	Total bits per subcarrier
N_{CBPS}	Number of coded bits per OFDM symbol
N_{DBPS}	Number of data bits per OFDM symbol
N_{ES}	Number of BCC encoders

Unequal MCS for 20 MHz

Table 12.7 defines the 20 MHz unequal MCS parameters and data rates for two spatial stream transmission. The data rate is calculated by dividing N_{DBPS} by the symbol time of 4 μs for 800 ns GI, and dividing by 3.6 μs for 400 ns GI.

Table 12.8 defines the 20 MHz unequal MCS parameters and data rates for three spatial stream transmission.

Table 12.9 defines the 20 MHz unequal MCS parameters and data rates for four spatial stream transmission. All 20 MHz MCSs use a single encoder.

Table 12.7 20 MHz unequal MCS parameters for two spatial streams (IEEE, 2007)

MCS Index	Modulation		R	N_{TBPS}	N_{CBPS}	N_{DBPS}	Data rate (Mbps)	
	Stream 1	Stream 2					800 ns GI	400 ns GI
33	16-QAM	QPSK	½	6	312	156	39	43.3
34	64-QAM	QPSK	½	8	416	208	52	57.8
35	64-QAM	16-QAM	½	10	520	260	65	72.2
36	16-QAM	QPSK	¾	6	312	234	58.5	65.0
37	64-QAM	QPSK	¾	8	416	312	78	86.7
38	64-QAM	16-QAM	¾	10	520	390	97.5	108.3

Table 12.8 20 MHz unequal MCS parameters for three spatial streams (IEEE, 2007)

MCS Index	Modulation			R	N_{TBPS}	N_{CBPS}	N_{DBPS}	Data rate (Mbps)	
	Stream 1	Stream 2	Stream 3					800 ns GI	400 ns GI
39	16-QAM	QPSK	QPSK	½	8	416	208	52	57.8
40	16-QAM	16-QAM	QPSK	½	10	520	260	65	72.2
41	64-QAM	QPSK	QPSK	½	10	520	260	65	72.2
42	64-QAM	16-QAM	QPSK	½	12	624	312	78	86.7
43	64-QAM	16-QAM	16-QAM	½	14	728	364	91	101.1
44	64-QAM	64-QAM	QPSK	½	14	728	364	91	101.1
45	64-QAM	64-QAM	16-QAM	½	16	832	416	104	115.6
46	16-QAM	QPSK	QPSK	¾	8	416	312	78	86.7
47	16-QAM	16-QAM	QPSK	¾	10	520	390	97.5	108.3
48	64-QAM	QPSK	QPSK	¾	10	520	390	97.5	108.3
49	64-QAM	16-QAM	QPSK	¾	12	624	468	117	130.0
50	64-QAM	16-QAM	16-QAM	¾	14	728	546	136.5	151.7
51	64-QAM	64-QAM	QPSK	¾	14	728	546	136.5	151.7
52	64-QAM	64-QAM	16-QAM	¾	16	832	624	156	173.3

Table 12.9 20 MHz unequal MCS parameters for four spatial streams (IEEE, 2007)

MCS Index	Modulation				R	N_{TBPS}	N_{CBPS}	N_{DBPS}	Data rate (Mbps)	
	Stream 1	Stream 2	Stream 3	Stream 4					800 ns GI	400 ns GI
53	16-QAM	QPSK	QPSK	QPSK	½	10	520	260	65	72.2
54	16-QAM	16-QAM	QPSK	QPSK	½	12	624	312	78	86.7
55	16-QAM	16-QAM	16-QAM	QPSK	½	14	728	364	91	101.1
56	64-QAM	QPSK	QPSK	QPSK	½	12	624	312	78	86.7
57	64-QAM	16-QAM	QPSK	QPSK	½	14	728	364	91	101.1
58	64-QAM	16-QAM	16-QAM	QPSK	½	16	832	416	104	115.6
59	64-QAM	16-QAM	16-QAM	16-QAM	½	18	936	468	117	130.0
60	64-QAM	64-QAM	QPSK	QPSK	½	16	832	416	104	115.6

Table 12.9 (*cont.*)

MCS Index	Modulation				R	N_{TBPS}	N_{CBPS}	N_{DBPS}	Data rate (Mbps)	
	Stream 1	Stream 2	Stream 3	Stream 4					800 ns GI	400 ns GI
61	64-QAM	64-QAM	16-QAM	QPSK	½	18	936	468	117	130.0
62	64-QAM	64-QAM	16-QAM	16-QAM	½	20	1040	520	130	144.4
63	64-QAM	64-QAM	64-QAM	QPSK	½	20	1040	520	130	144.4
64	64-QAM	64-QAM	64-QAM	16-QAM	½	22	1144	572	143	158.9
65	16-QAM	QPSK	QPSK	QPSK	¾	10	520	390	97.5	108.3
66	16-QAM	16-QAM	QPSK	QPSK	¾	12	624	468	117	130.0
67	16-QAM	16-QAM	16-QAM	QPSK	¾	14	728	546	136.5	151.7
68	64-QAM	QPSK	QPSK	QPSK	¾	12	624	468	117	130.0
69	64-QAM	16-QAM	QPSK	QPSK	¾	14	728	546	136.5	151.7
70	64-QAM	16-QAM	16-QAM	QPSK	¾	16	832	624	156	173.3
71	64-QAM	16-QAM	16-QAM	16-QAM	¾	18	936	702	175.5	195.0
72	64-QAM	64-QAM	QPSK	QPSK	¾	16	832	624	156	173.3
73	64-QAM	64-QAM	16-QAM	QPSK	¾	18	936	702	175.5	195.0
74	64-QAM	64-QAM	16-QAM	16-QAM	¾	20	1040	780	195	216.7
75	64-QAM	64-QAM	64-QAM	QPSK	¾	20	1040	780	195	216.7
76	64-QAM	64-QAM	64-QAM	16-QAM	¾	22	1144	858	214.5	238.3

Unequal MCS for 40 MHz

Table 12.10 defines the 40 MHz unequal MCS parameters and data rates for two spatial stream transmission. A single encoder is used for all two stream MCSs.

Table 12.10   40 MHz unequal MCS parameters for two spatial streams (IEEE, 2007)

MCS Index	Modulation		R	N_{BPSC}	N_{SD}	N_{SP}	N_{CBPS}	N_{DBPS}	Data rate (Mb/s)	
	Stream 1	Stream 2							800 ns GI	400 nsec GI
33	16-QAM	QPSK	½	6	108	6	648	324	81	90
34	64-QAM	QPSK	½	8	108	6	864	432	108	120
35	64-QAM	16-QAM	½	10	108	6	1080	540	135	150
36	16-QAM	QPSK	¾	6	108	6	648	486	121.5	135
37	64-QAM	QPSK	¾	8	108	6	864	648	162	180
38	64-QAM	16-QAM	¾	10	108	6	1080	810	202.5	225

Table 12.11 defines the 40 MHz unequal MCS parameters and data rates for three spatial stream transmission. A single encoder is used for every three stream MCS, except for MCS 52. Table 12.12 defines the 40 MHz unequal MCS parameters and data rates for two spatial stream transmission. A single encoder is used for MCS 53–69 and two encoders are used for MCS 70–76.

Table 12.11 40 MHz unequal MCS parameters for three spatial streams (IEEE, 2007)

MCS Index	Modulation			R	N_{BPSC}	N_{SD}	N_{SP}	N_{CBPS}	N_{DBPS}	N_{ES}	Data rate (Mbps)	
	Stream 1	Stream 2	Stream 3								800 ns GI	400 ns GI
39	16-QAM	QPSK	QPSK	$\frac{1}{2}$	8	108	6	864	432	1	108	120
40	16-QAM	16-QAM	QPSK	$\frac{1}{2}$	10	108	6	1080	540	1	135	150
41	64-QAM	QPSK	QPSK	$\frac{1}{2}$	10	108	6	1080	540	1	135	150
42	64-QAM	16-QAM	QPSK	$\frac{1}{2}$	12	108	6	1296	648	1	162	180
43	64-QAM	16-QAM	16-QAM	$\frac{1}{2}$	14	108	6	1512	756	1	189	210
44	64-QAM	64-QAM	QPSK	$\frac{1}{2}$	14	108	6	1512	756	1	189	210
45	64-QAM	64-QAM	16-QAM	$\frac{1}{2}$	16	108	6	1728	864	1	216	240
46	16-QAM	QPSK	QPSK	$\frac{3}{4}$	8	108	6	864	648	1	162	180
47	16-QAM	16-QAM	QPSK	$\frac{3}{4}$	10	108	6	1080	810	1	202.5	225
48	64-QAM	QPSK	QPSK	$\frac{3}{4}$	10	108	6	1080	810	1	202.5	225
49	64-QAM	16-QAM	QPSK	$\frac{3}{4}$	12	108	6	1296	972	1	243	270
50	64-QAM	16-QAM	16-QAM	$\frac{3}{4}$	14	108	6	1512	1134	1	283.5	315
51	64-QAM	64-QAM	QPSK	$\frac{3}{4}$	14	108	6	1512	1134	1	283.5	315
52	64-QAM	64-QAM	16-QAM	$\frac{3}{4}$	16	108	6	1728	1296	2	324	360

Table 12.12 40 MHz unequal MCS parameters for four spatial streams (IEEE, 2007)

MCS Index	Modulation				R	N_{BPSC}	N_{CBPS}	N_{DBPS}	N_{ES}	Data rate (Mbps)	
	Stream 1	Stream 2	Stream 3	Stream 4						800 ns GI	400 ns GI
53	16-QAM	QPSK	QPSK	QPSK	1/2	10	1080	540	1	135	150
54	16-QAM	16-QAM	QPSK	QPSK	1/2	12	1296	648	1	162	180
55	16-QAM	16-QAM	16-QAM	QPSK	1/2	14	1512	756	1	189	210
56	64-QAM	QPSK	QPSK	QPSK	1/2	12	1296	648	1	162	180
57	64-QAM	16-QAM	QPSK	QPSK	1/2	14	1512	756	1	189	210
58	64-QAM	16-QAM	16-QAM	QPSK	1/2	16	1728	864	1	216	240
59	64-QAM	16-QAM	16-QAM	16-QAM	1/2	18	1944	972	1	243	270
60	64-QAM	64-QAM	QPSK	QPSK	1/2	16	1728	864	1	216	240
61	64-QAM	64-QAM	16-QAM	QPSK	1/2	18	1944	972	1	243	270
62	64-QAM	64-QAM	16-QAM	16-QAM	1/2	20	2160	1080	1	270	300
63	64-QAM	64-QAM	64-QAM	QPSK	1/2	20	2160	1080	1	270	300
64	64-QAM	64-QAM	64-QAM	16-QAM	1/2	22	2376	1188	1	297	330
65	16-QAM	QPSK	QPSK	QPSK	3/4	10	1080	810	1	202.5	225
66	16-QAM	16-QAM	QPSK	QPSK	3/4	12	1296	972	1	243	270
67	16-QAM	16-QAM	16-QAM	QPSK	3/4	14	1512	1134	1	283.5	315
68	64-QAM	QPSK	QPSK	QPSK	3/4	12	1296	972	1	243	270
69	64-QAM	16-QAM	QPSK	QPSK	3/4	14	1512	1134	1	283.5	315
70	64-QAM	16-QAM	16-QAM	QPSK	3/4	16	1728	1296	2	324	360
71	64-QAM	16-QAM	16-QAM	16-QAM	3/4	18	1944	1458	2	364.5	405
72	64-QAM	64-QAM	QPSK	QPSK	3/4	16	1728	1296	2	324	360
73	64-QAM	64-QAM	16-QAM	16-QAM	3/4	18	1944	1458	2	364.5	405
74	64-QAM	64-QAM	16-QAM	16-QAM	3/4	20	2160	1620	2	405	450
75	64-QAM	64-QAM	64-QAM	QPSK	3/4	20	2160	1620	2	405	450
76	64-QAM	64-QAM	64-QAM	16-QAM	3/4	22	2376	1782	2	445.5	495

Index